Advances in Geophysical and Environmental Mechanics and Mathematics

Series Editor: Professor Kolumban Hutter

Board of Editors

Ralf Greve · Heinz Blatter

Dynamics of Ice Sheets and Glaciers

Springer

Prof. Dr. Ralf Greve
Hokkaido University
Inst. Low Temperature
Science
Kita-19, Nishi-8, Kita-ku
Sapporo 060-0819
Japan

Prof. Dr. Heinz Blatter
ETH Zürich
Inst. Atmospheric & Climate
Science
Universitätstr. 16
8092 Zürich
Switzerland

ISSN 1866-8348 e-ISSN 1866-8356
ISBN 978-3-642-26935-6 e-ISBN 978-3-642-03415-2
DOI 10.1007/978-3-642-03415-2
Springer Dordrecht Heidelberg London New York

Cover design: deblik, Berlin

Printed on acid-free paper

Springer is part of Springer Science+Business Media (www.springer.com)

To Professor Kolumban Hutter,
Professor Atsumu Ohmura
and
Professor Takeo Hondoh

Preface

Ice sheets, ice shelves, ice caps and glaciers are active, dynamic components of the climate system of the Earth, and they deserve the same scientific attention as the atmosphere and the oceans. However, while the dynamics of the atmosphere and the oceans have been studied intensively and literature on these topics abound, awareness of the importance of ice dynamics within the big picture has increased only recently. Just as an example, the widely acclaimed and valued book *Geophysical Fluid Dynamics* by Pedlosky (1987) states that "the subject has tended to focus on the dynamics of large-scale phenomena in the atmosphere and the oceans", and, consequently, only these are presented in the book. On the other hand, *glaciology* is an established field of research, and glacier dynamics has been dealt with in the literature to some extent; however, with a certain focus on smaller-scale phenomena. Treatments of the large-scale dynamics of ice sheets are mainly found in the specialist literature.

In this book, we try to bridge the gap between the conventional understandings of geophysical fluid dynamics and glacier dynamics. Chapter 1 puts the subject into the wider context of climate research. In Chapter 2, the mathematical properties of vectors and tensors are reviewed briefly. Chapter 3 presents a solid, continuum-mechanical background, which is the foundation for the subject matter of the remainder of the book. This chapter goes further than the immediate needs of ice sheet and glacier dynamics in order to provide a framework applicable to a great variety of related problems in geophysics (including the above-mentioned large-scale dynamics of the atmosphere and the ocean), physics and engineering sciences. The material properties of polycrystalline ice, as it occurs in land ice masses on Earth, are discussed in Chapter 4. The core of this book is made up by Chapters 5–7, devoted to the dynamics of ice sheets, ice shelves and glaciers, respectively. Special emphasis is put on systematically developing hierarchies of approximations for the different systems, and suitable numerical solution techniques are discussed. Chapter 8 is concerned with simple models for glacial isostasy, the reaction of the solid Earth to temporally varying ice loads. In Chapter 9 some more advanced and demanding topics of current research related to ice dynamics (induced anisotropy, compressible firn, polythermal glaciers) are treated. Chapter 10 concludes the book.

The content is based, largely, on lectures about ice-sheet/glacier dynamics and numerical models in glaciology developed by the authors over the past years. These lectures are offered jointly at the Hokkaido University, Sapporo, and the Swiss Federal Institute of Technology (ETH) Zurich, in association with the International Antarctic Institute (IAI), an international, multicampus programme in cryosphere science education (http://wwwearth.ees.hokudai.ac.jp/IAI/, http://www.iai.utas.edu.au/). The level of treatment caters mainly to graduate students, post-graduate students and researchers, but most of the material should also be understandable for motivated upper-level undergraduate students.

In order to eliminate one source of distraction especially for student readers, we have refrained from giving detailed references to original literature in the style of scientific articles. Instead, references have been kept at a reasonable minimum, and whenever possible, overview articles and textbooks have been given preference. An exception is Chapter 9, in which less well established topics are discussed, and which is more biased by the authors' own perspective than the material in the preceding chapters. The literature list at the end of the book is also understood as a suggestion for supplemental and/or further-reaching reading. We apologize to those colleagues whose publications are not quoted.

Hutter and Jöhnk (2004) end the preface of their book *Continuum Methods of Physical Modeling* with the statement that "writing a book can never be finished, a book has to be abandoned!" We cannot put it in better words. Abandoning this book is what we are now going to do, well knowing that it is not perfect, but nevertheless hoping that a variety of readers with backgrounds in glaciology, climate science, geophysical fluid dynamics, continuum mechanics, physics and applied mathematics will find it useful and inspiring in the years to come.

Sapporo, Japan; Zurich, Switzerland
June 2009

Ralf Greve
Heinz Blatter

Acknowledgements

First and foremost, we would like to express our sincere gratitude to Professor Kolumban Hutter (Zurich), who has invited us to submit this book to Springer's new series *Advances in Geophysical and Environmental Mechanics and Mathematics* (an offer which we have gladly accepted), provided many insightful comments and supervised the whole process from the idea to the final product as competent editor.

Many of our colleagues and students have read all or parts of various draft versions of the book. In particular, we would like to acknowledge the efforts by Professor Garry K. C. Clarke (Vancouver), Professor Leslie W. Morland (Norwich), Dr. Sérgio H. Faria (Göttingen), Dr. Andreas Aschwanden and Dr. Tracy Ewen (Zurich), whose contributions helped considerably to improve the content and the writing. Further, we would like to express our gratitude to Professor Takeo Hondoh, Dr. Shin Sugiyama, Ms. Hiromi Kimura and Ms. Tomoko Endo (Sapporo), whose commitment for the IAI Programme of Hokkaido University has greatly supported the development of the joint lectures and, ultimately, this book.

Grateful acknowledgement is made to the following publishers, organisations and authors for permission to use previously published, copyrighted figures:

American Geophysical Union:

Figure 6.12 – adapted from Humbert et al. (2005).
Figure 6.13 – reproduced from Humbert et al. (2005).
Figure 7.8 – adapted from Aschwanden and Blatter (2005).
Figure 7.9 – adapted from Aschwanden and Blatter (2005).
Figure 7.10 – adapted from Aschwanden and Blatter (2005).
Figure 9.27 – adapted from Aschwanden and Blatter (2005).

Darmstadt University of Technology:

Figure 4.2 – reproduced from Faria (2003).
Figure 5.15 – adapted from Calov (1994).

Elsevier:

Figure 4.1 – adapted from Paterson (1994).

GeoForschungsZentrum Potsdam:

Figure 8.1 – adapted from Thoma and Wolf (1999).

International Glaciological Society:

Figure 7.13 – adapted from Hubbard et al. (2000).
Figure 8.9 – reproduced from Le Meur and Huybrechts (1996).
Figure 9.13 – reproduced from Seddik et al. (2008).
Figure 9.14 – reproduced from Seddik et al. (2008).

Swiss Federal Institute of Technology Zurich:

Figure 9.15 – reproduced from Schwerzmann (2006).
Figure 9.16 – reproduced from Schwerzmann (2006).

Last but not least, we would like to say thanks to our dear families. Michiyo, Natsumi, Kenta, Susi, Samuel, Jonas, Annina, you make our lives richer, and your continued support and patience are invaluable and greatly appreciated.

Contents

1

Ice in the Climate System

1.1 The Terrestrial Cryosphere

The frozen part of the terrestrial climate system is referred to as the cryosphere. The cryosphere consists of several subsystems, namely ice sheets, ice shelves, ice caps, glaciers, sea ice, lake ice, river ice, ground ice and snow. Ice sheets are ice masses of continental size (area greater than 50,000 km^2) which rest on solid land, whereas ice shelves consist of floating ice nourished by the inflow from an adjacent ice sheet, typically stabilised by large bays. Extended land-based masses of ice covering less than 50,000 km^2 are termed ice caps, and smaller ice masses constrained by topographical features (for instance a mountain valley) are called glaciers. Sea ice floats on the ocean; however, in contrast to an ice shelf it forms directly by freezing sea water. Similarly, lake ice and river ice form directly on lake and river water, respectively. Ground ice occurs as permafrost, that is, soil that stays in a frozen state year-round. Snow is precipitation in the form of crystalline water ice, consisting of a multitude of snowflakes, which accumulate on the ground at a bulk density significantly less than that of ice.

Ice sheets (with their attached ice shelves), ice caps and glaciers, which are subsumed as land ice, are the focus of this book. As a common feature, these ice bodies show gravity-driven creep flow ("glacial flow"), sustained by the underlying land. This leads to thinning and horizontal spreading, which is essentially compensated by snow accumulation in the higher (interior) areas and melting and calving in the lower (marginal) areas. Any imbalance of this dynamic equilibrium leads to either growing or shrinking ice masses.

1.2 Land Ice on the Present-Day Earth

By far the largest single land ice body on the present-day Earth is the Antarctic Ice Sheet (Fig. 1.1), with a total ice volume of $25.7\times10^6\ \mathrm{km}^3$, and additionally $0.58\times10^6\ \mathrm{km}^3$ of the attached ice shelves (Ross Ice Shelf, Filchner-Rønne

R. Greve, H. Blatter, *Dynamics of Ice Sheets and Glaciers*, Advances in Geophysical and Environmental Mechanics and Mathematics,
DOI 10.1007/978-3-642-03415-2_1, © Springer-Verlag Berlin Heidelberg 2009

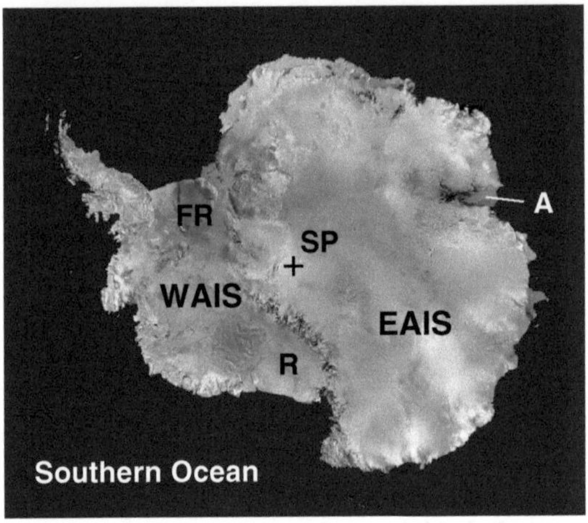

Fig. 1.1. Satellite composite image of the Antarctic Ice Sheet. EAIS: East Antarctic Ice Sheet, WAIS: West Antarctic Ice Sheet, R: Ross Ice Shelf, FR: Filchner-Rønne Ice Shelf, A: Amery Ice Shelf, SP: South Pole. Image size is approximately 6500 km × 5600 km. (Credit: NASA Goddard Space Flight Center; public domain.)

Ice Shelf, Amery Ice Shelf and others). This corresponds to a sea level rise equivalent of 61.1 m (Church et al. 2001). The ice sheet and the ice shelves cover an area of $12.4 \times 10^6\,\mathrm{km}^2$ and $1.1 \times 10^6\,\mathrm{km}^2$, respectively, so that the mean ice thickness is approximately 2 km. Additional extremes include the highest surface elevation of the ice sheet of 4.2 km AMSL (above mean sea level), an annual mean surface temperature which can be as low as −60°C in central East Antarctica, and the lowest temperature ever measured on the surface of the Earth, −89.2°C, at the Russian Vostok station. Due to these low temperatures, surface melting over the ice sheet is essentially non-existent, and the ice sheet loses its mass mainly by drainage into the surrounding ice shelves, from where it is ultimately released into the Southern Ocean by either calving (break-off of icebergs) or basal melting (Bentley 2004).

Compared to this, the second present-day ice sheet on Earth, the Greenland Ice Sheet, appears modest. Its ice volume amounts to $2.85 \times 10^6\,\mathrm{km}^3$ or 7.2 m sea level rise equivalent, with an ice-covered area of $1.71 \times 10^6\,\mathrm{km}^2$ (Church et al. 2001). Because of the absence of large bays, the ice sheet releases its outward mass flow directly into the ocean where it reaches the coast, and consequently ice shelves do not exist. An important difference to the Antarctic Ice Sheet is that, due to the higher surface temperatures, the regions close to the ice margin experience a considerable amount of melting during the summer season, so that the mass loss of the Greenland Ice Sheet is divided roughly equally between melting and calving (Thomas 2004).

Glaciers occur on every continent and in approximately 47 of the world's countries, including tropical locations like the Andes in northern South Amer-

Fig. 1.2. Satellite image of Vatnajökull, Iceland, in September 2002. Image size is approximately 150 km × 120 km. (Credit: Jacques Descloitres, MODIS Rapid Response Team, NASA Goddard Space Flight Center; public domain.)

ica, Mount Kilimanjaro (Tanzania) and Puncak Jaya (Western New Guinea, Indonesia). Examples for ice caps are the two largest European land ice bodies, Vatnajökull in Iceland (largest by volume; Fig. 1.2) and Austfonna in Svalbard, Norway (largest by area). More than 160,000 glaciers and approximately 70 ice caps of the world have a combined volume of $0.18 \times 10^6\,\mathrm{km}^3$ and cover an area of $0.68 \times 10^6\,\mathrm{km}^2$. Their total sea level rise equivalent is therefore estimated as 0.5 m (Church et al. 2001).

1.3 An Excursion into the Past

In the early Tertiary, the global climate was characterised by tropical-to-moderate worldwide temperatures and the complete absence of a cryosphere. However, in the course of the Tertiary, climates slowly cooled. Antarctica drifted to its current position at the South Pole, and in the early Oligocene (about 30 million years ago) the Antarctic Ice Sheet started to form as a small ice cap which retreated and advanced many times until the Pliocene, when it came to occupy almost all of Antarctica. The Greenland Ice Sheet did not form at all until the late Pliocene, but developed very rapidly with the onset of the Pleistocene Glacial Epoch about 2 million years ago.

The Pleistocene lasted until about 10,000 years ago and showed a sequence of advances ("ice ages" or "glacials") and retreats ("interglacials") of ice sheets and glaciers, known as glacial cycles. According to the now widely-accepted Milankovitch theory, the main mechanism at work is due to the periodic changes in the parameters of Earth's orbit around the sun (eccentricity, obliquity, precession), which affect the seasonal and latitudinal distribution of the solar insolation on Earth and, together with the effects of multiple positive and negative feedbacks (atmospheric CO_2 content, albedo, ice sheet dynamics etc.), govern the glacial cycles (e.g., Wilson et al. 2000). Until about 1 million years ago, their main period was 41,000 years (obliquity cycle), whereas thereafter the 100,000-year period (eccentricity cycle) prevailed.

About 21,000 years ago, at the Last Glacial Maximum, ice sheets covered large parts of North America, Greenland, the European Alps, northern Europe including Scandinavia and Britain, north-western Eurasia, Patagonia and Antarctica. Also, there were glaciers in the equatorial Andes, on Mauna Kea (Hawaii), in New Zealand and Tasmania, on several mountains in east and central Africa and in the Atlas Mountains. Owing to the additional water stored in these ice masses, the sea level was about 120-135 m lower than today, so that Great Britain was a part of continental Europe, the present Bering Strait was a land bridge between East Siberia and Alaska, and the northern Japanese island of Hokkaido was connected to Russian Sakhalin. After that, the ice retreated gradually, and at around 10,000 years ago the last ice age ended, marking the transition to the Holocene Epoch with its current, interglacial ice cover.

1.4 Ice Sheets, Glaciers and Global Warming

What will be the fate of the present-day ice sheets, ice caps and glaciers in a warming climate during the next decades and centuries? First of all, it is important to note that the smaller an ice body is, the faster it can respond to a change in the climatic conditions (surface temperature, precipitation). Therefore, the smaller glaciers and ice caps are much more vulnerable to global warming than the large ice sheets of Antarctica and Greenland. On the other hand, let us recall that the potential for sea level rise of the glaciers and ice caps is limited to 0.5 m, whereas that of the ice sheets is almost 70 m.

According to the Fourth Assessment Report of the Intergovernmental Panel on Climate Change (IPCC), Contribution of Working Group I, Chap. 5 (Bindoff et al. 2007), for the periods 1961-2003 and 1993-2003, global sea level rises of 1.8 ± 0.5 and 3.1 ± 0.7 mm a^{-1}, respectively, have been observed (the symbol "a" represents 1 year $\approx$ 31,556,926 s). Estimates of the various contributions suggest that, for the decade 1993-2003, 1.6 ± 0.5 mm a^{-1} can be attributed to ocean thermal expansion, 0.77 ± 0.22 mm a^{-1} to the melting of glaciers and ice caps, 0.21 ± 0.07 mm a^{-1} to changes of the Greenland Ice Sheet and 0.21 ± 0.35 mm a^{-1} to adjustments of the Antarctic Ice Sheet. Evidently, the largest contribution among the different types of land ice is from the small ice bodies, which becomes manifest in a significant trend towards retreat of glaciers all over the world (see Fig. 1.3 for an example). By contrast, the contribution of recent adjustments of the Antarctic Ice Sheet may even be negative. This surprising finding is due to the fact that the extremely low temperatures over Antarctica do not allow for significant increases in sur-

Fig. 1.3. Recession of McCarty Glacier in Kenai Fjords National Park, Alaska, 1909 – 2004. The glacier retreated approximately 20 km during this period and is not visible anymore in the 2004 photo. (Sources: 1909 photo by Ulysses Sherman Grant, U. S. Geological Survey Photo Library, public domain. 2004 photo by Bruce F. Molnia, U. S. Geological Survey, public domain.)

face melting, whereas increased precipitation rates as a consequence of global warming deposit more snow on the Antarctic Ice Sheet.

IPCC (Fourth Assessment Report 2007, Contribution of Working Group I, Chap. 10; Meehl et al. 2007) projections for climate change in the 21st century (more precisely, for the 2090-2099 average relative to the 1980-1999 average) give an increase of the globally averaged surface temperature in the range of 1.1-6.4°C, and a global-average sea level rise in the range of 0.18-0.59 m. These uncertainties are partly due to the assumption of a variety of greenhouse gas emission scenarios and partly due to model uncertainties themselves. The contribution to sea level rise from the glaciers and ice caps is estimated to be in the range of 0.07 to 0.17 m, the contribution from changes of the surface mass balance of the Greenland Ice Sheet as 0.01 to 0.12 m and that from changes of the surface mass balance of the Antarctic Ice Sheet as -0.14 to -0.02 m. Further contributions are due to ocean thermal expansion, ice sheet dynamics, thawing of permafrost and anthropogenic change in terrestrial water storage. Again, the largest contribution among the different types of land ice will likely be from the small ice bodies. Note that the upper end of the range for glaciers and ice caps (0.17 m) is approximately one third of their entire sea level rise equivalent (0.5 m). This illustrates the large vulnerability especially of small glaciers, many of which will probably have vanished by the end of the 21st century. In contrast, the large ice sheets are much more inert, and the positive contribution of Greenland (due to increased melting and runoff) is expected to be more or less compensated by the negative contribution of Antarctica (due to increased precipitation).

On a longer term, if global warming continues, glacier and ice cap retreat will go on, and the loss of a substantial fraction of their mass is likely within a few centuries. Also, the Greenland Ice Sheet will finally suffer a significant decay. For instance, Ridley et al. (2005) found that under climatic conditions resulting from a constant atmospheric CO_2 concentration four times the preindustrial level, after 1000 years only about 40% of the original ice volume remains. The impact on the Antarctic Ice Sheet as a whole will likely be limited; however, melting and runoff will finally outweigh increased precipitation, and due to its huge volume, a contribution to sea level rise of a few metres within 1000 years is possible.

Poorly understood internal ice flow dynamics make these predictions to some extent uncertain. For Antarctica, disintegration of attached ice shelves and accelerating ice streams and outlet glaciers may lead to an acceleration of the coastward mass flux of the ice sheet and therefore destabilise it. This possibility has been discussed, in particular, for the smaller part of the ice sheet in the western hemisphere (West Antarctic Ice Sheet, sea level rise equivalent of 6 m). For Greenland, surface meltwater percolating to the base may act as a lubricant on which parts of the ice sheet can glide off into the ocean. Firm predictions whether such ice-dynamic instabilities are likely for the next centuries are not possible at present due to inadequate understanding of the related processes.

Since the cryosphere is an integral part of the climate system, changes of its state will inevitably feed back on other subsystems. While for the smaller glaciers and ice caps such feedbacks are limited to local effects due to changes in albedo and hydrology, ice sheet decay can affect the climate on a global scale. For the 21st century, the greatest foreseeable problem is the increased freshwater discharge into the North Atlantic from the melting Greenland Ice Sheet. Together with increased precipitation rates, this meltwater reduces the salinity and density of the surface water in the North Atlantic and therefore hampers the formation of North Atlantic Deep Water (NADW). Since NADW plays a vital role in driving the North Atlantic drift (also known as Gulf Stream), this warm surface current may experience a weakening or even a complete shutdown, with severe consequences for the climate in Europe and the whole pattern of heat distribution by the Global Conveyor Belt (e.g., Alley 2000). On longer time-scales, albedo changes due to exposed ice-free land in Greenland feed back positively on surface temperatures, which can lead to an accelerated, irreversible disintegration of the ice sheet. Also, major orographic changes of the Greenland Ice Sheet disturb the atmospheric circulation by altering the stationary Rossby wave pattern. This process may entail a complex pattern of regional climate change in the Arctic and sub-Arctic areas, which is difficult, at this point, to assess in detail.

2

Vectors, Tensors and Their Representation

2.1 Definition of a Vector, Basic Properties

In mathematics, a *vector* is defined as an element of a vector space, and a vector space is a commutative (Abelian) group with a scalar multiplication. This is an abstract definition which has many possible realisations (numbers, functions, geometric objects and so on). For our purposes, it is sufficient to consider one of them, namely the geometric object of an *arrow* in the three-dimensional, Euclidian, physical space $\mathcal{E}$. Therefore, in our sense a vector $\mathbf{a} \in \mathcal{E}$ is an arrow which is characterised by a length and a direction. Physical quantities which can be described by such vectors are, for instance, velocity, acceleration, momentum and force. By contrast, *scalars* are simple numbers and characterise physical quantities without a direction, like mass, density, temperature etc.

We will usually denote vectors by bold-face symbols like $\mathbf{a}$, $\mathbf{b}$, $\mathbf{c}$, etc. The sum

$$\mathbf{s} = \mathbf{a} + \mathbf{b} \tag{2.1}$$

of two vectors is obtained by the parallelogram construction, and the scalar multiplication

$$\mathbf{p} = \lambda\mathbf{a}, \quad \lambda \in \mathbb{R} \tag{2.2}$$

($\mathbb{R}$ denotes the set of real numbers) is a vector parallel to $\mathbf{a}$ with length $\lambda|\mathbf{a}|$, where $|\mathbf{a}|$ is the length (absolute value, norm) of $\mathbf{a}$ (Fig. 2.1).

The direction of a vector can be characterised by the unit vector (length equal to one) $\mathbf{e}_a = \mathbf{a}/|\mathbf{a}|$. Further, the dot product (inner product)

$$\delta = \mathbf{a} \cdot \mathbf{b} \tag{2.3}$$

of two vectors is equal to the scalar given by $|\mathbf{a}|\,|\mathbf{b}|\cos\varphi$ (where φ is the angle between the two vectors), and the cross product (vector product)

$$\mathbf{c} = \mathbf{a} \times \mathbf{b} \tag{2.4}$$

R. Greve, H. Blatter, *Dynamics of Ice Sheets and Glaciers*, Advances in Geophysical and Environmental Mechanics and Mathematics,
DOI 10.1007/978-3-642-03415-2_2, © Springer-Verlag Berlin Heidelberg 2009

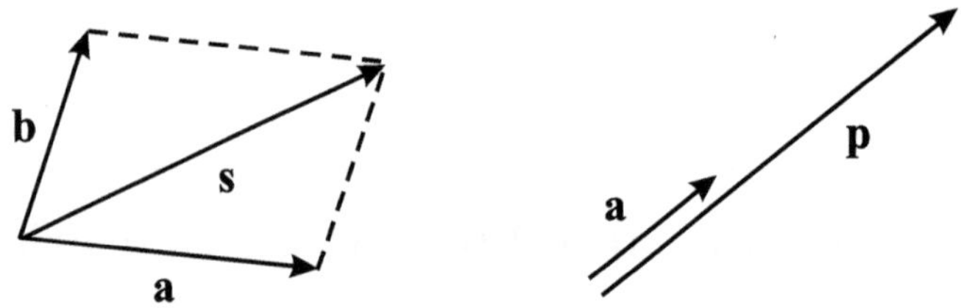

Fig. 2.1. Sum $\mathbf{s} = \mathbf{a} + \mathbf{b}$ and scalar multiplication $\mathbf{p} = \lambda\mathbf{a}$ of vectors.

is equal to the vector with length $|\mathbf{a}|\,|\mathbf{b}|\sin\varphi$ and direction perpendicular to the plane spanned by $\mathbf{a}$ and $\mathbf{b}$, such that $\mathbf{a}$, $\mathbf{b}$ and $\mathbf{c}$ form a right-handed system (Fig. 2.2). Note that

$$|\mathbf{a}| = \sqrt{\mathbf{a}\cdot\mathbf{a}}\,, \tag{2.5}$$

$$\mathbf{a}\cdot\mathbf{b} = 0 \quad \Leftrightarrow \quad \mathbf{a}\perp\mathbf{b} \;\text{ or }\; \mathbf{a} = \mathbf{0} \;\text{ or }\; \mathbf{b} = \mathbf{0}\,, \tag{2.6}$$

$$\mathbf{a}\times\mathbf{b} = -\mathbf{b}\times\mathbf{a} \tag{2.7}$$

and

$$\mathbf{a}\times\mathbf{a} = \mathbf{0}\,, \tag{2.8}$$

where $\mathbf{0}$ denotes the vector of length zero ("zero vector").

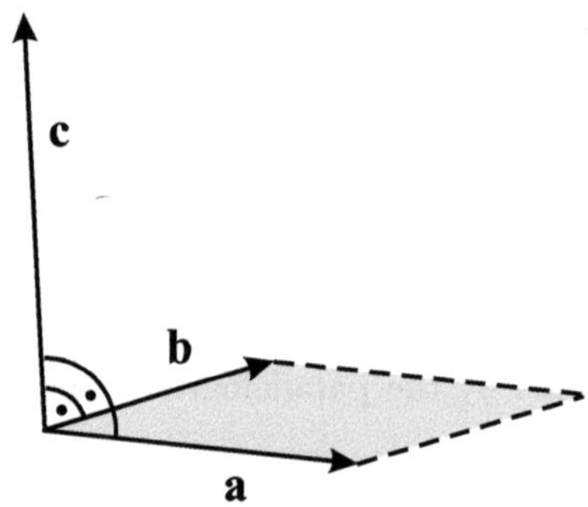

Fig. 2.2. Cross product $\mathbf{c} = \mathbf{a}\times\mathbf{b}$ of vectors.

Finally, the dyadic, outer or tensor product $\mathbf{a}\,\mathbf{b}$ (sometimes denoted as $\mathbf{a}\otimes\mathbf{b}$) is the linear transformation which, when applied to an arbitrary vector $\mathbf{x}$, obeys the relation

$$(\mathbf{a}\,\mathbf{b})\cdot\mathbf{x} = \mathbf{a}(\mathbf{b}\cdot\mathbf{x})\,, \tag{2.9}$$

where $(\mathbf{b}\cdot\mathbf{x})$ means the dot product (2.3). In other words, the transformation $\mathbf{a}\,\mathbf{b}$ maps the vector $\mathbf{x}$ on the vector which has the direction of $\mathbf{a}$ and length $(\mathbf{b}\cdot\mathbf{x})\,|\mathbf{a}|$.

2.2 Representation of Vectors as Number Triples

Let $\{\mathbf{e}_i\}_{i=1,2,3}$ be a set of unit vectors which are perpendicular to each other and form a right-handed system. In other words,

$$\mathbf{e}_i \cdot \mathbf{e}_j = \delta_{ij} , \tag{2.10}$$

where δ_{ij} is the Kronecker symbol defined as

$$\delta_{ij} = \begin{cases} 1 , & \text{for } i = j , \\ 0 , & \text{for } i \neq j , \end{cases} \tag{2.11}$$

and

$$\mathbf{e}_i \times \mathbf{e}_j = \mathbf{e}_k , \quad (i,j,k) \in \{(1,2,3),(2,3,1),(3,1,2)\} . \tag{2.12}$$

We will refer to such a set $\{\mathbf{e}_i\}$ as an *orthonormal basis* (also *Cartesian basis*). An arbitrary vector $\mathbf{a}$ can then be uniquely written as

$$\mathbf{a} = a_1\mathbf{e}_1 + a_2\mathbf{e}_2 + a_3\mathbf{e}_3 = \sum_{i=1}^{3} a_i\mathbf{e}_i , \tag{2.13}$$

where the a_i are real numbers. With Einstein's summation convention, which says that double indices (here i) automatically imply summation, this can be written in compact form as

$$\mathbf{a} = a_i\mathbf{e}_i . \tag{2.14}$$

Since the coefficients a_i are unique for a given basis $\{\mathbf{e}_i\}$, it is possible to represent the vector $\mathbf{a}$ by these coefficients. It is usual to arrange them in a column (number triple) and write

$$\mathbf{a}_{\{\mathbf{e}_i\}} = \begin{pmatrix} a_1 \\ a_2 \\ a_3 \end{pmatrix} , \tag{2.15}$$

which is to say, the vector $\mathbf{a}$ is represented by the components a_i with respect to the basis $\{\mathbf{e}_i\}$. Of course, when a different orthonormal basis $\{\mathbf{e}_i^\star\}$ is used, the representation of the vector $\mathbf{a}$ will change:

$$\mathbf{a} = a_i^\star\mathbf{e}_i^\star , \tag{2.16}$$

or

$$\mathbf{a}_{\{\mathbf{e}_i^\star\}} = \begin{pmatrix} a_1^\star \\ a_2^\star \\ a_3^\star \end{pmatrix} . \tag{2.17}$$

Note that the vector $\mathbf{a}$ is still the same object (arrow in space), whereas its components have changed. It is therefore of great importance to distinguish between vectors themselves and their representation as number triples. Mixing up these two different things is a notorious source of confusion. Only when a single basis $\{\mathbf{e}_i\}$ is defined from the outset, is $\mathbf{a}$ uniquely expressed by

$$\mathbf{a} = \begin{pmatrix} a_1 \\ a_2 \\ a_3 \end{pmatrix} . \tag{2.18}$$

In components with respect to a given basis $\{\mathbf{e}_i\}$, the dot product (2.3) can be evaluated as

$$\mathbf{a} \cdot \mathbf{b} = a_i b_i \, , \tag{2.19}$$

and the ith component of the cross product (2.4) is

$$(\mathbf{a} \times \mathbf{b})_i = \varepsilon_{ijk} \, a_j b_k \tag{2.20}$$

(summation over j and k). In the latter expression, ε_{ijk} is called the Levi-Civita symbol or alternator, defined as

$$\varepsilon_{ijk} = \begin{cases} 1 \, , & \text{for } (i,j,k) \in \{(1,2,3),\, (2,3,1),\, (3,1,2)\} \, , \\ -1 \, , & \text{for } (i,j,k) \in \{(1,3,2),\, (3,2,1),\, (2,1,3)\} \, , \\ 0 \, , & \text{otherwise (at least two indices are equal)} \, . \end{cases} \tag{2.21}$$

The dyadic product defined in Eq. (2.9) is expressed as

$$\mathbf{a}\,\mathbf{b} = (a_i \mathbf{e}_i)\,(b_j \mathbf{e}_j) = a_i b_j \, \mathbf{e}_i \, \mathbf{e}_j \, , \tag{2.22}$$

(summation over i and j), where $\mathbf{e}_i \, \mathbf{e}_j$ is the dyadic product of the respective basis vectors.

2.3 Tensors of Order 2

A *tensor* A of order 2 (often simply called a tensor) is defined as a linear transformation which maps vectors on vectors:

$$\mathbf{y} = \mathsf{A} \cdot \mathbf{x} \, . \tag{2.23}$$

Tensors will generally be denoted by sans-serif symbols like A, B, C, etc. We have already encountered special tensors of order 2, namely the dyadic products between two vectors introduced in Eq. (2.9). Their expression with respect to an orthonormal basis $\{\mathbf{e}_i\}$ was given by Eq. (2.22), and similarly a general tensor of order 2 can be written as

$$\mathsf{A} = A_{ij} \, \mathbf{e}_i \, \mathbf{e}_j \, . \tag{2.24}$$

Evidently, the tensor A is represented by the components A_{ij}, and analogous to Eq. (2.15) this can be denoted by

$$\mathsf{A}_{\{\mathbf{e}_i\}} = \begin{pmatrix} A_{11} & A_{12} & A_{13} \\ A_{21} & A_{22} & A_{23} \\ A_{31} & A_{32} & A_{33} \end{pmatrix} , \tag{2.25}$$

where the components have been arranged into a square matrix. Again, if a different basis $\{\mathbf{e}_i^\star\}$ is used, the representation will change,

$$\mathsf{A}_{\{\mathbf{e}_i^\star\}} = \begin{pmatrix} A_{11}^\star & A_{12}^\star & A_{13}^\star \\ A_{21}^\star & A_{22}^\star & A_{23}^\star \\ A_{31}^\star & A_{32}^\star & A_{33}^\star \end{pmatrix} , \tag{2.26}$$

so that tensors and matrices must be distinguished in the same way as vectors and number triples. Only when a single, fixed basis $\{\mathbf{e}_i\}$ is used, is A uniquely expressed by the square matrix

$$\mathsf{A} = \begin{pmatrix} A_{11} & A_{12} & A_{13} \\ A_{21} & A_{22} & A_{23} \\ A_{31} & A_{32} & A_{33} \end{pmatrix} . \tag{2.27}$$

With the representation of Eq. (2.24), the linear transformation (2.23) is given by

$$\begin{aligned} \mathbf{y} &= (A_{ij}\,\mathbf{e}_i\,\mathbf{e}_j) \cdot (x_k\,\mathbf{e}_k) = A_{ij}x_k\,\mathbf{e}_i\,(\mathbf{e}_j \cdot \mathbf{e}_k) \\ &= A_{ij}x_k\,\mathbf{e}_i\,\delta_{jk} = A_{ij}x_j\,\mathbf{e}_i\,, \end{aligned} \tag{2.28}$$

or

$$y_i = A_{ij}x_j\,. \tag{2.29}$$

Evidently, this is nothing else but the matrix-column product

$$\begin{pmatrix} y_1 \\ y_2 \\ y_3 \end{pmatrix} = \begin{pmatrix} A_{11} & A_{12} & A_{13} \\ A_{21} & A_{22} & A_{23} \\ A_{31} & A_{32} & A_{33} \end{pmatrix} \cdot \begin{pmatrix} x_1 \\ x_2 \\ x_3 \end{pmatrix} \tag{2.30}$$

expressed in (Cartesian) *index notation.* Index notation is a very efficient method of carrying out computations in vector/tensor algebra and analysis [see also the expressions (2.19) and (2.20) for the dot product and the cross product, respectively], and we will use it frequently.

The transpose of a tensor A is the unique tensor A^{T} defined by

$$\mathsf{A}^{\mathrm{T}} = A_{ji}\,\mathbf{e}_i\,\mathbf{e}_j\,. \tag{2.31}$$

In matrix form, this reads

$$\mathsf{A}^{\mathrm{T}}_{\{\mathbf{e}_i\}} = \begin{pmatrix} A_{11} & A_{21} & A_{31} \\ A_{12} & A_{22} & A_{32} \\ A_{13} & A_{23} & A_{33} \end{pmatrix} ; \tag{2.32}$$

that is, the elements of the original matrix $\mathsf{A}_{\{\mathbf{e}_i\}}$ [Eq. (2.25)] have been mirrored along the main diagonal A_{11}-A_{22}-A_{33}. In index notation, this is expressed by the relation

$$A^{\mathrm{T}}_{ij} = A_{ji} \,. \tag{2.33}$$

A symmetric tensor is defined by

$$\mathsf{A}^{\mathrm{T}} = \mathsf{A} \qquad (A_{ij} = A_{ji})\,, \tag{2.34}$$

whereas for an antisymmetric tensor

$$\mathsf{A}^{\mathrm{T}} = -\mathsf{A} \qquad (A_{ij} = -A_{ji}) \tag{2.35}$$

holds. The latter case implies that all main-diagonal elements (A_{11}, A_{22}, A_{33}) are equal to zero.

An important example for a tensor of order 2 is the unit tensor I, which provides the identity transformation $\mathbf{x} = \mathsf{I}\cdot\mathbf{x}$. Its components in *any* orthonormal basis $\{\mathbf{e}_i\}$ are given by the Kronecker symbol δ_{ij}, that is,

$$\mathsf{I} = \delta_{ij}\,\mathbf{e}_i\,\mathbf{e}_j\,, \tag{2.36}$$

so that its matrix representation is given by the unit matrix,

$$\mathsf{I}_{\{\mathbf{e}_i\}} = \begin{pmatrix} 1 & 0 & 0 \\ 0 & 1 & 0 \\ 0 & 0 & 1 \end{pmatrix} \,. \tag{2.37}$$

The multiplication of two tensors A and B yields the tensor $\mathsf{A}\cdot\mathsf{B}$ defined by successive application of first B and then A on an arbitrary vector $\mathbf{x}$,

$$(\mathsf{A}\cdot\mathsf{B})\cdot\mathbf{x} = \mathsf{A}\cdot(\mathsf{B}\cdot\mathbf{x})\,. \tag{2.38}$$

By expressing the two tensors according to Eq. (2.24), we find

$$\begin{aligned} \mathsf{A}\cdot(\mathsf{B}\cdot\mathbf{x}) &= (A_{ij}\,\mathbf{e}_i\,\mathbf{e}_j)\cdot[(B_{kl}\,\mathbf{e}_k\,\mathbf{e}_l)\cdot(x_m\,\mathbf{e}_m)] \\ &= (A_{ij}\,\mathbf{e}_i\,\mathbf{e}_j)\cdot(B_{kl}x_l\,\mathbf{e}_k) \\ &= A_{ij}B_{kl}x_l\,\mathbf{e}_i\,(\mathbf{e}_j\cdot\mathbf{e}_k) \\ &= A_{ij}B_{kl}x_l\,\mathbf{e}_i\,\delta_{jk} \\ &= A_{ij}B_{jl}x_l\,\mathbf{e}_i \end{aligned}$$

$$\Rightarrow\ \mathsf{A}\cdot\mathsf{B} = A_{ij}B_{jl}\,\mathbf{e}_i\,\mathbf{e}_l\,, \quad \text{or} \quad (\mathsf{A}\cdot\mathsf{B})_{il} = A_{ij}B_{jl}\,. \tag{2.39}$$

This represents the familiar matrix multiplication

$$(\mathsf{A}\cdot\mathsf{B})_{\{\mathbf{e}_i\}} = \begin{pmatrix} A_{11} & A_{12} & A_{13} \\ A_{21} & A_{22} & A_{23} \\ A_{31} & A_{32} & A_{33} \end{pmatrix} \cdot \begin{pmatrix} B_{11} & B_{12} & B_{13} \\ B_{21} & B_{22} & B_{23} \\ B_{31} & B_{32} & B_{33} \end{pmatrix} \,. \tag{2.40}$$

The above rules can be extended to the multiplication of more than two tensors (e.g., $\mathsf{A} \cdot \mathsf{B} \cdot \mathsf{C}$), and powers of a tensor A are defined by

$$\mathsf{A}^2 = \mathsf{A} \cdot \mathsf{A} \,, \quad \mathsf{A}^3 = \mathsf{A} \cdot \mathsf{A} \cdot \mathsf{A} \,, \quad \text{etc.} \tag{2.41}$$

Note that the multiplication of tensors is associative, but in general not commutative.

An order 2 tensor A in three-dimensional space has three independent scalar invariants. If A is represented by the matrix $\mathsf{A}_{\{\mathbf{e}_i\}}$ [Eq. (2.25)], the invariants are

$$\begin{aligned} I_{\mathsf{A}} &= \operatorname{tr} \mathsf{A} = A_{11} + A_{22} + A_{33} \,, \\ II_{\mathsf{A}} &= \tfrac{1}{2} \left[\operatorname{tr} (\mathsf{A}^2) - (\operatorname{tr} \mathsf{A})^2 \right] \\ &= A_{12}A_{21} + A_{13}A_{31} + A_{23}A_{32} - A_{11}A_{22} - A_{11}A_{33} - A_{22}A_{33} \,, \\ III_{\mathsf{A}} &= \det \mathsf{A} = \tfrac{1}{6} \left[2 \operatorname{tr} (\mathsf{A}^3) - 3 \operatorname{tr} \mathsf{A} \operatorname{tr} (\mathsf{A}^2) + (\operatorname{tr} \mathsf{A})^3 \right] \\ &\qquad = A_{11}A_{22}A_{33} + A_{12}A_{23}A_{31} + A_{13}A_{21}A_{32} \\ &\qquad - A_{11}A_{23}A_{32} - A_{12}A_{21}A_{33} - A_{13}A_{22}A_{31} \,. \end{aligned} \tag{2.42}$$

The important point is that these scalar invariants do not depend on the particular basis; that is, the same values are obtained when they are computed with a different matrix representation $\mathsf{A}_{\{\mathbf{e}_i^\star\}}$. The first ($I_{\mathsf{A}}$) and third ($III_{\mathsf{A}}$) invariants are also called the trace and determinant of the tensor A, respectively. For the determinant, a useful expression in index notation is

$$\det \mathsf{A} = \tfrac{1}{6} \varepsilon_{ijk} \varepsilon_{lmn} A_{il} A_{jm} A_{kn} \,, \tag{2.43}$$

which involves a sixfold summation and the Levi-Civita symbol defined in Eq. (2.21).

The inverse A^{-1} of the tensor A is defined by the relations

$$\mathsf{A} \cdot \mathsf{A}^{-1} = \mathsf{A}^{-1} \cdot \mathsf{A} = \mathsf{I} \,. \tag{2.44}$$

It only exists if the determinant of A is not equal to zero:

$$\det \mathsf{A} \neq 0 \quad \Leftrightarrow \quad \exists \mathsf{A}^{-1} : \text{Eq. (2.44) fulfilled} \,. \tag{2.45}$$

When expressed in component form, a tensor of order 2 is a quantity with two indices [see Eq. (2.24)]. As we have seen in Sect. 2.2, the expression of a vector in component form leads to a quantity with one index (for instance, a_i), and a scalar quantity, of course, does not have any indices at all. Therefore, vectors and scalars are also referred to as tensors of order 1 and 0, respectively.

2.4 Higher Order Tensors

As a generalisation of Eq. (2.23), tensors $\mathsf{A}^{[r]}$ of order $r > 2$ can be defined inductively as linear transformations which map vectors $\mathbf{x}$ on tensors $\mathsf{Y}^{[r-1]}$ of order $r - 1$,

$$\mathsf{Y}^{[r-1]} = \mathsf{A}^{[r]} \cdot \mathbf{x} . \qquad (2.46)$$

Such tensors can be written in component form as

$$\mathsf{A}^{[r]} = A_{i_1 i_2 \ldots i_r} \, \mathbf{e}_{i_1} \, \mathbf{e}_{i_2} \ldots \mathbf{e}_{i_r} \qquad (2.47)$$

(summation over the r indices i_1, i_2, ..., i_r). As an example, the Levi-Civita symbol (2.21) can be interpreted as the components of an order 3 tensor $\epsilon^{[3]}$,

$$\epsilon^{[3]} = \varepsilon_{ijk} \, \mathbf{e}_i \, \mathbf{e}_j \, \mathbf{e}_k \, , \qquad (2.48)$$

which is known as the epsilon or permutation tensor. Tensors of order 4 play a role in the theories of elasticity and visco-elasticity.

The tensor multiplication introduced in Eq. (2.39) can be generalised to tensors of arbitrary orders $\mathsf{A}^{[r]}$ and $\mathsf{B}^{[s]}$, and the result is a tensor of order $r + s - 2$,

$$\begin{aligned} \mathsf{A}^{[r]} \cdot \mathsf{B}^{[s]} &= (A_{i_1 i_2 \ldots i_r} \, \mathbf{e}_{i_1} \, \mathbf{e}_{i_2} \ldots \mathbf{e}_{i_r}) \cdot (B_{j_1 j_2 \ldots j_s} \, \mathbf{e}_{j_1} \, \mathbf{e}_{j_2} \ldots \mathbf{e}_{j_s}) \\ &= A_{i_1 i_2 \ldots i_r} B_{j_1 j_2 \ldots j_s} \, \mathbf{e}_{i_1} \, \mathbf{e}_{i_2} \ldots \mathbf{e}_{i_{r-1}} \, (\mathbf{e}_{i_r} \cdot \mathbf{e}_{j_1}) \, \mathbf{e}_{j_2} \ldots \mathbf{e}_{j_s} \\ &= A_{i_1 i_2 \ldots i_r} B_{j_1 j_2 \ldots j_s} \delta_{i_r j_1} \, \mathbf{e}_{i_1} \, \mathbf{e}_{i_2} \ldots \mathbf{e}_{i_{r-1}} \, \mathbf{e}_{j_2} \ldots \mathbf{e}_{j_s} \\ &= A_{i_1 i_2 \ldots i_r} B_{i_r j_2 \ldots j_s} \, \mathbf{e}_{i_1} \, \mathbf{e}_{i_2} \ldots \mathbf{e}_{i_{r-1}} \, \mathbf{e}_{j_2} \ldots \mathbf{e}_{j_s} \, . \end{aligned} \qquad (2.49)$$

This operation is called tensor contraction. The dot product (2.3) between two vectors, the tensor-vector multiplications (2.23) and (2.46) as well as the tensor multiplication (2.39) can all be considered as special cases of the general tensor contraction (2.49).

Further, the dyadic product between two vectors [Eqs. (2.9), (2.22)] can be readily generalised to arbitrary tensors $\mathsf{A}^{[r]}$ and $\mathsf{B}^{[s]}$,

$$\begin{aligned} \mathsf{A}^{[r]} \, \mathsf{B}^{[s]} &= (A_{i_1 i_2 \ldots i_r} \, \mathbf{e}_{i_1} \, \mathbf{e}_{i_2} \ldots \mathbf{e}_{i_r}) \, (B_{j_1 j_2 \ldots j_s} \, \mathbf{e}_{j_1} \, \mathbf{e}_{j_2} \ldots \mathbf{e}_{j_s}) \\ &= A_{i_1 i_2 \ldots i_r} B_{j_1 j_2 \ldots j_s} \, \mathbf{e}_{i_1} \, \mathbf{e}_{i_2} \ldots \mathbf{e}_{i_r} \, \mathbf{e}_{j_1} \, \mathbf{e}_{j_2} \ldots \mathbf{e}_{j_s} \, , \end{aligned} \qquad (2.50)$$

which is called tensor product. The result is a tensor of order $r + s$.

2.5 Vector and Tensor Analysis

In physical applications like the continuum-mechanical modelling of ice sheets and glaciers, we are often concerned with scalar, vector or tensor *fields*, in which the respective quantities depend on space and time. Let us assume a fixed orthonormal basis $\{\mathbf{e}_1, \mathbf{e}_2, \mathbf{e}_3\}$ (or $\{\mathbf{e}_x, \mathbf{e}_y, \mathbf{e}_z\}$), then space can be described by the Cartesian coordinates x_1, x_2, x_3 (or x, y, z), and time is designated by the variable t.

Partial derivatives of a scalar field $\lambda(x_1, x_2, x_3, t)$ will be denoted by the alternative notations

$$\frac{\partial \lambda}{\partial t} = \lambda_{,t} \, , \qquad \frac{\partial \lambda}{\partial x_1} = \frac{\partial \lambda}{\partial x} = \lambda_{,1} = \lambda_{,x} \, , \qquad \text{etc.} \qquad (2.51)$$

For a vector field $\mathbf{a}(x_1, x_2, x_3, t)$,

$$\begin{aligned} \frac{\partial \mathbf{a}}{\partial t} = \mathbf{a}_{,t} &= \frac{\partial (a_i\,\mathbf{e}_i)}{\partial t} = \frac{\partial a_i}{\partial t}\,\mathbf{e}_i = a_{i,t}\,\mathbf{e}_i\,, \\ \frac{\partial \mathbf{a}}{\partial x_1} = \frac{\partial \mathbf{a}}{\partial x} = \mathbf{a}_{,1} = \mathbf{a}_{,x} &= \frac{\partial (a_i\,\mathbf{e}_i)}{\partial x_1} = \frac{\partial a_i}{\partial x_1}\,\mathbf{e}_i = a_{i,1}\,\mathbf{e}_i\,, \\ &\text{etc.}\,, \end{aligned} \tag{2.52}$$

and analogous for order 2 and higher order tensor fields.

The nabla operator ∇ is defined in terms of spatial partial derivative operators as

$$\nabla = \mathbf{e}_x \frac{\partial}{\partial x} + \mathbf{e}_y \frac{\partial}{\partial y} + \mathbf{e}_z \frac{\partial}{\partial z} = \mathbf{e}_i \frac{\partial}{\partial x_i}\,. \tag{2.53}$$

It is useful in order to introduce the gradient of a scalar field,

$$\operatorname{grad}\lambda = \nabla\lambda = \frac{\partial\lambda}{\partial x}\,\mathbf{e}_x + \frac{\partial\lambda}{\partial y}\,\mathbf{e}_y + \frac{\partial\lambda}{\partial z}\,\mathbf{e}_z = \frac{\partial\lambda}{\partial x_i}\,\mathbf{e}_i = \lambda_{,i}\,\mathbf{e}_i\,, \tag{2.54}$$

the divergence of a vector field,

$$\operatorname{div}\mathbf{a} = \nabla\cdot\mathbf{a} = \frac{\partial a_x}{\partial x} + \frac{\partial a_y}{\partial y} + \frac{\partial a_z}{\partial z} = \frac{\partial a_i}{\partial x_i} = a_{i,i}\,, \tag{2.55}$$

the curl of a vector field,

$$\begin{aligned} \operatorname{curl}\mathbf{a} &= \nabla\times\mathbf{a} \\ &= \left(\frac{\partial a_z}{\partial y} - \frac{\partial a_y}{\partial z}\right)\mathbf{e}_x + \left(\frac{\partial a_x}{\partial z} - \frac{\partial a_z}{\partial x}\right)\mathbf{e}_y + \left(\frac{\partial a_y}{\partial x} - \frac{\partial a_x}{\partial y}\right)\mathbf{e}_z \\ &= \varepsilon_{ijk}\,\frac{\partial a_k}{\partial x_j}\,\mathbf{e}_i = \varepsilon_{ijk}\,a_{k,j}\,\mathbf{e}_i\,, \end{aligned} \tag{2.56}$$

and the Laplacian of a scalar field,

$$\begin{aligned} \Delta\lambda &= \operatorname{div}\operatorname{grad}\lambda = \nabla^2\lambda \\ &= \frac{\partial^2\lambda}{\partial x^2} + \frac{\partial^2\lambda}{\partial y^2} + \frac{\partial^2\lambda}{\partial z^2} = \frac{\partial}{\partial x_i}\left(\frac{\partial\lambda}{\partial x_i}\right) = \lambda_{,ii}\,. \end{aligned} \tag{2.57}$$

These expressions can be generalised to tensor fields

$$\mathsf{A}^{[r]}(x_1, x_2, x_3, t) = A_{i_1 i_2 \ldots i_r}(x_1, x_2, x_3, t)\,\mathbf{e}_{i_1}\,\mathbf{e}_{i_2}\ldots\mathbf{e}_{i_r} \tag{2.58}$$

of order $r \geq 1$ as follows:

$$\operatorname{grad}\mathsf{A}^{[r]} = \frac{\partial A_{i_1 i_2 \ldots i_r}}{\partial x_j}\,\mathbf{e}_{i_1}\,\mathbf{e}_{i_2}\ldots\mathbf{e}_{i_r}\,\mathbf{e}_j$$

$$= A_{i_1 i_2 \ldots i_r, j}\, \mathbf{e}_{i_1}\, \mathbf{e}_{i_2} \ldots \mathbf{e}_{i_r}\, \mathbf{e}_j \qquad \text{(order } r+1\text{)}, \tag{2.59}$$

$$\operatorname{div} \mathsf{A}^{[r]} = \frac{\partial A_{i_1 i_2 \ldots i_{r-1} j}}{\partial x_j}\, \mathbf{e}_{i_1}\, \mathbf{e}_{i_2} \ldots \mathbf{e}_{i_{r-1}}$$
$$= A_{i_1 i_2 \ldots i_{r-1} j, j}\, \mathbf{e}_{i_1}\, \mathbf{e}_{i_2} \ldots \mathbf{e}_{i_{r-1}} \qquad \text{(order } r-1\text{)}, \tag{2.60}$$

$$\operatorname{curl} \mathsf{A}^{[r]} = \varepsilon_{ijk} \frac{\partial A_{i_1 i_2 \ldots i_{r-1} k}}{\partial x_j}\, \mathbf{e}_{i_1}\, \mathbf{e}_{i_2} \ldots \mathbf{e}_{i_{r-1}}\, \mathbf{e}_i$$
$$= \varepsilon_{ijk}\, A_{i_1 i_2 \ldots i_{r-1} k, j}\, \mathbf{e}_{i_1}\, \mathbf{e}_{i_2} \ldots \mathbf{e}_{i_{r-1}}\, \mathbf{e}_i \quad \text{(order } r\text{)}, \tag{2.61}$$

$$\Delta \mathsf{A}^{[r]} = \frac{\partial}{\partial x_j} \left(\frac{\partial A_{i_1 i_2 \ldots i_r}}{\partial x_j} \right) \mathbf{e}_{i_1}\, \mathbf{e}_{i_2} \ldots \mathbf{e}_{i_r}$$
$$= A_{i_1 i_2 \ldots i_r, jj}\, \mathbf{e}_{i_1}\, \mathbf{e}_{i_2} \ldots \mathbf{e}_{i_r} \qquad \text{(order } r\text{)}. \tag{2.62}$$

For the sake of simplicity, we refrain from giving the corresponding expressions for curvilinear coordinates like cylindrical coordinates, spherical coordinates etc., and refer the interested reader to the mathematical literature (e.g., Heinbockel 1996) instead.

Let us finally note two important integral theorems. The divergence theorem relates the integral of the divergence of a vector field $\mathbf{a}$ over a volume ω and the integral of the "flux" of $\mathbf{a}$ through the surface $\partial\omega$,

$$\int_\omega (\operatorname{div} \mathbf{a})\, \mathrm{d}v = \oint_{\partial\omega} \mathbf{a} \cdot \mathbf{n}\, \mathrm{d}a\,, \tag{2.63}$$

where $\mathrm{d}v$ is the volume increment, $\mathrm{d}a$ the surface increment and $\mathbf{n}$ the outer unit normal vector on $\partial\omega$. The curl theorem states that the integral of the curl of a vector field $\mathbf{a}$ over a surface σ equals the line integral of $\mathbf{a}$ over the curve $\partial\sigma$ bounding the surface,

$$\int_\sigma (\operatorname{curl} \mathbf{a}) \cdot \mathbf{n}\, \mathrm{d}a = \oint_{\partial\sigma} \mathbf{a} \cdot \mathrm{d}\mathbf{l}\,, \tag{2.64}$$

where $\mathrm{d}\mathbf{l}$ is the vectorial line increment along the curve $\partial\sigma$.

3

Elements of Continuum Mechanics

3.1 Bodies and Configurations

Continuum mechanics is concerned with the motion and deformation of continuous *bodies* (for instance, a glacier). A body consists of an infinite number of material elements, called *particles*. For any time t, each particle is identified by a position vector $\mathbf{x}$ (relative to a prescribed origin O) in the physical space $\mathcal{E}$, and the continuous set of position vectors for all particles of the body is called a *configuration* κ of the body. If t is the actual time, the corresponding configuration is called the *present configuration* κ_t. In addition, we define a *reference configuration* κ_r which refers to a fixed (or initial) time t_0. Position vectors in the reference configuration will be written in capitals, for example as $\mathbf{X}$; they can be used for identifying the individual particles of the body, independent of the actual time. Note that different sets of basis vectors ($\{\mathbf{E}_A\}_{A=1,2,3}$, $\{\mathbf{e}_i\}_{i=1,2,3}$) and different origins may be used in the two configurations (Fig. 3.1).

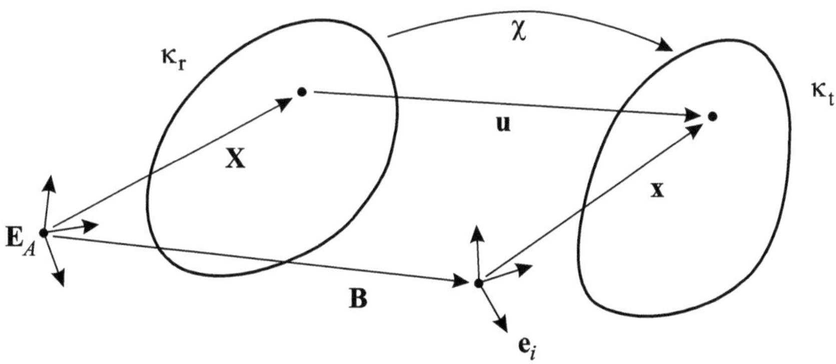

Fig. 3.1. Bodies, reference configuration κ_r and present configuration κ_t.

R. Greve, H. Blatter, *Dynamics of Ice Sheets and Glaciers*, Advances in Geophysical and Environmental Mechanics and Mathematics,
DOI 10.1007/978-3-642-03415-2_3,

The mapping $\boldsymbol{\chi}$ which provides the position $\mathbf{x}$ of each particle at time t as a function of its reference position $\mathbf{X}$ is called the *motion* of the body,

$$\begin{aligned} \boldsymbol{\chi}: \kappa_r &\rightarrow \kappa_t \\ \mathbf{X} &\rightarrow \mathbf{x} = \mathbf{x}(\mathbf{X}, t) \,. \end{aligned} \tag{3.1}$$

It is assumed that the motion $\mathbf{x}(\mathbf{X}, t)$ is continuously differentiable in the entire body (with the possible exception of discrete singular lines or surfaces), and that the inverse mapping $\boldsymbol{\chi}^{-1}$ exists:

$$\begin{aligned} \boldsymbol{\chi}^{-1}: \kappa_t &\rightarrow \kappa_r \\ \mathbf{x} &\rightarrow \mathbf{X} = \mathbf{X}(\mathbf{x}, t) \,. \end{aligned} \tag{3.2}$$

The *displacement* is defined as the connecting vector between a given particle in the reference and present configuration. If the connecting vector between the two origins of the basis systems is denoted by $\mathbf{B}$, then

$$\mathbf{u} = \mathbf{x} - \mathbf{X} + \mathbf{B} \tag{3.3}$$

holds. The above relations are illustrated in Fig. 3.1.

Of course, in a deformable body the displacement at time t will in general be different for different particles, so that it can be written as the vector field $\mathbf{u} = \mathbf{u}(\mathbf{X}, t)$. However, this is not the only possibility. Equation (3.2) shows that $\mathbf{X}$ can be expressed in terms of $\mathbf{x}$ and t, so that we can also assume the displacement field as a function of $\mathbf{x}$ and t, that is, $\mathbf{u} = \mathbf{u}(\mathbf{x}, t)$. These two possibilities also hold for other field quantities ψ (density, temperature, velocity etc.), and we call $\psi(\mathbf{X}, t)$ the *Lagrangian* or *material description*, whereas $\psi(\mathbf{x}, t)$ is referred to as the *Eulerian* or *spatial description*. Most frequently, for solid bodies the Lagrangian description is used, whereas the Eulerian description is more appropriate for problems of fluid dynamics (like glacier flow).

3.2 Kinematics

3.2.1 Deformation Gradient, Stretch Tensors

The *deformation gradient* F is defined as the material gradient (gradient with respect to $\mathbf{X}$) of the motion (3.1),

$$\mathsf{F} = \operatorname{Grad} \mathbf{x}(\mathbf{X}, t) \,, \tag{3.4}$$

or in components

$$F_{iA} = \frac{\partial x_i(\mathbf{X}, t)}{\partial X_A} = x_{i,A} \,, \tag{3.5}$$

where $\mathsf{F} = F_{iA}\, \mathbf{e}_i\, \mathbf{E}_A$, the operator Grad $(\cdot)$ is the material gradient, and the notation $(\cdot)_{,A}$ means the partial derivative $\partial(\cdot)/\partial X_A$. Note that small indices

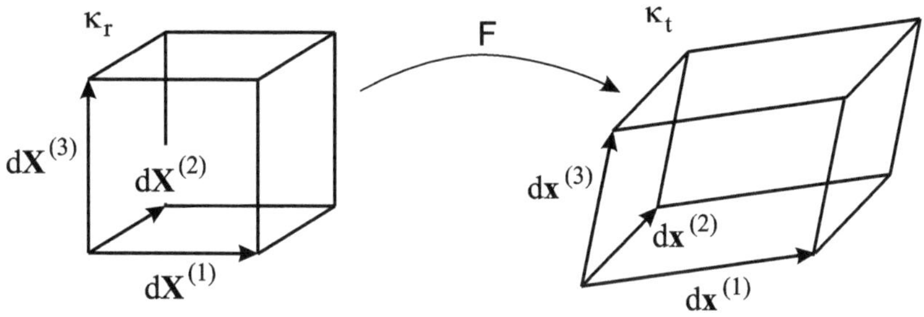

Fig. 3.2. Deformation gradient: Transformation between line and volume elements in the reference and present configuration.

generally refer to the present configuration and capital indices to the reference configuration. The deformation gradient is a tensor field of order 2.

According to definition (3.4), the deformation gradient F can be interpreted as the functional matrix of the motion function (3.1). It transforms line elements from the reference configuration ($\mathrm{d}\mathbf{X}$) to the present configuration ($\mathrm{d}\mathbf{x}$),

$$\mathrm{d}\mathbf{x} = \mathsf{F} \cdot \mathrm{d}\mathbf{X}\,, \quad \text{or} \quad \mathrm{d}x_i = F_{iA}\,\mathrm{d}X_A\,, \tag{3.6}$$

which is illustrated in Fig. 3.2.

The determinant of the deformation gradient, called the *Jacobian*, is given by

$$J = \det \mathsf{F}\,, \quad \text{or} \quad J = \tfrac{1}{6}\varepsilon_{ijk}\varepsilon_{ABC}F_{iA}F_{jB}F_{kC}\,; \tag{3.7}$$

for the component form on the right see Eq. (2.43). Since we have ensured that the motion function is invertible, J must be different from zero, and the inverse deformation gradient F^{-1} exists. Further, real motions cannot invert the orientation, so that

$$J > 0 \tag{3.8}$$

must hold. The Jacobian determines the local volume change due to the motion,

$$\mathrm{d}v = J\,\mathrm{d}V\,, \tag{3.9}$$

where $\mathrm{d}V$ is the volume element in the reference configuration which may be spanned by three line elements $\mathrm{d}\mathbf{X}^{(1)}$, $\mathrm{d}\mathbf{X}^{(2)}$, $\mathrm{d}\mathbf{X}^{(3)}$, and $\mathrm{d}v$ is the volume element in the present configuration spanned by $\mathrm{d}\mathbf{x}^{(1)}$, $\mathrm{d}\mathbf{x}^{(2)}$, $\mathrm{d}\mathbf{x}^{(3)}$ (Fig. 3.2).

The *theorem of polar decomposition* tells us that, like any tensor with positive determinant, the deformation gradient F can be uniquely decomposed according to

$$\mathsf{F} = \mathsf{R} \cdot \mathsf{U} = \mathsf{V} \cdot \mathsf{R}\,, \tag{3.10}$$

where R is a proper orthogonal tensor ($\mathsf{R} \cdot \mathsf{R}^{\mathrm{T}} = \mathsf{R}^{\mathrm{T}} \cdot \mathsf{R} = \mathsf{I}$ and $\det \mathsf{R} = +1$), and the tensors U and V are symmetric ($\mathsf{U} = \mathsf{U}^{\mathrm{T}}$, $\mathsf{V} = \mathsf{V}^{\mathrm{T}}$) and positive definite ($\forall \mathbf{x} \neq \mathbf{0}$: $\mathbf{x} \cdot \mathsf{U} \cdot \mathbf{x} > 0$, $\mathbf{x} \cdot \mathsf{V} \cdot \mathbf{x} > 0$) [see, e.g., Liu (2002), Hutter and Jöhnk (2004)]. The tensors U and V are called the *right* and *left stretch tensor*, respectively, and R is the *rotation tensor*.

The polar decomposition of F can be obtained as follows. From Eq. (3.10) we compute

$$\mathsf{F}^{\mathrm{T}} \cdot \mathsf{F} = \mathsf{U}^{\mathrm{T}} \cdot \mathsf{R}^{\mathrm{T}} \cdot \mathsf{R} \cdot \mathsf{U} = \mathsf{U} \cdot \mathsf{I} \cdot \mathsf{U} \quad \Rightarrow \quad \mathsf{U}^2 = \mathsf{F}^{\mathrm{T}} \cdot \mathsf{F} \tag{3.11}$$

and

$$\mathsf{F} \cdot \mathsf{F}^{\mathrm{T}} = \mathsf{V} \cdot \mathsf{R} \cdot \mathsf{R}^{\mathrm{T}} \cdot \mathsf{V}^{\mathrm{T}} = \mathsf{V} \cdot \mathsf{I} \cdot \mathsf{V} \quad \Rightarrow \quad \mathsf{V}^2 = \mathsf{F} \cdot \mathsf{F}^{\mathrm{T}}, \tag{3.12}$$

which determine the stretch tensors U and V in terms of F. The rotation tensor R follows then from Eq. (3.10) as

$$\mathsf{R} = \mathsf{F} \cdot \mathsf{U}^{-1}, \quad \text{or} \quad \mathsf{R} = \mathsf{V}^{-1} \cdot \mathsf{F}. \tag{3.13}$$

Note that Eq. (3.10) also implies the relation

$$\mathsf{V} = \mathsf{R} \cdot \mathsf{U} \cdot \mathsf{R}^{\mathrm{T}}, \tag{3.14}$$

which means that the two stretch tensors are connected by a similarity transformation (e.g., Jänich 1994).

The tensors $\mathsf{U}^2 = \mathsf{F}^{\mathrm{T}} \cdot \mathsf{F}$ and $\mathsf{V}^2 = \mathsf{F} \cdot \mathsf{F}^{\mathrm{T}}$ which appear in Eqs. (3.11) and (3.12) are referred to as the *right Cauchy Green tensor* C and *left Cauchy Green tensor* B, respectively. They are related by the same similarity transformation as the stretch tensors U and V,

$$\begin{aligned} \mathsf{B} = \mathsf{V}^2 &\overset{(3.14)}{=} \mathsf{R} \cdot \mathsf{U} \cdot \mathsf{R}^{\mathrm{T}} \cdot \mathsf{R} \cdot \mathsf{U} \cdot \mathsf{R}^{\mathrm{T}} \\ &= \mathsf{R} \cdot \mathsf{U} \cdot \mathsf{I} \cdot \mathsf{U} \cdot \mathsf{R}^{\mathrm{T}} \\ &= \mathsf{R} \cdot \mathsf{U}^2 \cdot \mathsf{R}^{\mathrm{T}} = \mathsf{R} \cdot \mathsf{C} \cdot \mathsf{R}^{\mathrm{T}}, \end{aligned} \tag{3.15}$$

and play a role in the description of solid bodies which undergo large deformations (such as rubber).

The polar decomposition of the deformation gradient F allows the interpretation of an arbitrary deformation as a sequence of a stretching followed by a local rigid body rotation, or vice versa. With Eq. (3.6) we can write

$$\mathrm{d}\mathbf{x} = \mathsf{R} \cdot \mathsf{U} \cdot \mathrm{d}\mathbf{X} = \mathsf{V} \cdot \mathsf{R} \cdot \mathrm{d}\mathbf{X}. \tag{3.16}$$

Since U is symmetric, there exists a special set of orthonormal basis vectors $\{\bar{\mathbf{e}}_i\}$, called the *principal axes*, for which the matrix $\mathsf{U}_{\{\bar{\mathbf{e}}_i\}}$ is diagonal, that is,

$$\mathsf{U}_{\{\bar{\mathbf{e}}_i\}} = \mathrm{diag}(\lambda_1, \lambda_2, \lambda_3) = \begin{pmatrix} \lambda_1 & 0 & 0 \\ 0 & \lambda_2 & 0 \\ 0 & 0 & \lambda_3 \end{pmatrix} \tag{3.17}$$

(e.g., Jänich 1994). The λ_i, $i = 1 \ldots 3$, are the eigenvalues of U, and due to the positive definiteness they are all positive. This holds also for V, and because of the similarity transformation (3.14) U and V have the same eigenvalues.

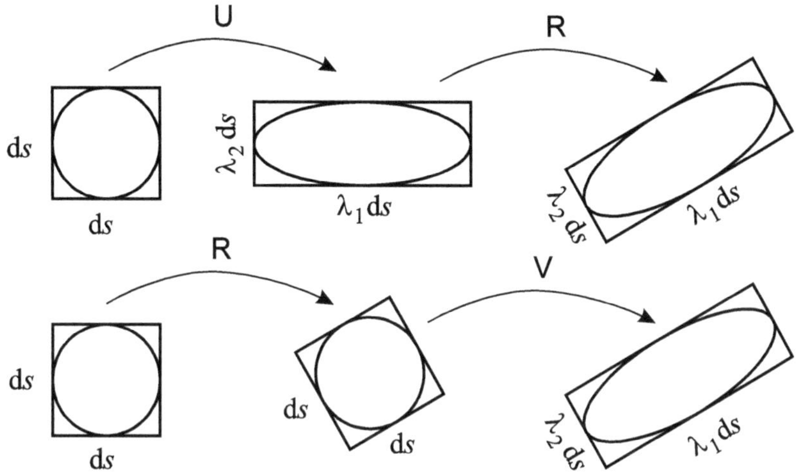

Fig. 3.3. Polar decomposition of the deformation gradient.

Let us now consider an infinitesimal cube (volume element) in the reference configuration, of which the edges $\mathrm{d}s$ are aligned with the principal axes of U (Fig. 3.3). The sequence of transformations $\mathrm{d}\mathbf{x} = \mathsf{R}\cdot\mathsf{U}\cdot\mathrm{d}\mathbf{X}$ stretches, in the first step, the edges of the cube by the factors λ_i (eigenvalues of U), which deforms the cube to a rectangular cuboid. In the second step, this element is rotated by the orthogonal transformation R. The alternative sequence $\mathrm{d}\mathbf{x} = \mathsf{V}\cdot\mathsf{R}\cdot\mathrm{d}\mathbf{X}$ leads to the same result, but here the initial cube is first rotated by R and then stretched by V.

Of course, since F, U, V and R are, in general, functions of $\mathbf{X}$ and t (or $\mathbf{x}$ and t), this decomposition is only local. In other words, a volume element at a different position will experience a different stretching and a different rotation.

3.2.2 Velocity, Acceleration, Velocity Gradient

As usual, we define the *velocity* $\mathbf{v}$ as the first time derivative of the position $\mathbf{x}$ [motion (3.1)],

$$\mathbf{v} = \mathbf{v}(\mathbf{X},t) = \frac{\partial\mathbf{x}(\mathbf{X},t)}{\partial t}\,, \tag{3.18}$$

and the *acceleration* $\mathbf{a}$ as its second time derivative,

$$\mathbf{a} = \mathbf{a}(\mathbf{X},t) = \frac{\partial^2\mathbf{x}(\mathbf{X},t)}{\partial t^2} = \frac{\partial\mathbf{v}(\mathbf{X},t)}{\partial t}\,. \tag{3.19}$$

Evidently, this yields the velocity and acceleration fields in Lagrangian description. By inserting the inverse motion (3.2) one can readily obtain the corresponding Eulerian descriptions $\mathbf{v}(\mathbf{x},t)$ and $\mathbf{a}(\mathbf{x},t)$.

The time derivatives in Eqs. (3.18) and (3.19) are taken for fixed material position vectors $\mathbf{X}$. We will call this the *material time derivative*, denoted briefly by the operators $\mathrm{d}(\bullet)/\mathrm{d}t$ or $(\bullet)^{\cdot}$. Therefore, for any field quantity ψ,

$$\dot{\psi} = \frac{\mathrm{d}\psi}{\mathrm{d}t} = \frac{\partial \psi(\mathbf{X},t)}{\partial t} \,. \tag{3.20}$$

For example, Eqs. (3.18) and (3.19) read

$$\mathbf{v} = \dot{\mathbf{x}} = \frac{\mathrm{d}\mathbf{x}}{\mathrm{d}t} \,, \qquad \mathbf{a} = \ddot{\mathbf{x}} = \frac{\mathrm{d}^2\mathbf{x}}{\mathrm{d}t^2} = \dot{\mathbf{v}} = \frac{\mathrm{d}\mathbf{v}}{\mathrm{d}t} \,. \tag{3.21}$$

By contrast, the operator $\partial(\bullet)/\partial t$ denotes the *local time derivative* for fixed spatial position vector $\mathbf{x}$,

$$\frac{\partial \psi}{\partial t} = \frac{\partial \psi(\mathbf{x},t)}{\partial t} \,. \tag{3.22}$$

With the chain rule, the relation between the material and local time derivative is

$$\begin{aligned} \frac{\mathrm{d}\psi}{\mathrm{d}t} &= \frac{\mathrm{d}}{\mathrm{d}t}\psi(\mathbf{x}(\mathbf{X},t),t) \\ &= \frac{\partial \psi(\mathbf{x},t)}{\partial t} + \operatorname{grad}\psi(\mathbf{x},t) \cdot \frac{\mathrm{d}\mathbf{x}(\mathbf{X},t)}{\mathrm{d}t} \\ &= \frac{\partial \psi}{\partial t} + (\operatorname{grad}\psi) \cdot \mathbf{v} \,, \end{aligned} \tag{3.23}$$

where the operator $\operatorname{grad}(\cdot)$ is the spatial gradient, of which the components are the partial derivatives $(\cdot)_{,i} = \partial(\cdot)/\partial x_i$. One says that the material time derivative $\mathrm{d}\psi/\mathrm{d}t$ is composed of a local part $\partial\psi/\partial t$ and an advective part $(\operatorname{grad}\psi) \cdot \mathbf{v}$.

Relation (3.23) is equally valid if ψ is a vector or tensor field. Therefore, for the acceleration expressed by Eq. $(3.21)_2$,

$$\mathbf{a} = \frac{\partial \mathbf{v}}{\partial t} + (\operatorname{grad}\mathbf{v}) \cdot \mathbf{v} \,. \tag{3.24}$$

The tensor quantity

$$\mathsf{L} = \operatorname{grad}\mathbf{v} = \frac{\partial v_i}{\partial x_j}\,\mathbf{e}_i\,\mathbf{e}_j = v_{i,j}\,\mathbf{e}_i\,\mathbf{e}_j \tag{3.25}$$

which appears in Eq. (3.24) is called the *velocity gradient.* It is related to the material time derivative of the deformation gradient as follows,

$$\begin{aligned} &\dot{F}_{iA} = \frac{\partial^2 x_i(\mathbf{X},t)}{\partial t\,\partial X_A} = \frac{\partial v_i(\mathbf{X},t)}{\partial X_A} = \frac{\partial v_i(\mathbf{x},t)}{\partial x_j}\,\frac{\partial x_j(\mathbf{X},t)}{\partial X_A} = \frac{\partial v_i(\mathbf{x},t)}{\partial x_j}\,F_{jA} \\ &\Rightarrow\ \dot{\mathsf{F}} = \mathsf{L}\cdot\mathsf{F}\,, \quad \text{or} \quad \mathsf{L} = \dot{\mathsf{F}}\cdot\mathsf{F}^{-1}\,. \end{aligned} \tag{3.26}$$

Without proof, we note that the material time derivative of the Jacobian is

$$\dot{J} = J \operatorname{div} \mathbf{v} = J \operatorname{tr} \mathsf{L} \tag{3.27}$$

[$\operatorname{div} \mathbf{v} = v_{i,i} = L_{ii} = \operatorname{tr} \mathsf{L}$; for the definition of the trace of a tensor see Eq. $(2.42)_1$]. In words, the divergence of the velocity field determines local volume changes [see also Eq. (3.9)], which is a very intuitive result.

Like any arbitrary tensor, the velocity gradient can be additively decomposed into unique symmetric and antisymmetric parts,

$$\mathsf{L} = \mathsf{D} + \mathsf{W}, \tag{3.28}$$

with

$$\begin{aligned} \mathsf{D} &= \tfrac{1}{2}(\mathsf{L} + \mathsf{L}^{\mathrm{T}}) \quad (\text{"strain-rate tensor" or "stretching tensor"}), \\ \mathsf{W} &= \tfrac{1}{2}(\mathsf{L} - \mathsf{L}^{\mathrm{T}}) \quad (\text{"spin tensor"}). \end{aligned} \tag{3.29}$$

Evidently, $\mathsf{D} = \mathsf{D}^{\mathrm{T}}$ (symmetry) and $\mathsf{W} = -\mathsf{W}^{\mathrm{T}}$ (antisymmetry) are fulfilled.

In order to give an interpretation of the elements of the matrix of the strain-rate tensor D (with respect to a given orthonormal basis $\{\mathbf{e}_i\}$), we now compute the material time derivative of line elements $\mathrm{d}\mathbf{x}$ in the present configuration:

$$(\mathrm{d}\mathbf{x})^{\cdot} = \dot{\mathsf{F}} \cdot \mathrm{d}\mathbf{X} = \mathsf{L} \cdot \mathsf{F} \cdot \mathrm{d}\mathbf{X} = \mathsf{L} \cdot \mathrm{d}\mathbf{x} \tag{3.30}$$

[where Eqs. (3.6), (3.26) and $(\mathrm{d}\mathbf{X})^{\cdot} = \mathbf{0}$ were used]. For the scalar product between two line elements $\mathrm{d}\mathbf{x}^{(1)}$ and $\mathrm{d}\mathbf{x}^{(2)}$, this yields

$$\begin{aligned} (\mathrm{d}\mathbf{x}^{(1)} \cdot \mathrm{d}\mathbf{x}^{(2)})^{\cdot} &= (\mathrm{d}\mathbf{x}^{(1)})^{\cdot} \cdot \mathrm{d}\mathbf{x}^{(2)} + \mathrm{d}\mathbf{x}^{(1)} \cdot (\mathrm{d}\mathbf{x}^{(2)})^{\cdot} \\ &= (\mathsf{L} \cdot \mathrm{d}\mathbf{x}^{(1)}) \cdot \mathrm{d}\mathbf{x}^{(2)} + \mathrm{d}\mathbf{x}^{(1)} \cdot (\mathsf{L} \cdot \mathrm{d}\mathbf{x}^{(2)}) \\ &= \mathrm{d}\mathbf{x}^{(1)} \cdot \mathsf{L}^{\mathrm{T}} \cdot \mathrm{d}\mathbf{x}^{(2)} + \mathrm{d}\mathbf{x}^{(1)} \cdot \mathsf{L} \cdot \mathrm{d}\mathbf{x}^{(2)} \\ &= 2\,\mathrm{d}\mathbf{x}^{(1)} \cdot \mathsf{D} \cdot \mathrm{d}\mathbf{x}^{(2)}. \end{aligned} \tag{3.31}$$

Let us assume

$$\begin{aligned} \mathrm{d}\mathbf{x}^{(1)} &= \mathbf{n}^{(1)} \mathrm{d}s^{(1)}, \\ \mathrm{d}\mathbf{x}^{(2)} &= \mathbf{n}^{(2)} \mathrm{d}s^{(2)}, \end{aligned} \qquad \mathbf{n}^{(1)} \cdot \mathbf{n}^{(2)} = \cos((\pi/2) - \gamma) = \sin\gamma, \tag{3.32}$$

where $\mathbf{n}^{(1)}$, $\mathbf{n}^{(2)}$ are unit vectors, and γ is the deviation of the angle between $\mathbf{n}^{(1)}$ and $\mathbf{n}^{(2)}$ from a right angle. Equation (3.31) then reads

$$\begin{aligned} &(\sin\gamma\, \mathrm{d}s^{(1)} \mathrm{d}s^{(2)})^{\cdot} = 2 \mathrm{d}s^{(1)} \mathrm{d}s^{(2)}\, \mathbf{n}^{(1)} \cdot \mathsf{D} \cdot \mathbf{n}^{(2)} \\ &\Rightarrow\ \dot{\gamma} \cos\gamma + \sin\gamma \left(\frac{(\mathrm{d}s^{(1)})^{\cdot}}{\mathrm{d}s^{(1)}} + \frac{(\mathrm{d}s^{(2)})^{\cdot}}{\mathrm{d}s^{(2)}} \right) = 2\mathbf{n}^{(1)} \cdot \mathsf{D} \cdot \mathbf{n}^{(2)}. \end{aligned} \tag{3.33}$$

We first make the special choice $\mathbf{n}^{(1)} = \mathbf{n}^{(2)} = \mathbf{e}_x$ ($\gamma = 90°$) and $\mathrm{d}s^{(1)} = \mathrm{d}s^{(2)} = \mathrm{d}s$ (Fig. 3.4, left). Then,

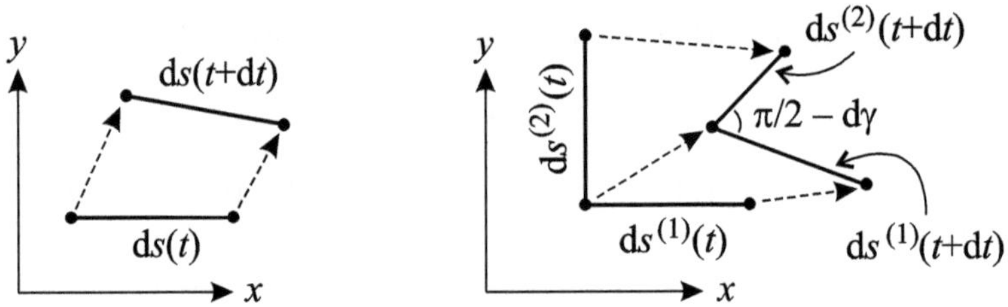

Fig. 3.4. Dilatation (left) and shear (right) of line elements in the present configuration.

$$2\frac{(\mathrm{d}s)^{\cdot}}{\mathrm{d}s} = 2\mathbf{e}_x \cdot \mathsf{D} \cdot \mathbf{e}_x \quad \Rightarrow \quad D_{xx} = \frac{(\mathrm{d}s)^{\cdot}}{\mathrm{d}s} . \tag{3.34}$$

Analogous results are found for the y- and z-directions. Therefore, the elements D_{xx}, D_{yy}, D_{zz} on the main diagonal of the matrix of D are equal to the *dilatation rates* in the x-, y- and z-direction, respectively.

Second, we choose $\mathbf{n}^{(1)} = \mathbf{e}_x$ and $\mathbf{n}^{(2)} = \mathbf{e}_y$, so that $\gamma = 0$ (Fig. 3.4, right). This yields

$$\dot{\gamma} = 2\mathbf{e}_x \cdot \mathsf{D} \cdot \mathbf{e}_y \quad \Rightarrow \quad D_{xy} = \frac{\dot{\gamma}}{2} . \tag{3.35}$$

Analogous relations can be obtained for the two other off-diagonal elements D_{xz} and D_{yz}. Therefore, $2D_{xy}$, $2D_{xz}$ and $2D_{yz}$ denote the *shear rates* $\dot{\gamma}_{xy}$, $\dot{\gamma}_{xz}$ and $\dot{\gamma}_{yz}$, that is, the temporal changes of right angles formed by the respective coordinate directions.

As for the spin tensor W, we note that its matrix has only three independent elements (this holds for any antisymmetric tensor). Without loss of generality, it can therefore be written as

$$\mathsf{W} = \begin{pmatrix} 0 & -w_3 & w_2 \\ w_3 & 0 & -w_1 \\ -w_2 & w_1 & 0 \end{pmatrix} \tag{3.36}$$

[see also Eq. (2.18) and the discussion there]. The w_i arranged in the above form are the components of the *dual vector*

$$\mathbf{w} = \operatorname{dual} \mathsf{W} = \begin{pmatrix} w_1 \\ w_2 \\ w_3 \end{pmatrix} , \tag{3.37}$$

with which the linear transformation $\mathsf{W} \cdot \mathbf{a}$ (arbitrary vector $\mathbf{a}$) can be expressed as a cross product,

$$\mathsf{W} \cdot \mathbf{a} = \mathbf{w} \times \mathbf{a} . \tag{3.38}$$

Thus, Eq. (3.30) yields

$$(\mathrm{d}\mathbf{x})^{\cdot} = \mathsf{D} \cdot \mathrm{d}\mathbf{x} + \mathbf{w} \times \mathrm{d}\mathbf{x} \,. \tag{3.39}$$

The first summand on the right-hand side describes the strain (deformation) part of the motion, the second summand the local rigid body rotation with angular velocity $\mathbf{w}$. This justifies the names "strain-rate tensor" and "spin tensor" for D and W, respectively.

3.3 Balance Equations

3.3.1 Reynolds' Transport Theorem

We consider a material volume $\omega \subset \kappa_t$ in the present configuration. "Material" means that the volume consists of the same particles for all times, $\partial\omega$ denotes the boundary of ω, $\mathbf{v}$ the velocity field of the body and $\mathbf{n}$ the unit normal vector on $\partial\omega$ (see Fig. 3.5). For an arbitrary scalar, vector or tensor field quantity $\psi(\mathbf{x}, t)$, we now calculate the term $(\mathrm{d}/\mathrm{d}t) \int_\omega \psi \, \mathrm{d}v$, that is, the temporal change of the field quantity integrated over the volume ω.

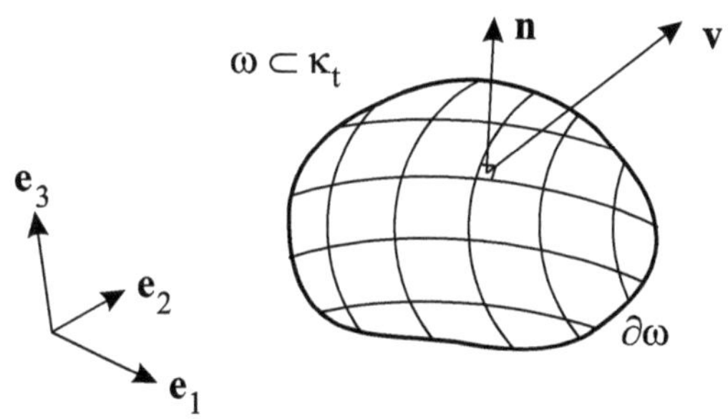

Fig. 3.5. On the Reynolds' transport theorem: Material volume ω with boundary $\partial\omega$ in the present configuration κ_t.

To this end, we transform the integration variable to material coordinates $\mathbf{X}$, which changes the integration domain ω to the volume $\Omega \subset \kappa_r$ in the reference configuration as

$$\frac{\mathrm{d}}{\mathrm{d}t} \int_\omega \psi(\mathbf{x}, t) \, \mathrm{d}v = \frac{\mathrm{d}}{\mathrm{d}t} \int_\Omega \psi(\mathbf{x}(\mathbf{X}, t), t) \, J(\mathbf{X}, t) \mathrm{d}V \,. \tag{3.40}$$

For the transformation of the volume element Eq. (3.9) was used. Since Ω, as a material volume in the reference configuration, is time-independent, differentiation and integration can be exchanged on the right-hand side, provided that the fields ψ, J and $\mathbf{v}$ are sufficiently smooth:

$$\begin{aligned}\frac{\mathrm{d}}{\mathrm{d}t}\int_\omega \psi(\mathbf{x},t)\,\mathrm{d}v &= \int_\Omega (\dot{\psi}J + \psi\dot{J})\,\mathrm{d}V \\ &= \int_\Omega (\dot{\psi} + \psi\,\mathrm{div}\,\mathbf{v})\,J\mathrm{d}V \\ &= \int_\omega (\dot{\psi} + \psi\,\mathrm{div}\,\mathbf{v})\,\mathrm{d}v\,. \end{aligned} \tag{3.41}$$

In the second step, Eq. (3.27) was used, and in the last step the integral was transformed back to spatial coordinates. The result can be further rewritten as

$$\begin{aligned}\frac{\mathrm{d}}{\mathrm{d}t}\int_\omega \psi(\mathbf{x},t)\,\mathrm{d}v &= \int_\omega \left(\frac{\partial\psi}{\partial t} + (\mathrm{grad}\,\psi)\cdot\mathbf{v} + \psi\,\mathrm{div}\,\mathbf{v}\right)\mathrm{d}v \\ &= \int_\omega \left(\frac{\partial\psi}{\partial t} + \mathrm{div}\,(\psi\mathbf{v})\right)\mathrm{d}v\,. \end{aligned} \tag{3.42}$$

By using the divergence theorem (2.63), we now obtain

$$\frac{\mathrm{d}}{\mathrm{d}t}\int_\omega \psi(\mathbf{x},t)\,\mathrm{d}v = \int_\omega \frac{\partial\psi}{\partial t}\,\mathrm{d}v + \oint_{\partial\omega} \psi\mathbf{v}\cdot\mathbf{n}\,\mathrm{d}a\,, \tag{3.43}$$

which is known as *Reynolds' transport theorem*. It says that the temporal change of the integral $\int_\omega \psi\,\mathrm{d}v$ over the material volume ω is composed of two parts, (i) the local change $\partial\psi/\partial t$ within ω, and (ii) the advective flux $\psi\mathbf{v}$ in the normal direction $\mathbf{n}$ across the boundary $\partial\omega$. Note that, if ψ is a tensor field of order $r \geq 1$, $\psi\mathbf{v}$ is a tensor product which yields a tensor of order $r+1$.

3.3.2 General Balance Equation

Let $\mathcal{G}(\omega,t)$ be a physical quantity of the entire material volume ω, which is supposed to be additive over subsets of ω (e.g., mass, momentum or internal energy, but not temperature or velocity). We assume that the change of $\mathcal{G}$ with time may be due to three different processes, namely

- the *flux* $\mathcal{F}(\partial\omega,t)$ of $\mathcal{G}$ across the boundary $\partial\omega$,
- the *production* $\mathcal{P}(\omega,t)$ of $\mathcal{G}$ within the volume ω,
- the *supply* $\mathcal{S}(\omega,t)$ of $\mathcal{G}$ within the volume ω.

Therefore, we can balance $\mathrm{d}\mathcal{G}/\mathrm{d}t$ as follows:

$$\frac{\mathrm{d}}{\mathrm{d}t}\mathcal{G}(\omega,t) = -\mathcal{F}(\partial\omega,t) + \mathcal{P}(\omega,t) + \mathcal{S}(\omega,t)\,, \tag{3.44}$$

where positive fluxes have been defined as outflows from the volume, so that the flux term has a negative sign.

The idea behind distinguishing between the mathematically equivalent quantities production and supply is that production is due to internal processes (within the volume ω) only, whereas supply has an external source. *Conserved quantities* are characterised by a vanishing production.

In order to reformulate the statement (3.44), we assume that $\mathcal{G}$, $\mathcal{P}$ and $\mathcal{S}$ can be expressed as volume integrals of corresponding *densities* g, p and s,

$$\begin{aligned}
\mathcal{G}(\omega,t) &= \textstyle\int_\omega g(\mathbf{x},t)\,\mathrm{d}v\,, \quad && g: \text{ density of the quantity } \mathcal{G}\,, \\
\mathcal{P}(\omega,t) &= \textstyle\int_\omega p(\mathbf{x},t)\,\mathrm{d}v\,, \quad && p: \text{ production density of } \mathcal{G}\,, \\
\mathcal{S}(\omega,t) &= \textstyle\int_\omega s(\mathbf{x},t)\,\mathrm{d}v\,, \quad && s: \text{ supply density of } \mathcal{G}\,,
\end{aligned} \tag{3.45}$$

and that $\mathcal{F}$ can be obtained as the surface integral of a *flux density* $\boldsymbol{\phi}$,

$$\mathcal{F}(\partial\omega,t) = \oint_{\partial\omega} \boldsymbol{\phi}(\mathbf{x},t)\cdot\mathbf{n}\,\mathrm{d}a\,, \tag{3.46}$$

where $\mathrm{d}a$ is the scalar surface element. Note that, if $\mathcal{G}$ is a tensor quantity of order $r \geq 0$ (scalar, vector etc.), then the order of g, p and s is also equal to r, whereas the order of $\boldsymbol{\phi}$ is $r+1$.

Inserting the expressions (3.45) and (3.46) in Eq. (3.44) yields the *general balance equation in integral form*,

$$\begin{aligned}
\frac{\mathrm{d}}{\mathrm{d}t}\int_\omega g(\mathbf{x},t)\,\mathrm{d}v = &-\oint_{\partial\omega} \boldsymbol{\phi}(\mathbf{x},t)\cdot\mathbf{n}\,\mathrm{d}a \\
&+ \int_\omega p(\mathbf{x},t)\,\mathrm{d}v + \int_\omega s(\mathbf{x},t)\,\mathrm{d}v\,.
\end{aligned} \tag{3.47}$$

Provided that all fields in this equation are sufficiently smooth, it can be localised as follows. We apply Reynolds' transport theorem (3.43) to the left-hand side (with $\psi = g$), transform all surface integrals to volume integrals with the divergence theorem (2.63) and assemble all terms on the left-hand side:

$$\int_\omega \left(\frac{\partial g}{\partial t} + \operatorname{div}(g\mathbf{v}) + \operatorname{div}\boldsymbol{\phi} - p - s \right) \mathrm{d}v = 0\,. \tag{3.48}$$

This relation must hold for any arbitrary material volume ω, which is only possible if the integrand itself vanishes. Thus,

$$\frac{\partial g}{\partial t} = -\operatorname{div}(\boldsymbol{\phi} + g\mathbf{v}) + p + s\,, \tag{3.49}$$

which is the *general balance equation in local form*. It balances the local change of the density g with the production and supply densities and the negative divergence of two flux terms, the actual flux density $\boldsymbol{\phi}$ and the advective (or convective) flux density $g\mathbf{v}$.

3.3.3 General Jump Condition

The local balance equation (3.49) is only valid for those parts of the material volume ω for which all fields are sufficiently smooth. We now consider the case that there exists an oriented *singular surface* σ within ω for which this is not fulfilled. In particular, the density g may be discontinuous on σ. Denote the unit normal vector on σ by $\mathbf{n}$, the side of ω into which $\mathbf{n}$ points by ω^+ and the other side by ω^-. The singular surface need not be material, that is to say, it may travel with its own velocity $\mathbf{w}$ which may differ from the particle velocity $\mathbf{v}$ (Fig. 3.6).

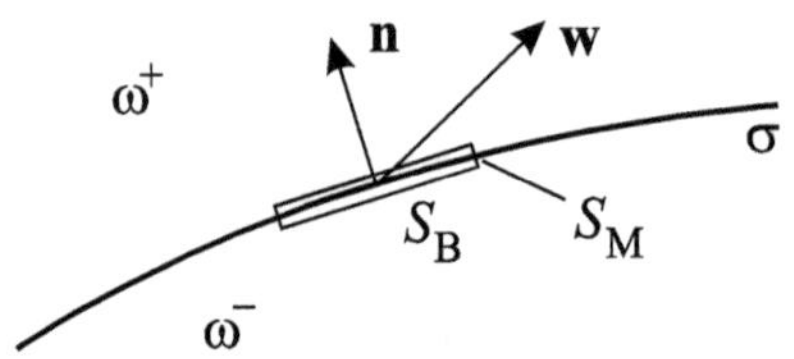

Fig. 3.6. Singular surface σ within the material volume ω.

The values which a field quantity $\psi(\mathbf{x}, t)$ assumes when the point $\mathbf{x} \in \sigma$ (on the singular surface) is approached on an arbitrary path in ω^- or ω^+ are denoted by ψ^- and ψ^+, respectively:

$$\begin{aligned} \forall \mathbf{x} \in \sigma : \quad & \psi^-(\mathbf{x}, t) = \lim_{\mathbf{y} \to \mathbf{x},\, \mathbf{y} \in \omega^-} \psi(\mathbf{y}, t)\,, \\ & \psi^+(\mathbf{x}, t) = \lim_{\mathbf{y} \to \mathbf{x},\, \mathbf{y} \in \omega^+} \psi(\mathbf{y}, t)\,. \end{aligned} \tag{3.50}$$

Of course, this requires that the limits exist and are finite. We define the *jump* $[\![\psi]\!]$ of ψ on σ as

$$\forall \mathbf{x} \in \sigma : \quad [\![\psi]\!]\,(\mathbf{x}, t) = \psi^+(\mathbf{x}, t) - \psi^-(\mathbf{x}, t)\,. \tag{3.51}$$

If $[\![\psi]\!] \neq 0$, the quantity ψ experiences a discontinuity on the singular surface.

We now motivate a balance equation similar to Eq. (3.47) for the pill-box volume ν around the singular surface σ with basal area S_B und mantle area S_M, which is also indicated in Fig. 3.6. It is hereby assumed that S_B and S_M are very small, so that the curvature of σ can be neglected. This entails a very small volume of the pill-box, so that the three volume integrals in Eq. (3.47) are negligible compared to the surface integral of the flux density $\boldsymbol{\phi}$ (provided that all integrands are bounded). So the first guess for the balance equation of the pill-box volume ν would be $\oint_{\partial\nu} \boldsymbol{\phi}(\mathbf{x}, t) \cdot \mathbf{n}\, da = 0$. However, in general we are concerned with a non-material volume here, so that, in addition to the actual flux density $\boldsymbol{\phi}$ an advective flux density $g(\mathbf{v} - \mathbf{w})$ due to the particle motion (velocity $\mathbf{v}$) relative to the motion of the singular surface (velocity $\mathbf{w}$) must be taken into account. Thus, the correct form of the balance equation is

$$\oint_{\partial\nu} (\boldsymbol{\phi} + g(\mathbf{v} - \mathbf{w})) \cdot \mathbf{n}\, da = 0\,. \tag{3.52}$$

The geometry of the pill-box volume is such that $S_\mathrm{M} \ll S_\mathrm{B}$, so that all fluxes through S_M can be neglected in comparison with fluxes through S_B. Therefore, only the basal surfaces with area S_B on the ω^+ and ω^- side of σ (denoted as S_B^+ and S_B^-, respectively) contribute to the surface integral (3.52):

$$\int_{S_\mathrm{B}^+} (\boldsymbol{\phi} + g(\mathbf{v} - \mathbf{w})) \cdot \mathbf{n}\, da + \int_{S_\mathrm{B}^-} (\boldsymbol{\phi} + g(\mathbf{v} - \mathbf{w})) \cdot (-\mathbf{n})\, da = 0 \tag{3.53}$$

(note that the outer unit normal vector is $\mathbf{n}$ on S_B^+ and $-\mathbf{n}$ on S_B^-). Since all field quantities are virtually constant on the very small surfaces S_B^+ and S_B^-, this can be written as

$$\begin{aligned} & S_\mathrm{B}\,(\boldsymbol{\phi}^+ + g^+(\mathbf{v}^+ - \mathbf{w})) \cdot \mathbf{n} + S_\mathrm{B}\,(\boldsymbol{\phi}^- + g^-(\mathbf{v}^- - \mathbf{w})) \cdot (-\mathbf{n}) = 0 \\ \Rightarrow\ & \boldsymbol{\phi}^+ \cdot \mathbf{n} - \boldsymbol{\phi}^- \cdot \mathbf{n} + (g^+(\mathbf{v}^+ - \mathbf{w})) \cdot \mathbf{n} - (g^-(\mathbf{v}^- - \mathbf{w})) \cdot \mathbf{n} = 0\,, \end{aligned} \tag{3.54}$$

and with definition (3.51) we obtain

$$[\![\boldsymbol{\phi} \cdot \mathbf{n}]\!] + [\![g\,((\mathbf{v} - \mathbf{w}) \cdot \mathbf{n})]\!] = 0\,. \tag{3.55}$$

This is the *general jump condition on singular surfaces*. For a formal derivation of this jump condition see, e.g., Liu (2002), Hutter and Jöhnk (2004).

3.3.4 Mass Balance

If the physical quantity $\mathcal{G}$ is identified with the total mass M of the material volume ω, it is clear that $\mathrm{d}M/\mathrm{d}t = 0$, because the mass of a material volume cannot change. With the (mass) density ρ this can be expressed as

$$\frac{\mathrm{d}}{\mathrm{d}t} \int_\omega \rho\, \mathrm{d}v = 0\,. \tag{3.56}$$

By comparison with the general balance equation (3.47) we find immediately

$$g = \rho\,, \quad \boldsymbol{\phi} = \mathbf{0}\,, \quad p = 0\,, \quad s = 0\,. \tag{3.57}$$

With these densities, the local balance equation (3.49) reads

$$\frac{\partial \rho}{\partial t} + \mathrm{div}\,(\rho \mathbf{v}) = 0\,. \tag{3.58}$$

This is the *mass balance*, also known as the *continuity equation*. An equivalent form can be derived by differentiating the product,

$$\begin{aligned} & \frac{\partial \rho}{\partial t} + (\mathrm{grad}\,\rho) \cdot \mathbf{v} + \rho\,\mathrm{div}\,\mathbf{v} = 0 \\ \Rightarrow\ & \dot{\rho} + \rho\,\mathrm{div}\,\mathbf{v} = 0\,. \end{aligned} \tag{3.59}$$

An important special case is that of an *incompressible* material, defined by a constant density, that is, $\rho = \text{const}$ or $\dot{\rho} = 0$. For this case, Eq. (3.59) simplifies to

$$\operatorname{div} \mathbf{v} = 0\,, \tag{3.60}$$

which is the mass balance or continuity equation for incompressible materials. Evidently, the corresponding velocity field is source-free (solenoidal).

By inserting the densities (3.57) in the general jump condition (3.55), we obtain the *mass jump condition* on singular surfaces as

$$[\![\rho\,((\mathbf{v}-\mathbf{w})\cdot\mathbf{n})]\!] = 0\,. \tag{3.61}$$

It simply states that the mass inflow at one side of the singular surface must equal the mass outflow at the other side.

Let us come back to the general balance equation (3.49) and write the density g as

$$g = \rho g_{\text{s}}\,, \tag{3.62}$$

where g_{s} denotes the physical quantity under consideration *per unit mass* (which, of course, only makes sense if the quantity is not the mass itself). This yields

$$\begin{aligned}
&\frac{\partial(\rho g_s)}{\partial t} + \operatorname{div}(\boldsymbol{\phi} + \rho g_s \mathbf{v}) = p + s \\
&\Rightarrow\ \rho\frac{\partial g_s}{\partial t} + g_s\frac{\partial \rho}{\partial t} + \operatorname{div}\boldsymbol{\phi} + g_s \operatorname{div}(\rho\mathbf{v}) + (\operatorname{grad} g_s)\cdot\rho\mathbf{v} = p + s \\
&\Rightarrow\ \rho\left\{\frac{\partial g_s}{\partial t} + (\operatorname{grad} g_s)\cdot\mathbf{v}\right\} + g_s\left\{\frac{\partial \rho}{\partial t} + \operatorname{div}(\rho\mathbf{v})\right\} \\
&\qquad\qquad = -\operatorname{div}\boldsymbol{\phi} + p + s\,.
\end{aligned} \tag{3.63}$$

The first term in curly brackets is the material time derivative of g [see Eq. (3.23)], the second term vanishes because of the mass balance (3.58). It remains

$$\rho\frac{\mathrm{d}g_s}{\mathrm{d}t} = -\operatorname{div}\boldsymbol{\phi} + p + s \tag{3.64}$$

as an alternative representation of the general balance equation, which can be used for any quantity except mass.

3.3.5 Momentum Balance

Let us now identify the physical quantity $\mathcal{G}$ with the total momentum $\mathbf{P}$ (vector!) of the material volume ω. Since momentum is equal to mass times velocity, the momentum density can be expressed as mass density times velocity,

$$g = \rho\mathbf{v}\,, \tag{3.65}$$

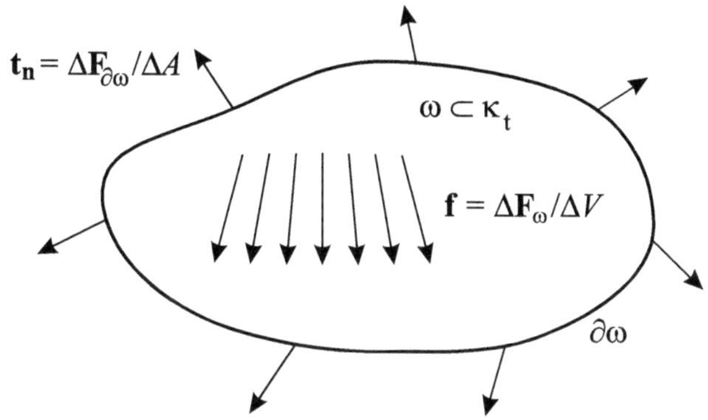

Fig. 3.7. Volume forces $\mathbf{f}$ and surface forces $\mathbf{t_n}$ which act on a material volume ω and its surface $\partial\omega$ in the present configuration κ_t.

and the total momentum $\mathbf{P}$ is equal to $\int_\omega \rho\mathbf{v}\,\mathrm{d}v$. Following Newton's Second Law, its temporal change $\mathrm{d}\mathbf{P}/\mathrm{d}t$ must be given by the sum of all forces $\mathbf{F}$ which act on the volume ω. These forces can either be external *volume forces* $\mathbf{f}(\mathbf{x},t)$ acting on any volume element within ω (e.g., the gravity field), or internal *stresses* (surface forces) $\mathbf{t_n}(\mathbf{x},t)$ acting on the boundary surface $\partial\omega$ (Fig. 3.7). The latter do not only depend on position $\mathbf{x}$ and time t, but also on the orientation of the surface, expressed by the unit normal vector $\mathbf{n}$.

The total force acting on the material volume ω is therefore

$$\mathbf{F} = \oint_{\partial\omega} \mathbf{t_n}(\mathbf{x},t)\,\mathrm{d}a + \int_\omega \mathbf{f}(\mathbf{x},t)\,\mathrm{d}v\,, \tag{3.66}$$

and Newton's Second Law reads

$$\frac{\mathrm{d}}{\mathrm{d}t}\int_\omega \rho(\mathbf{x},t)\mathbf{v}(\mathbf{x},t)\,\mathrm{d}v = \oint_{\partial\omega} \mathbf{t_n}(\mathbf{x},t)\,\mathrm{d}a + \int_\omega \mathbf{f}(\mathbf{x},t)\,\mathrm{d}v\,. \tag{3.67}$$

Except for the surface integral (flux term), this has the form of the general balance equation (3.47). By comparing the flux terms, we infer that the stress vector $\mathbf{t_n}$ must be a *linear* function of $\mathbf{n}$, that is,

$$\mathbf{t_n}(\mathbf{x},t) = \mathsf{t}(\mathbf{x},t)\cdot\mathbf{n}\,, \tag{3.68}$$

where $\mathsf{t}(\mathbf{x},t)$ is a tensor field of order 2 which is called the *Cauchy stress tensor*. Now we can identify

$$g = \rho\mathbf{v}\,,\ \ \boldsymbol{\phi} = -\mathsf{t}\,,\ \ p = \mathbf{0}\,,\ \ s = \mathbf{f} \tag{3.69}$$

(the volume force is interpreted as a supply term and not a production term because it is assumed to have an external source), and from Eq. (3.49) we obtain the local form of the *momentum balance* as

$$\frac{\partial(\rho\mathbf{v})}{\partial t} + \operatorname{div}(\rho\mathbf{v}\,\mathbf{v}) = \operatorname{div}\mathsf{t} + \mathbf{f}\,. \tag{3.70}$$

With the specific momentum (momentum per unit mass)

$$g_{\mathrm{s}} = \frac{g}{\rho} = \mathbf{v} \tag{3.71}$$

and the representation (3.64) of the general balance equation, an equivalent form of the momentum balance is

$$\rho\frac{\mathrm{d}\mathbf{v}}{\mathrm{d}t} = \operatorname{div}\mathsf{t} + \mathbf{f}\,. \tag{3.72}$$

Note that, if the momentum balance is formulated in a non-inertial system (for instance, the rotating Earth), the volume force $\mathbf{f}$ contains contributions from inertial forces (centrifugal force, Coriolis force etc.).

The *momentum jump condition* on singular surfaces is readily obtained from Eqs. (3.55) and (3.69),

$$[\![\mathsf{t}\cdot\mathbf{n}]\!] - [\![\rho\mathbf{v}\,((\mathbf{v}-\mathbf{w})\cdot\mathbf{n})]\!] = \mathbf{0}\,. \tag{3.73}$$

It relates the jump of the stress vector $(\mathsf{t}\cdot\mathbf{n})$ to the jump of the advective momentum flux across the interface. In the case of a *material singular surface* $(\mathbf{v}^{+}\cdot\mathbf{n} = \mathbf{v}^{-}\cdot\mathbf{n} = \mathbf{w}\cdot\mathbf{n})$ the stress vector is continuous,

$$[\![\mathsf{t}\cdot\mathbf{n}]\!] = \mathbf{0}\,. \tag{3.74}$$

With respect to a given orthonormal basis $\{\mathbf{e}_i\}$, the matrix of the Cauchy stress tensor t defined by Eq. (3.68) is

$$\mathsf{t} = \begin{pmatrix} t_{xx} & t_{xy} & t_{xz} \\ t_{yx} & t_{yy} & t_{yz} \\ t_{zx} & t_{zy} & t_{zz} \end{pmatrix}. \tag{3.75}$$

The elements of this matrix can be interpreted as follows. For a cut along the xy plane, that is, with unit normal vector $\mathbf{n} = \pm\mathbf{e}_z$ (the sign depends on the orientation of the plane), the stress vector

$$\mathbf{t}_{\pm\mathbf{e}_z} = \pm\mathsf{t}\cdot\mathbf{e}_z = \pm\begin{pmatrix} t_{xz} \\ t_{yz} \\ t_{zz} \end{pmatrix} \tag{3.76}$$

is obtained. Evidently, the diagonal element t_{zz} is perpendicular to the cut plane, whereas the off-diagonal elements t_{xz} and t_{yz} are parallel to the plane. The same result is found for cuts along the xz and yz planes. Therefore, the three diagonal elements (t_{xx}, t_{yy}, t_{zz}) are referred to as *normal stresses,* and the six off-diagonal elements $(t_{xy}, t_{yx}, t_{xz}, t_{zx}, t_{yz}, t_{zy})$ are called *shear stresses.* The meaning of these components is illustrated in Fig. 3.8.

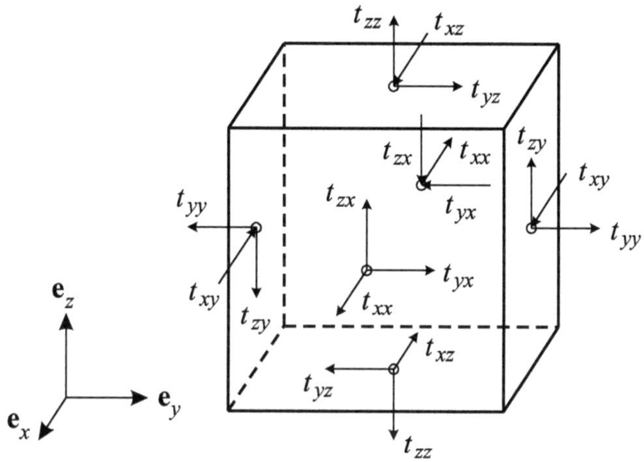

Fig. 3.8. Components of the Cauchy stress tensor.

3.3.6 Balance of Angular Momentum

In point mechanics, the angular momentum $\mathbf{L}$ of a mass point is given by $\mathbf{L} = \mathbf{x} \times \mathbf{P}$ (cross product of position vector and momentum), and the torque $\mathbf{M}$ is defined as $\mathbf{M} = \mathbf{x} \times \mathbf{F}$ (cross product of position vector and force acting on the mass point). In view of the momentum densities (3.69) for the continuous body, this motivates the following identities for the densities of angular momentum:

$$g = \mathbf{x} \times \rho\mathbf{v}\,, \quad \phi = -\mathbf{x} \times \mathrm{t}\,, \quad p = \mathbf{0}\,, \quad s = \mathbf{x} \times \mathbf{f}\,. \tag{3.77}$$

Inserting these identities in Eq. (3.49) yields the *balance of angular momentum*,

$$\frac{\partial(\mathbf{x} \times \rho\mathbf{v})}{\partial t} + \operatorname{div}\left[(\mathbf{x} \times \rho\mathbf{v})\,\mathbf{v}\right] = \operatorname{div}(\mathbf{x} \times \mathrm{t}) + \mathbf{x} \times \mathbf{f}\,, \tag{3.78}$$

or in index notation

$$\frac{\partial}{\partial t}(\rho\,\varepsilon_{ijk}x_j v_k) + (\rho\,\varepsilon_{ijk}x_j v_k v_l)_{,l} = (\varepsilon_{ijk}x_j t_{kl})_{,l} + \varepsilon_{ijk}x_j f_k\,. \tag{3.79}$$

Using the momentum balance (3.70), this can be drastically simplified. We compute $\mathbf{x} \times (3.70)$ in index notation,

$$\begin{aligned}
&\frac{\partial}{\partial t}(\rho v_k) + (\rho v_k v_l)_{,l} = t_{kl,l} + f_k \quad | \cdot \varepsilon_{ijk}x_j \\
\Rightarrow\ &\frac{\partial}{\partial t}(\rho\,\varepsilon_{ijk}x_j v_k) + x_j(\rho\,\varepsilon_{ijk}v_k v_l)_{,l} = x_j(\varepsilon_{ijk}t_{kl})_{,l} + \varepsilon_{ijk}x_j f_k \\
\Rightarrow\ &\frac{\partial}{\partial t}(\rho\,\varepsilon_{ijk}x_j v_k) + (\rho\,\varepsilon_{ijk}x_j v_k v_l)_{,l} - \rho\,\varepsilon_{ijk}v_k v_l x_{j,l} \\
&\qquad = (\varepsilon_{ijk}x_j t_{kl})_{,l} - \varepsilon_{ijk}t_{kl}x_{j,l} + \varepsilon_{ijk}x_j f_k\,,
\end{aligned} \tag{3.80}$$

and subtract this from the balance of angular momentum (3.79), which leaves

$$\begin{aligned}
&\rho\,\varepsilon_{ijk} v_k v_l \delta_{jl} = \varepsilon_{ijk} t_{kl} \delta_{jl} \\
&\Rightarrow\ \rho\,\varepsilon_{ijk} v_k v_j = \varepsilon_{ijk} t_{kj} \\
&\Rightarrow\ \tfrac{1}{2}(\rho\,\varepsilon_{ijk} v_k v_j + \rho\,\varepsilon_{ikj} v_j v_k) = \tfrac{1}{2}(\varepsilon_{ijk} t_{kj} + \varepsilon_{ikj} t_{jk}) \\
&\Rightarrow\ \rho\,\varepsilon_{ijk}(v_k v_j - v_j v_k) = \varepsilon_{ijk}(t_{kj} - t_{jk}) \\
&\Rightarrow\ \varepsilon_{ijk}(t_{kj} - t_{jk}) = 0\,.
\end{aligned} \tag{3.81}$$

Evaluation of this result for $i = 1$ yields

$$\begin{aligned}
&\varepsilon_{123}(t_{32} - t_{23}) + \varepsilon_{132}(t_{23} - t_{32}) = 0 \\
&\Rightarrow\ (t_{zy} - t_{yz}) - (t_{yz} - t_{zy}) = 0 \quad \Rightarrow \quad t_{yz} = t_{zy}\,.
\end{aligned} \tag{3.82}$$

Similarly, for $i = 2$ and 3 one finds $t_{xz} = t_{zx}$ and $t_{xy} = t_{yx}$, respectively. The balance of angular momentum thus reduces to the statement that the *Cauchy stress tensor is symmetric*,

$$\mathsf{t} = \mathsf{t}^{\mathrm{T}}, \quad \text{or} \quad t_{ij} = t_{ji}\,. \tag{3.83}$$

By contrast, an independent jump condition of angular momentum does not exist; it is equivalent to the momentum jump condition (3.73).

3.3.7 Energy Balance

Balance of Kinetic Energy

We now compute the dot product of the momentum balance (3.72) and the velocity $\mathbf{v}$. Expressed in index notation, this is

$$\begin{aligned}
\rho v_k \frac{\mathrm{d}v_k}{\mathrm{d}t} &= t_{kl,l} v_k + f_k v_k \\
\Rightarrow\ \rho \frac{\mathrm{d}}{\mathrm{d}t}\left(\frac{v_k v_k}{2}\right) &= (t_{kl} v_k)_{,l} - t_{kl} v_{k,l} + f_k v_k \\
&= (\mathsf{t}\cdot\mathbf{v})_{l,l} - (\mathsf{t}\cdot\mathsf{L})_{ll} + f_k v_k \\
\Rightarrow\ \rho \frac{\mathrm{d}}{\mathrm{d}t}\left(\frac{v^2}{2}\right) &= \mathrm{div}\,(\mathsf{t}\cdot\mathbf{v}) - \mathrm{tr}\,(\mathsf{t}\cdot\mathsf{L}) + \mathbf{f}\cdot\mathbf{v}
\end{aligned} \tag{3.84}$$

[in the step from line 2 to 3 the symmetry of t, Eq. (3.83), has been used, and in the last line we have introduced the *speed* $v = |\mathbf{v}|$, which is the absolute value of the velocity]. For the second summand on the right-hand side, $\mathrm{tr}\,(\mathsf{t}\cdot\mathsf{L})$, we apply the decomposition (3.28) of L, the symmetry of t and the antisymmetry of W,

$$\begin{aligned}\mathrm{tr}\,(\mathsf{t}\cdot\mathsf{L}) &= \mathrm{tr}\,(\mathsf{t}\cdot\mathsf{D}) + \mathrm{tr}\,(\mathsf{t}\cdot\mathsf{W}) \\ &= \mathrm{tr}\,(\mathsf{t}\cdot\mathsf{D}) + t_{ij}W_{ji} \\ &= \mathrm{tr}\,(\mathsf{t}\cdot\mathsf{D}) + \tfrac{1}{2}(t_{ij}W_{ji} + t_{ji}W_{ij}) \\ &= \mathrm{tr}\,(\mathsf{t}\cdot\mathsf{D}) + \tfrac{1}{2}(t_{ij}W_{ji} - t_{ij}W_{ji}) = \mathrm{tr}\,(\mathsf{t}\cdot\mathsf{D})\,, \end{aligned} \tag{3.85}$$

so that

$$\rho\frac{\mathrm{d}}{\mathrm{d}t}\left(\frac{v^2}{2}\right) = \mathrm{div}\,(\mathsf{t}\cdot\mathbf{v}) - \mathrm{tr}\,(\mathsf{t}\cdot\mathsf{D}) + \mathbf{f}\cdot\mathbf{v}\,. \tag{3.86}$$

Since the kinetic energy of a mass m is given by $mv^2/2$, the term $v^2/2$ denotes the specific kinetic energy (per unit mass) of a continuous body. Comparison of the above result with the general balance equation (3.64) shows that it can be interpreted as the *balance of kinetic energy*, where

$$\begin{aligned} g &= \rho v^2/2 && \text{(kinetic energy density)}\,, \\ g_s &= v^2/2 && \text{(specific kinetic energy)}\,, \\ \boldsymbol{\phi} &= -\mathsf{t}\cdot\mathbf{v} && \text{(power of stresses)}\,, \\ p &= -\mathrm{tr}\,(\mathsf{t}\cdot\mathsf{D}) && (-p\text{: dissipation power})\,, \\ s &= \mathbf{f}\cdot\mathbf{v} && \text{(power of volume forces)}\,. \end{aligned} \tag{3.87}$$

The attribution of the dissipation power as a production term and the power of volume forces as a supply term was done because the former is only due to intrinsic quantities, whereas in the latter the volume force occurs which has an external source. Thus, in contrast to mass, momentum and angular momentum, the kinetic energy has a non-zero production density, which means that it is not a conserved quantity.

Energy Balance, Balance of Internal Energy

The balance of kinetic energy, Eq. (3.86), is not an independent statement, but a mere consequence of the momentum balance (3.72). However, classical mechanics and thermodynamics tells us that the kinetic energy is only one part of the total energy of a system (here: continuous body), and that the total energy is a conserved quantity (no production). In order to formulate the (total) energy balance, we thus extend Eq. (3.87) by introducing an internal energy, a heat flux and a radiation power and setting the production to zero:

$$\begin{aligned} g &= \rho(u + v^2/2) && (u\text{: specific internal energy})\,, \\ g_s &= u + v^2/2\,, \\ \boldsymbol{\phi} &= \mathbf{q} - \mathsf{t}\cdot\mathbf{v} && (\mathbf{q}\text{: heat flux})\,, \\ p &= 0\,, \\ s &= \rho r + \mathbf{f}\cdot\mathbf{v} && (r\text{: specific radiation power})\,. \end{aligned} \tag{3.88}$$

By inserting these densities in Eq. (3.49), the *energy balance*

$$\begin{aligned}\frac{\partial}{\partial t}\left[\rho\left(u + \frac{v^2}{2}\right)\right] &+ \mathrm{div}\left[\rho\left(u + \frac{v^2}{2}\right)\mathbf{v}\right] \\ &= -\mathrm{div}\,\mathbf{q} + \mathrm{div}\,(\mathsf{t}\cdot\mathbf{v}) + \rho r + \mathbf{f}\cdot\mathbf{v} \end{aligned} \tag{3.89}$$

is obtained, which is also known as the *First Law of Thermodynamics.* An alternative form follows from Eq. (3.64),

$$\rho \frac{\mathrm{d}}{\mathrm{d}t}\left(u + \frac{v^2}{2}\right) = -\operatorname{div}\mathbf{q} + \operatorname{div}(\mathsf{t}\cdot\mathbf{v}) + \rho r + \mathbf{f}\cdot\mathbf{v}\,. \tag{3.90}$$

This can be simplified further:

$$\begin{aligned} \rho\frac{\mathrm{d}u}{\mathrm{d}t} + \rho v_k \frac{\mathrm{d}v_k}{\mathrm{d}t} &= -q_{k,k} + (t_{kl}v_k)_{,l} + \rho r + f_k v_k \\ &= -q_{k,k} + t_{kl,l}v_k + t_{kl}v_{k,l} + \rho r + f_k v_k \\ \Rightarrow\ \rho\frac{\mathrm{d}u}{\mathrm{d}t} + v_k\left\{\rho\frac{\mathrm{d}v_k}{\mathrm{d}t} - t_{kl,l} - f_k\right\} &= -q_{k,k} + t_{lk}L_{kl} + \rho r\,. \end{aligned} \tag{3.91}$$

The term in curly brackets vanishes because of the momentum balance (3.72), and the term $t_{lk}L_{kl} = \operatorname{tr}(\mathsf{t}\cdot\mathsf{L})$ can again be replaced by $\operatorname{tr}(\mathsf{t}\cdot\mathsf{D})$ [see Eq. (3.85)], so that we obtain

$$\rho\frac{\mathrm{d}u}{\mathrm{d}t} = -\operatorname{div}\mathbf{q} + \operatorname{tr}(\mathsf{t}\cdot\mathsf{D}) + \rho r\,. \tag{3.92}$$

Evidently, this is the *balance of internal energy* in the form (3.64), with the corresponding densities

$$\begin{aligned} g &= \rho u\,, & & \\ g_s &= u & &\text{(specific internal energy)}\,, \\ \boldsymbol{\phi} &= \mathbf{q} & &\text{(heat flux)}\,, \\ p &= \operatorname{tr}(\mathsf{t}\cdot\mathsf{D}) & &\text{(dissipation power)}\,, \\ s &= \rho r & &(r\text{: specific radiation power})\,. \end{aligned} \tag{3.93}$$

In contrast to the total energy, the internal energy is not a conserved quantity. Its production density is equal to the dissipation power, which already appeared in the balance of kinetic energy (3.86) with a negative sign. The name "dissipation power" results from the fact that it annihilates kinetic energy and changes it into internal energy. In other words, macroscopic mechanical energy is transformed into heat (microscopic, unordered motion). Therefore, the dissipation power can also be interpreted as heat production due to internal friction.

From Eqs. (3.55) and (3.88) we obtain the *energy jump condition*

$$[\![\mathbf{q}\cdot\mathbf{n}]\!] - [\![\mathbf{v}\cdot\mathsf{t}\cdot\mathbf{n}]\!] + \left[\!\!\left[\rho\left(u + \tfrac{1}{2}v^2\right)\left((\mathbf{v}-\mathbf{w})\cdot\mathbf{n}\right)\right]\!\!\right] = 0\,. \tag{3.94}$$

In case of a *material singular surface* ($\mathbf{v}^+\cdot\mathbf{n} = \mathbf{v}^-\cdot\mathbf{n} = \mathbf{w}\cdot\mathbf{n}$) the third summand vanishes, and because of the continuity of the stress vector $\mathsf{t}\cdot\mathbf{n}$ [Eq. (3.74)] it can be factored out in the second summand:

$$\begin{aligned} &[\![\mathbf{q}\cdot\mathbf{n}]\!] - [\![\mathbf{v}]\!]\cdot\mathsf{t}\cdot\mathbf{n} = 0 \\ \Rightarrow\ &[\![\mathbf{q}\cdot\mathbf{n}]\!] - [\![\mathbf{v}_\perp]\!]\cdot\mathsf{t}\cdot\mathbf{n} - [\![\mathbf{v}_\parallel]\!]\cdot\mathsf{t}\cdot\mathbf{n} = 0 \\ \Rightarrow\ &[\![\mathbf{q}\cdot\mathbf{n}]\!] - [\![\mathbf{v}_\parallel]\!]\cdot\mathsf{t}\cdot\mathbf{n} = 0 \end{aligned} \tag{3.95}$$

[where $\mathbf{v} = \mathbf{v}_\perp + \mathbf{v}_\parallel$, with the normal component $\mathbf{v}_\perp = (\mathbf{v} \cdot \mathbf{n})\mathbf{n}$ and the tangential component $\mathbf{v}_\parallel = \mathbf{v} - (\mathbf{v} \cdot \mathbf{n})\mathbf{n}$; the jump of $\mathbf{v}_\perp$ vanishes because of $\mathbf{v}^+ \cdot \mathbf{n} = \mathbf{v}^- \cdot \mathbf{n}$]. Only under the additional assumption of a *no-slip condition*, that is, $[\![\mathbf{v}_\parallel]\!] = \mathbf{0}$, does the normal component of the heat flux $(\mathbf{q} \cdot \mathbf{n})$ become continuous.

3.4 Constitutive Equations

3.4.1 Homogeneous Viscous Thermoelastic Bodies

The balance equations of mass, momentum and internal energy derived in Sect. 3.3 read

$$\frac{\mathrm{d}\rho}{\mathrm{d}t} = -\rho\,\mathrm{div}\,\mathbf{v}\,, \tag{3.96}$$

$$\rho\frac{\mathrm{d}\mathbf{v}}{\mathrm{d}t} = \mathrm{div}\,\mathsf{t} + \mathbf{f}\,, \tag{3.97}$$

$$\rho\frac{\mathrm{d}u}{\mathrm{d}t} = -\mathrm{div}\,\mathbf{q} + \mathrm{tr}\,(\mathsf{t} \cdot \mathsf{D}) + \rho r \tag{3.98}$$

[see Eqs. (3.59), (3.72) and (3.92); the balance of angular momentum is implicitly included in the symmetry of t]. They constitute *evolution equations* for the unknown fields ρ, $\mathbf{v}$ and u; however, on the right-hand sides the fields t and $\mathbf{q}$ are also unknown. The supply terms $\mathbf{f}$ and r are assumed to be prescribed as external forcings. Thus, in component form we have $1+3+1 = 5$ equations (mass balance: scalar equation, momentum balance: vector equation, energy balance: scalar equation) for the $1 + 3 + 6 + 1 + 3 = 14$ unknown fields ρ (scalar), $\mathbf{v}$ (vector), t (symmetric tensor), u (scalar) and $\mathbf{q}$ (vector), and the system is highly under-determined. Therefore, additional *closure relations* between the field quantities are required. These closure relations describe the specific behaviour of the different materials (whereas the balance equations are universally valid), and they are called *constitutive equations* or *material equations*.

The general theory of constitutive equations is beyond the scope of this text [see e.g. Liu (2002), Hutter and Jöhnk (2004)]. Here, we confine ourselves to a simple class of materials, the so-called *homogeneous viscous thermoelastic bodies*. This will be sufficient for our purpose of describing ice-dynamic processes.

A homogeneous viscous thermoelastic body is defined as a material whose constitutive equations are functions of the form

$$\begin{aligned} \mathsf{t} &= \mathsf{t}\,(\mathsf{F},\, \dot{\mathsf{F}},\, T,\, \mathrm{grad}\,T)\,, \\ \mathbf{q} &= \mathbf{q}\,(\mathsf{F},\, \dot{\mathsf{F}},\, T,\, \mathrm{grad}\,T)\,, \\ u &= u\,(\mathsf{F},\, \dot{\mathsf{F}},\, T,\, \mathrm{grad}\,T)\,, \end{aligned} \tag{3.99}$$

where the *temperature* $T(\mathbf{x}, t)$ has been introduced as an additional scalar field quantity. Hence, the Cauchy stress tensor t, the heat flux $\mathbf{q}$ and the specific internal energy u are understood as the dependent *material quantities*, and the material shows neither non-local nor memory effects. Note that due to Eq. (3.26) the dependency on $\dot{\mathsf{F}}$ can also be expressed as a dependency on the velocity gradient L. In the following, we will discuss two examples of homogeneous viscous thermoelastic bodies relevant for ice dynamics, namely the linear elastic solid (Hookean body) and the Newtonian fluid.

3.4.2 Linear Elastic Solid

Hooke's Law

An *elastic body* is defined as a material for which the stress tensor depends on the deformation gradient only,

$$\mathsf{t} = \mathsf{t}(\mathsf{F})\,. \tag{3.100}$$

In particular, this excludes any temperature dependencies, so that the problem is purely mechanical, and the energy balance (3.98) need not be taken into account.

For many practical applications, it is sufficient to assume *small deformations*, that is, $\mathsf{F} \approx \mathsf{I}$. If we use the same origins ($\mathbf{B} = \mathbf{0}$) and bases ($\mathbf{e}_i = \delta_{iA}\mathbf{E}_A$) for the reference and the present configurations (see Fig. 3.1), then the displacements $\mathbf{u} = \mathbf{x} - \mathbf{X}$ will be small, that is, $\mathbf{x} \approx \mathbf{X}$. The reference configuration and the present configuration virtually fall together. It is then no longer necessary to distinguish between material and spatial derivatives ($\partial/\partial x_i \approx \partial/\partial X_A$ for $i = A$). For this situation, the *displacement gradient* H is defined as

$$\mathsf{H} = \mathrm{Grad}\,\mathbf{u} = \mathsf{F} - \mathsf{I}\,, \tag{3.101}$$

and the *infinitesimal strain tensor* ϵ is the symmetric part,

$$\epsilon = \tfrac{1}{2}(\mathsf{H} + \mathsf{H}^\mathrm{T})\,, \quad \text{or} \quad \varepsilon_{ij} = \tfrac{1}{2}(u_{i,j} + u_{j,i})\,. \tag{3.102}$$

Without proof let us note that the diagonal elements ε_{xx}, ε_{yy}, ε_{zz} correspond to the dilatations (relative length changes of line elements) in x-, y- and z-directions respectively. Also, the off-diagonal elements ε_{xy}, ε_{xz}, ε_{yz} are equal to one half of the shear angles in the x-y, x-z and y-z planes respectively; that is,

$$\varepsilon_{ij} = \tfrac{1}{2}\gamma_{ij}\,, \quad (i \neq j)\,. \tag{3.103}$$

These interpretations are analogous to those of the components of the strain-rate tensor D in terms of dilatation rates and shear rates [see Eqs. (3.34) and (3.35)].

The constitutive equation of an isotropic (identical properties in all directions), linear elastic solid for small deformations, also known as a *Hookean body*, is now

$$\mathsf{t} = (\lambda \operatorname{tr} \epsilon) \mathsf{I} + 2\mu \epsilon . \tag{3.104}$$

This material equation is called *Hooke's law*, and the two coefficients λ, μ are the *Lamé parameters.*

An alternative formulation is found by splitting up the infinitesimal strain tensor into an isotropic part and a traceless deviator ϵ^{D},

$$\epsilon = \left(\tfrac{1}{3} \operatorname{tr} \epsilon\right) \mathsf{I} + \epsilon^{\mathrm{D}} , \tag{3.105}$$

with $\operatorname{tr} \epsilon^{\mathrm{D}} = 0$. Inserting this into Hooke's law (3.104) yields

$$\begin{aligned} \mathsf{t} &= (\lambda \operatorname{tr} \epsilon) \mathsf{I} + \left(\tfrac{2}{3}\mu \operatorname{tr} \epsilon\right) \mathsf{I} + 2\mu \epsilon^{\mathrm{D}} \\ &= \left[(\lambda + \tfrac{2}{3}\mu) \operatorname{tr} \epsilon\right] \mathsf{I} + 2\mu \epsilon^{\mathrm{D}} \\ &= (\kappa \operatorname{tr} \epsilon) \mathsf{I} + 2\mu \epsilon^{\mathrm{D}} , \end{aligned} \tag{3.106}$$

where $\kappa = \lambda + 2\mu/3$ is the elastic *bulk modulus.* In this form of Hooke's law, volume-changing compressions or expansions are described by the first summand, whereas the second summand accounts for volume-preserving distortions.

Phenomenological Introduction of Hooke's Law

Let us consider a small cube of a linear elastic solid, which is subjected to the normal stress t_{xx} (Fig. 3.9, left). A linear relation between this stress and the resulting dilatation ε_{xx} will be observed,

$$t_{xx} = E \, \varepsilon_{xx} ; \tag{3.107}$$

the parameter E is called *Young's modulus.* Further, in the perpendicular directions y and z negative dilatations (compressions) ε_{yy} and ε_{zz} will occur, for which the relations

$$\varepsilon_{yy} = -\nu \, \varepsilon_{xx} = -\frac{\nu}{E} t_{xx} , \quad \varepsilon_{zz} = -\nu \, \varepsilon_{xx} = -\frac{\nu}{E} t_{xx} \tag{3.108}$$

hold. The factor ν is known as *Poisson's ratio.*

Now subject the same cube to a shear stress t_{xy} (Fig. 3.9, right). It will then suffer a proportional shear by the angle γ_{xy}, that is,

$$t_{xy} = \mu \, \gamma_{xy} . \tag{3.109}$$

The coefficient μ is the *shear modulus*, and, as we will see below, it is identical to the 2nd Lamé parameter and therefore denoted by the same symbol.

By superposition of relations (3.107) – (3.109) we obtain, for an arbitrary state of stress,

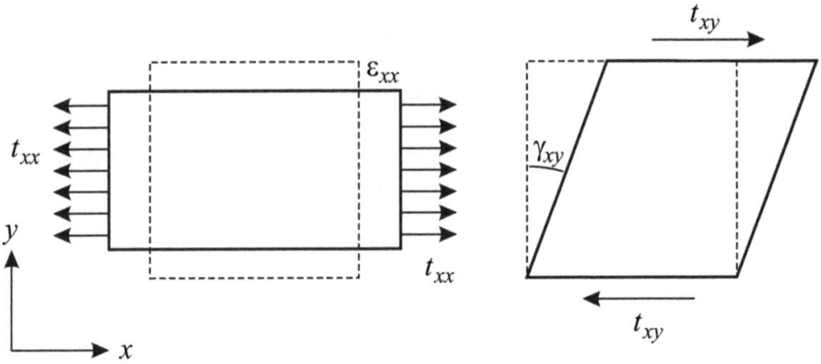

Fig. 3.9. Tension and shear experiment for a linear elastic solid (Hookean body).

$$
\begin{aligned}
\varepsilon_{xx} &= \frac{1}{E}t_{xx} - \frac{\nu}{E}(t_{yy} + t_{zz})\,, \\
\varepsilon_{yy} &= \frac{1}{E}t_{yy} - \frac{\nu}{E}(t_{xx} + t_{zz})\,, \\
\varepsilon_{zz} &= \frac{1}{E}t_{zz} - \frac{\nu}{E}(t_{xx} + t_{yy})\,, \\
\gamma_{xy} &= \frac{1}{\mu}t_{xy}\,, \\
\gamma_{xz} &= \frac{1}{\mu}t_{xz}\,, \\
\gamma_{yz} &= \frac{1}{\mu}t_{yz}\,.
\end{aligned}
\tag{3.110}
$$

Equations $(3.110)_{4,5,6}$, using (3.103), can be inverted straightforwardly to yield the stress expressions

$$
t_{xy} = 2\mu\,\varepsilon_{xy}\,, \quad t_{xz} = 2\mu\,\varepsilon_{xz}\,, \quad t_{yz} = 2\mu\,\varepsilon_{yz}\,. \tag{3.111}
$$

Equations $(3.110)_{1,2,3}$ can be rewritten as

$$
\begin{aligned}
\varepsilon_{xx} &= \frac{1+\nu}{E}t_{xx} - \frac{\nu}{E}\mathrm{tr}\,\mathbf{t}\,, \\
\varepsilon_{yy} &= \frac{1+\nu}{E}t_{yy} - \frac{\nu}{E}\mathrm{tr}\,\mathbf{t}\,, \\
\varepsilon_{zz} &= \frac{1+\nu}{E}t_{zz} - \frac{\nu}{E}\mathrm{tr}\,\mathbf{t}\,,
\end{aligned}
\tag{3.112}
$$

and by summation,

$$
\mathrm{tr}\,\boldsymbol{\epsilon} = \frac{1+\nu}{E}\mathrm{tr}\,\mathbf{t} - 3\frac{\nu}{E}\mathrm{tr}\,\mathbf{t} = \frac{1-2\nu}{E}\mathrm{tr}\,\mathbf{t}\,, \tag{3.113}
$$

or

$$\mathrm{tr}\,\mathsf{t} = \frac{E}{1-2\nu}\mathrm{tr}\,\epsilon\,. \tag{3.114}$$

This yields

$$\begin{aligned}
\varepsilon_{xx} &= \frac{1+\nu}{E}t_{xx} - \frac{\nu}{1-2\nu}\mathrm{tr}\,\epsilon\,,\\
\varepsilon_{yy} &= \frac{1+\nu}{E}t_{yy} - \frac{\nu}{1-2\nu}\mathrm{tr}\,\epsilon\,,\\
\varepsilon_{zz} &= \frac{1+\nu}{E}t_{zz} - \frac{\nu}{1-2\nu}\mathrm{tr}\,\epsilon\,,
\end{aligned} \tag{3.115}$$

and therefore, together with Eq. (3.111),

$$\begin{aligned}
t_{xx} &= \frac{E}{1+\nu}\varepsilon_{xx} + \frac{E\nu}{(1+\nu)(1-2\nu)}\mathrm{tr}\,\epsilon\,,\\
t_{yy} &= \frac{E}{1+\nu}\varepsilon_{yy} + \frac{E\nu}{(1+\nu)(1-2\nu)}\mathrm{tr}\,\epsilon\,,\\
t_{zz} &= \frac{E}{1+\nu}\varepsilon_{zz} + \frac{E\nu}{(1+\nu)(1-2\nu)}\mathrm{tr}\,\epsilon\,,\\
t_{xy} &= 2\mu\,\varepsilon_{xy}\,,\\
t_{xz} &= 2\mu\,\varepsilon_{xz}\,,\\
t_{yz} &= 2\mu\,\varepsilon_{yz}\,.
\end{aligned} \tag{3.116}$$

This result is identical to the six components of Hooke's law (3.104). The last three equations show that the shear modulus is indeed identical to the 2nd Lamé parameter, and from the first three equations we can infer the relations

$$\lambda = \frac{E\nu}{(1+\nu)(1-2\nu)}\,, \qquad \mu = \frac{E}{2(1+\nu)} \tag{3.117}$$

between the four parameters, of which only two are independent. By inverting Eq. (3.117), E and ν can also be expressed as functions of the Lamé parameters, which yields

$$E = \frac{\mu(3\lambda+2\mu)}{\lambda+\mu}\,, \qquad \nu = \frac{\lambda}{2(\lambda+\mu)}\,. \tag{3.118}$$

Navier Equation

The mass balance (3.96) can be integrated directly. With Eq. (3.27) and the assumption of small deformations, $\mathsf{F} \approx \mathsf{I} \Rightarrow J \approx 1$, we find

$$\begin{aligned}
&\dot{\rho} + \rho \operatorname{div} \mathbf{v} = 0 \\
&\Rightarrow \frac{\dot{\rho}}{\rho} + \frac{\dot{J}}{J} = 0 \quad \Rightarrow \int_{t_0}^{t} \left(\frac{\dot{\rho}}{\rho} + \frac{\dot{J}}{J} \right) \mathrm{d}t = 0 \\
&\Rightarrow \ln \frac{\rho}{\rho_0} + \ln J = 0 \quad \Rightarrow \frac{\rho}{\rho_0} J = 1 \quad \Rightarrow \rho = \frac{\rho_0}{J} \approx \rho_0 \,,
\end{aligned} \tag{3.119}$$

where t_0 is the initial time which defines the reference configuration, and ρ_0 is the constant density in the reference configuration.

We now insert Hooke's law (3.104) in the momentum balance (3.97). For the divergence of the stress tensor we find

$$\begin{aligned}
(\operatorname{div} \mathsf{t})_i &= t_{ij,j} = (\lambda\, \varepsilon_{kk}\, \delta_{ij} + 2\mu\, \varepsilon_{ij})_{,j} \\
&= (\lambda\, u_{k,k} \delta_{ij})_{,j} + \mu\, u_{i,jj} + \mu\, u_{j,ij} \\
&= \lambda\, u_{k,kj} \delta_{ij} + \mu\, u_{i,jj} + \mu\, u_{j,ji} \\
&= (\lambda + \mu)\, u_{k,ki} + \mu\, u_{i,jj} \\
\Rightarrow \operatorname{div} \mathsf{t} &= (\lambda + \mu) \operatorname{grad} \operatorname{div} \mathbf{u} + \mu\, \Delta \mathbf{u} \,,
\end{aligned} \tag{3.120}$$

where Δ is the Laplacian introduced in Eqs. (2.57) and (2.62). With this, Eq. (3.119) and $\mathbf{v} = \dot{\mathbf{x}} = (\mathbf{u} + \mathbf{X})^{\cdot} = \dot{\mathbf{u}}$ we obtain

$$\rho_0 \frac{\mathrm{d}^2 \mathbf{u}}{\mathrm{d}t^2} = (\lambda + \mu) \operatorname{grad} \operatorname{div} \mathbf{u} + \mu\, \Delta \mathbf{u} + \mathbf{f} \,. \tag{3.121}$$

This is the equation of motion for the Hookean body, and it is known as the *Navier equation.* It consists of three component equations for the three unknown displacement components u_x, u_y and u_z, which is a closed system.

Thin Elastic Plate

An important problem of linear elasticity is that of a thin elastic plate, loaded perpendicular to its plane. Let the plate be oriented in the horizontal x-y plane, its thickness H be much smaller than the horizontal extent L, and its load (force) per unit area be given by $q(x, y)$ (Fig. 3.10).

The problem is assumed to be static or quasi-static, so that the acceleration term on the left-hand side of the Navier equation (3.121) can be omitted. The displacement field is then approximately given by $\mathbf{u} = u_z(x, y)\, \mathbf{e}_z$, that is, a vertical displacement independent of z, with negligible horizontal displacement. A rather lengthy derivation, which is not carried out here [see, e.g., Marguerre and Woernle (1969)], shows that the vertical displacement $u_z(x, y)$ is approximately given by the *biharmonic equation*

$$K\, \Delta^2 u_z(x, y) = q(x, y) \,. \tag{3.122}$$

The parameter K is the *flexural stiffness*, defined as

$$K = \frac{E H^3}{12(1 - \nu^2)} = \frac{\mu H^3}{6(1 - \nu)} \,, \tag{3.123}$$

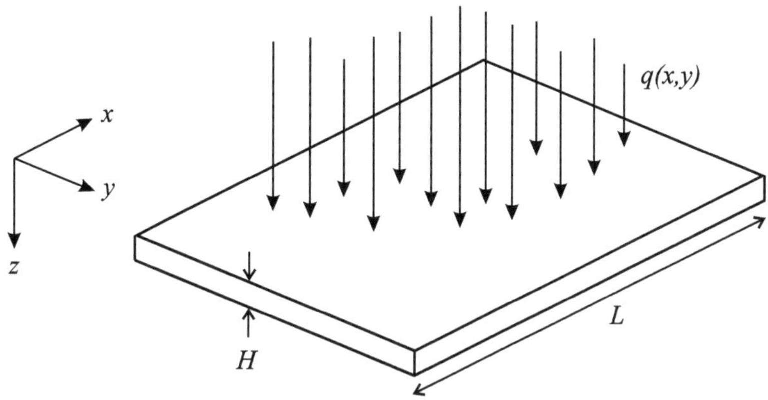

Fig. 3.10. Thin elastic plate with area load q. The thickness H is assumed to be much smaller than the horizontal extent L.

and Δ^2 is the biharmonic operator, which in Cartesian coordinates in the x-y plane takes the form

$$\Delta^2 = \frac{\partial^4}{\partial x^4} + 2\frac{\partial^4}{\partial x^2 \partial y^2} + \frac{\partial^4}{\partial y^4}\,. \tag{3.124}$$

Of course, in order to provide a unique solution, the biharmonic equation (3.122) must be complemented by suitable boundary conditions for the displacement at the rim of the plate, which depend on the actual problem.

3.4.3 Newtonian Fluid

Compressible Newtonian Fluid

A material is called *viscous* if the material function $(3.99)_1$ for the stress tensor t contains an explicit dependency on $\dot{\mathsf{F}}$ or L. The most important realisation of a viscous material is the *Newtonian fluid* (also called *linear viscous fluid*), which can either be compressible or incompressible. For the compressible case, t depends linearly on the strain-rate tensor D (symmetric part of L), the density ρ and the temperature T, through the following material function,

$$\mathsf{t} = -p(\rho, T)\,\mathsf{I} + (\lambda \operatorname{tr} \mathsf{D})\,\mathsf{I} + 2\eta\,\mathsf{D}\,, \tag{3.125}$$

where p is the *thermodynamic pressure*, which is a function of the density ρ and the temperature T ("thermal equation of state"), and λ and η are the *coefficients of viscosity*. In principle, λ and η can also depend on ρ and T, but for simplicity we assume that they are constant.

Analogous to Hooke's law [see Eqs. (3.105) and (3.106)], an alternative form results from splitting the strain-rate tensor into an isotropic part and a traceless deviator D^{D},

$$\mathsf{D} = \left(\tfrac{1}{3}\operatorname{tr}\mathsf{D}\right)\mathsf{I} + \mathsf{D}^{\mathrm{D}}\,, \tag{3.126}$$

with $\operatorname{tr}\mathsf{D}^{\mathrm{D}} = 0$. This yields

$$\begin{aligned} \mathsf{t} &= -p(\rho,T)\,\mathsf{I} + (\lambda\operatorname{tr}\mathsf{D})\,\mathsf{I} + \left(\tfrac{2}{3}\eta\operatorname{tr}\mathsf{D}\right)\mathsf{I} + 2\eta\,\mathsf{D}^{\mathrm{D}} \\ &= -p(\rho,T)\,\mathsf{I} + \left[(\lambda + \tfrac{2}{3}\eta)\operatorname{tr}\mathsf{D}\right]\mathsf{I} + 2\eta\,\mathsf{D}^{\mathrm{D}} \\ &= -p(\rho,T)\,\mathsf{I} + (\zeta\operatorname{tr}\mathsf{D})\,\mathsf{I} + 2\eta\,\mathsf{D}^{\mathrm{D}}\,, \end{aligned} \tag{3.127}$$

where $\zeta = \lambda + 2\eta/3$ is the *bulk viscosity*, and η is also known as the *shear viscosity*. We can combine the first two terms on the right-hand side to $-p_{\mathrm{tot}}\,\mathsf{I}$, where

$$p_{\mathrm{tot}} = p(\rho,T) - \zeta\operatorname{tr}\mathsf{D} = p(\rho,T) + p_{\mathrm{visc}} \tag{3.128}$$

is the *total pressure*, which consists of the thermodynamic pressure $p(\rho,T)$ and the *viscous pressure* p_{visc}, defined by

$$p_{\mathrm{visc}} = -\zeta\operatorname{tr}\mathsf{D}\,. \tag{3.129}$$

Except for the pressure term, the material function (3.125) corresponds largely to Hooke's law (3.104), and by a computation similar to (3.120) we solve for the divergence of the stress tensor

$$\operatorname{div}\mathsf{t} = -\operatorname{grad}p(\rho,T) + (\lambda+\eta)\operatorname{grad}\operatorname{div}\mathbf{v} + \eta\,\Delta\mathbf{v}\,. \tag{3.130}$$

With this result, the momentum balance (3.97) yields the equation of motion

$$\rho\frac{\mathrm{d}\mathbf{v}}{\mathrm{d}t} = -\operatorname{grad}p(\rho,T) + (\lambda+\eta)\operatorname{grad}\operatorname{div}\mathbf{v} + \eta\,\Delta\mathbf{v} + \mathbf{f}\,, \tag{3.131}$$

which is the *Navier-Stokes equation* for the case of a compressible Newtonian fluid. Note the formal similarity to the Navier equation (3.121) for the Hookean body. If we assume that the temperature of the system is known (for instance, nearly isothermal conditions), then, together with the mass balance (3.96), we have four component equations for the four unknown field components v_x, v_y, v_z and ρ, which is again a closed system.

Incompressible Newtonian Fluid

For the incompressible Newtonian fluid, $\rho = \mathrm{const}$ holds, so that the mass balance reduces to

$$\operatorname{div}\mathbf{v} = 0 \tag{3.132}$$

[see Eq. (3.60)]. It is then convenient to split the stress tensor as

$$\mathsf{t} = -p\,\mathsf{I} + \mathsf{t}^{\mathrm{D}}\,, \tag{3.133}$$

where

$$p = -\tfrac{1}{3}\operatorname{tr}\mathsf{t} \tag{3.134}$$

denotes the pressure, and t^{D} is the traceless *stress deviator* ($\mathrm{tr}\,\mathsf{t}^{\mathrm{D}} = 0$). Now the material function $(3.99)_1$ only determines the stress deviator t^{D} and reads

$$\mathsf{t}^{\mathrm{D}} = 2\eta\,\mathsf{D}\,, \tag{3.135}$$

where the coefficient η is again the *shear viscosity* (or simply the *viscosity*). Note also that the mass balance (3.132) is equivalent to $\mathrm{tr}\,\mathsf{D} = 0$, so that the strain-rate tensor is equal to its deviatoric part, that is, $\mathsf{D} = \mathsf{D}^{\mathrm{D}}$.

In order to derive the equation of motion for the incompressible case, we compute the divergence of the stress tensor with the decomposition (3.133), the material function (3.135) and the mass balance (3.132),

$$\mathrm{div}\,\mathsf{t} = -\mathrm{grad}\,p + \mathrm{div}\,\mathsf{t}^{\mathrm{D}}\,, \tag{3.136}$$

where

$$\begin{aligned}(\mathrm{div}\,\mathsf{t}^{\mathrm{D}})_i &= 2\eta\,D_{ij,j} = \eta\,(v_{i,jj} + v_{j,ij}) = \eta\,(v_{i,jj} + v_{j,ji}) \\ &= \eta\,[v_{i,jj} + (\mathrm{div}\,\mathbf{v})_{,i}] = \eta\,v_{i,jj} = \eta\,(\Delta\mathbf{v})_i\,.\end{aligned} \tag{3.137}$$

Insertion of these results in the momentum balance (3.97) yields the *Navier-Stokes equation* for the incompressible Newtonian fluid,

$$\rho\frac{\mathrm{d}\mathbf{v}}{\mathrm{d}t} = -\mathrm{grad}\,p + \eta\,\Delta\mathbf{v} + \mathbf{f}\,. \tag{3.138}$$

Note that, in contrast to the compressible case, there is only a single pressure p involved. It appears as a free field, so that we have the four component equations (3.132) and (3.138) for the four unknown field components v_x, v_y, v_z and p.

If the viscosity η is temperature-dependent and the temperature is not known *a priori*, then a *thermo-mechanically coupled problem* is obtained, for which the energy balance (3.98) must additionally be solved. This requires that the material functions $(3.99)_2$ and $(3.99)_3$ for the heat flux and the internal energy be specified. Insertion in the energy balance yields the missing evolution equation for the temperature. For instance, let the heat flow be given by *Fourier's law of heat conduction,*

$$\mathbf{q} = -\kappa\,\mathrm{grad}\,T\,, \tag{3.139}$$

and the internal energy depend linearly on temperature,

$$u = u_0 + c(T - T_0)\,, \tag{3.140}$$

where κ is the heat conductivity, c the specific heat, u_0 a fixed reference value for u and T_0 a fixed reference value for T [such that $u(T_0) = u_0$]. Then the energy balance (3.98) results in

$$\rho c \frac{\mathrm{d}T}{\mathrm{d}t} = \kappa \operatorname{div} \operatorname{grad} T + \operatorname{tr}\left[(-p\mathsf{I} + 2\eta\mathsf{D}) \cdot \mathsf{D}\right] + \rho r$$
$$= \kappa \Delta T - p \operatorname{tr} \mathsf{D} + 2\eta \operatorname{tr} \mathsf{D}^2 + \rho r$$

$$\Rightarrow \rho c \frac{\mathrm{d}T}{\mathrm{d}t} = \kappa \Delta T + 2\eta \operatorname{tr} \mathsf{D}^2 + \rho r \tag{3.141}$$

(note that $\operatorname{tr} \mathsf{D} = \operatorname{div} \mathbf{v} = 0$). Now we have the five equations (3.132), (3.138) and (3.141) which govern the evolution of the five fields v_x, v_y, v_z, p and T.

Gravity-Driven Thin Film Flow

Let us consider a thin film (thickness H) of an incompressible Newtonian fluid (density ρ, viscosity η), which flows down an impenetrable plane (inclination angle α) under the influence of gravity (acceleration due to gravity $\mathbf{g}$); see Fig. 3.11. The film is uniform and of infinite extent in the x (downhill) and y (lateral) directions. At the contact between the fluid and the underlying plane, no-slip conditions prevail, and the free surface is stress-free. Further, steady-state conditions are assumed.

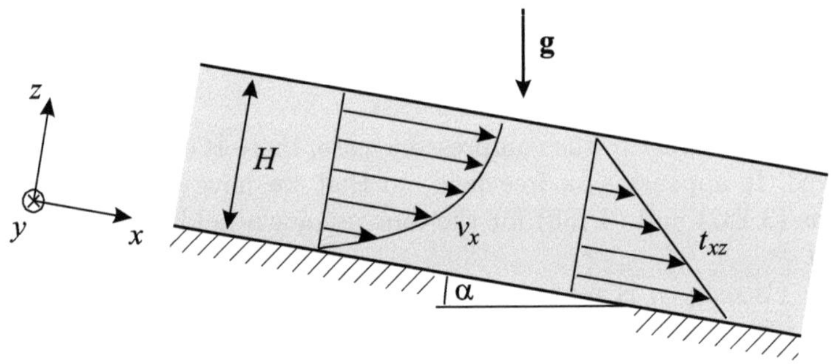

Fig. 3.11. Gravity-driven thin film flow of an incompressible Newtonian fluid.

This problem is a realisation of *plane strain*: due to the uniformity in the y (lateral) direction, the velocity component v_y and the strain-rate components D_{xy}, D_{yy} and D_{yz} vanish, and all field quantities are independent of y. Further, the uniformity in the x (downhill) direction and the steady-state assumption imply that dependencies on x and t will not occur either, so that only dependence on the vertical coordinate z remains. Moreover, due to the impenetrable basal plane, there will be no vertical velocity component v_z, and the only remaining velocity component is $v_x(z)$.

Taking into account $\mathbf{f} = \rho\mathbf{g}$ and $\mathbf{g} = g \sin\alpha\, \mathbf{e}_x - g\cos\alpha\, \mathbf{e}_z$, we note the x-component of the Navier-Stokes equation (3.138),

$$\rho\frac{\mathrm{d}v_x}{\mathrm{d}t} = \rho\left(\frac{\partial v_x}{\partial t} + v_x\frac{\partial v_x}{\partial x} + v_y\frac{\partial v_x}{\partial y} + v_z\frac{\partial v_x}{\partial z}\right)$$
$$= -\frac{\partial p}{\partial x} + \eta\left(\frac{\partial^2 v_x}{\partial x^2} + \frac{\partial^2 v_x}{\partial y^2} + \frac{\partial^2 v_x}{\partial z^2}\right) + \rho g \sin\alpha\,. \tag{3.142}$$

Owing to the above arguments, all terms except the last two vanish, and the equation simplifies to

$$\eta\frac{\partial^2 v_x}{\partial z^2} = -\rho g \sin\alpha\,. \tag{3.143}$$

This can readily be integrated,

$$\eta\frac{\partial v_x}{\partial z} = C_1 - \rho g z \sin\alpha\,, \tag{3.144}$$

where C_1 is an integration constant. Its value can be determined by noting that, due to the material function (3.135), the left-hand side is equal to the shear stress t_{xz},

$$t_{xz} = \eta\frac{\partial v_x}{\partial z} = C_1 - \rho g z \sin\alpha\,, \tag{3.145}$$

which vanishes at the free surface ($z = H$) as a consequence of the stress-free boundary condition. Thus,

$$t_{xz}\big|_{z=H} = C_1 - \rho g H \sin\alpha = 0 \quad\Rightarrow\quad C_1 = \rho g H \sin\alpha\,, \tag{3.146}$$

and we obtain for the shear stress the linear profile

$$t_{xz} = \eta\frac{\partial v_x}{\partial z} = \rho g (H - z) \sin\alpha\,. \tag{3.147}$$

A further integration yields the velocity,

$$v_x = \frac{\rho g}{\eta}\left(Hz - \frac{z^2}{2}\right)\sin\alpha + C_2\,, \tag{3.148}$$

and the integration constant C_2 is evidently equal to zero due to the no-slip condition $v_x|_{z=0} = 0$. Therefore, the solution for the downhill velocity is the parabolic profile

$$v_x = \frac{\rho g H \sin\alpha}{\eta}\left(z - \frac{z^2}{2H}\right)\,. \tag{3.149}$$

The solutions (3.147) and (3.149) are also shown in Fig. 3.11.

Analogous to Eq. (3.142), the z-component of the Navier-Stokes equation (3.138) reads

$$\rho\frac{\mathrm{d}v_z}{\mathrm{d}t} = \rho\left(\frac{\partial v_z}{\partial t} + v_x\frac{\partial v_z}{\partial x} + v_y\frac{\partial v_z}{\partial y} + v_z\frac{\partial v_z}{\partial z}\right)$$
$$= -\frac{\partial p}{\partial z} + \eta\left(\frac{\partial^2 v_z}{\partial x^2} + \frac{\partial^2 v_z}{\partial y^2} + \frac{\partial^2 v_z}{\partial z^2}\right) - \rho g \cos\alpha\,, \tag{3.150}$$

and simplifies for the thin film problem to

$$\frac{\partial p}{\partial z} = -\rho g \cos\alpha \,. \tag{3.151}$$

The integral of this equation is

$$p = C_3 - \rho g z \cos\alpha \,, \tag{3.152}$$

and the integration constant C_3 follows from the stress-free boundary condition at the surface,

$$p|_{z=H} = C_3 - \rho g H \cos\alpha = 0 \quad \Rightarrow \quad C_3 = \rho g H \cos\alpha \,. \tag{3.153}$$

Thus, we obtain

$$p = \rho g (H - z) \cos\alpha \,, \tag{3.154}$$

which is a hydrostatic pressure profile, that is, the pressure at any point in the thin film equals the weight of the overburden fluid.

While one may think first of an oil film ($\eta \sim 0.1\,\mathrm{Pa\,s}$) thinner than one millimetre flowing down some substrate as a realisation of gravity-driven thin film flow, we can also make the film 100 metres thick and assume a viscosity as large as $\eta \sim 10^{14}\,\mathrm{Pa\,s}$. Then we already have a very simple model of a flowing glacier. However, for a realistic description of glacier ice the incompressible Newtonian fluid is not sufficient. In the next chapter we will formulate more appropriate constitutive equations for glacier ice.

4

Constitutive Equations for Polycrystalline Ice

4.1 Microstructure of Ice

The phase of H_2O ice which exists at pressure and temperature conditions encountered in ice sheets and glaciers is called *ice Ih*. It forms hexagonal crystals, that is, the water molecules are arranged in layers of hexagonal rings (Fig. 4.1). The plane of such a layer is called the *basal plane*, which actually consists of two planes shifted slightly (by 0.0923 nm) against each other. The direction perpendicular to the basal planes is the *optic axis* or *c-axis*, and the distance between two adjacent basal planes is 0.276 nm.

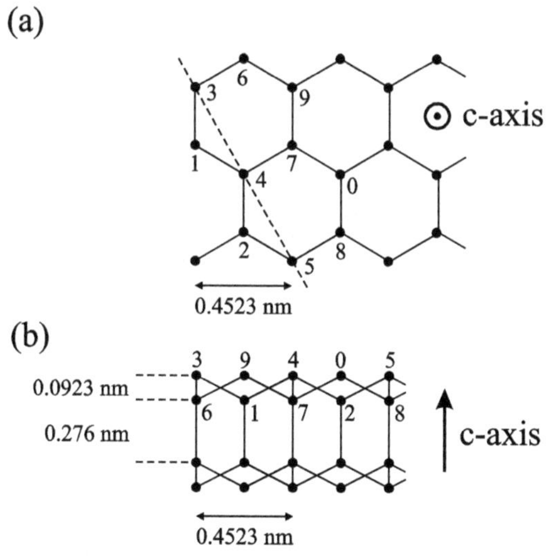

Fig. 4.1. Structure of an ice crystal. The circles denote the oxygen atoms of the H_2O molecules. (a) Projection on the basal plane. (b) Projection on plane indicated by the broken line in (a). Adapted from Paterson (1994), © Elsevier.

R. Greve, H. Blatter, *Dynamics of Ice Sheets and Glaciers*, Advances in Geophysical and Environmental Mechanics and Mathematics,
DOI 10.1007/978-3-642-03415-2_4, © Springer-Verlag Berlin Heidelberg 2009

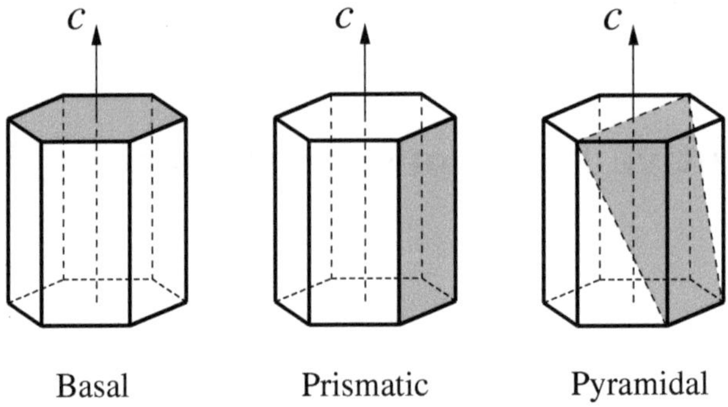

Fig. 4.2. Basal, prismatic and pyramidal glide planes in the hexagonal ice Ih crystal. Reproduced from Faria (2003), © S. H. Faria.

Owing to this relatively large distance, the basal planes can glide on each other when a shear stress is applied, comparable to the deformation of a deck of cards. To a much lesser extent, gliding is also possible in the prismatic and pyramidal planes (see Fig. 4.2). This means that the ice crystal responds to an applied shear stress with a continuous deformation, which goes on as long as the stress is applied (*creep*, fluid-like behaviour) and depends on the direction of the stress relative to the crystal planes (*anisotropy*).

Measurements have shown that ice crystals show some creep even for very low stresses. In a perfect crystal such a behaviour would not be expected. However, in real crystals *dislocations* occur, which are defects in its structure. These imperfections make the crystal much more easily deformable, and this is enhanced even more by the fact that during creep additional dislocations are generated. This creep mechanism is consequently called *dislocation creep*.

4.2 Creep of Polycrystalline Ice

Naturally, ice which occurs in ice sheets and glaciers does not consist of a single ice crystal. Rather, it is composed of a vast number of *crystallites* (also called *grains*), the typical size of which is of the order of millimetres to centimetres. Such a compound is called *polycrystalline ice*. An example is shown in Fig. 4.3.

The *c*-axis orientations of the crystallites in polycrystalline ice differ from one another. In the following, we will assume that the orientation distribution is completely random. In this case, the anisotropy of the crystallites averages out in the compound, so that its macroscopic behaviour will be *isotropic*. In other words, the material properties of polycrystalline ice do not show any directional dependence.

Let us assume to conduct a shear experiment with a small sample of polycrystalline ice as sketched in Fig. 4.4 (left panel). The shear stress τ is assumed

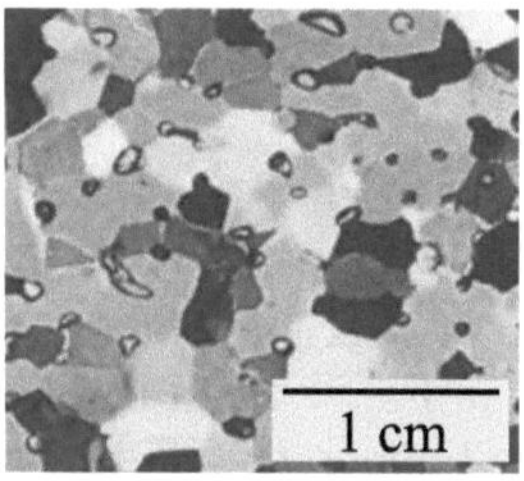

Fig. 4.3. Thin-section of polycrystalline glacier ice regarded between crossed polarisation filters. The crystallites (grains) are clearly visible, and their apparent colours depend on the c-axis orientation.

to be held constant, and the shear angle γ is measured as a function of time. The resulting creep curve $\gamma(t)$ is shown schematically in Fig. 4.4 (right panel). An initial, instantaneous *elastic deformation* of the polycrystalline aggregate is followed by a phase called *primary creep* during which the shear rate $\dot{\gamma}$ decreases continuously. This behaviour is related to the increasing geometric incompatibilities of the deforming crystallites with different orientations. After some time, a minimum shear rate is reached which remains constant subsequently, so that the shear angle increases linearly with time. This phase is known as *secondary creep*. In the case of rather high temperatures and/or high stresses, at a later stage *dynamic recrystallisation* (nucleation and growth of crystallites which are favourably oriented for deformation; also known as *migration recrystallisation*) sets in, which leads to accelerated creep and finally a constant shear rate (linear increase of the shear angle with time) significantly larger than that of the secondary creep. This is called *tertiary creep*.

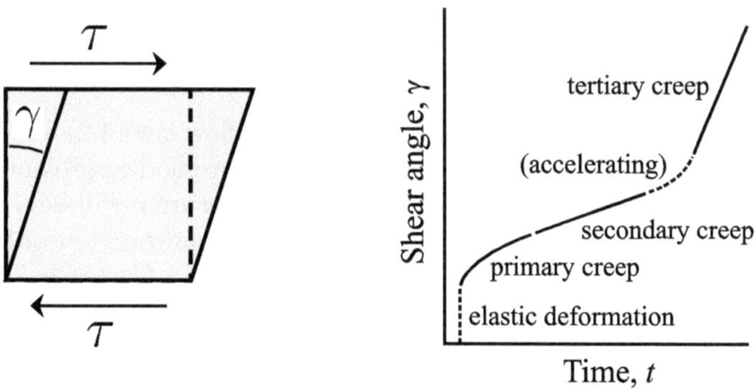

Fig. 4.4. Shear experiment for a sample of polycrystalline ice. τ denotes the applied shear stress, γ the shear angle and t the time.

4.3 Flow Relation

4.3.1 Glen's Flow Law

From the above considerations it becomes clear that the shear rate $\dot{\gamma}$ for secondary (minimum) creep of isotropic polycrystalline ice under the simple-shear conditions illustrated in Fig. 4.4 can be expressed as a unique function of the shear stress τ, the ice temperature T and the pressure p,

$$\dot{\gamma} = \dot{\gamma}(\tau, T, p)\,. \tag{4.1}$$

Numerous laboratory experiments and field measurements suggest that the concrete relation is that of a non-linearly viscous fluid,

$$\dot{\gamma} = \frac{1}{\eta(T, p, |\tau|)}\,\tau\,, \tag{4.2}$$

where η denotes the shear viscosity. The inverse $1/\eta$ is called *fluidity*, and its dependence on the temperature T, the pressure p and the absolute value of the shear stress $|\tau|$ can be factorised as

$$\frac{1}{\eta(T, p, |\tau|)} = 2A(T, p)\, f(|\tau|)\,, \tag{4.3}$$

where $A(T, p)$ is the *rate factor* and $f(|\tau|)$ the *creep function*. These are usually expressed in the form of an *Arrhenius law*

$$A(T, p) = A_0\, e^{-(Q+pV)/RT}\,, \tag{4.4}$$

(A_0: pre-exponential constant, Q: activation energy, V: activation volume, $R = 8.314\ \mathrm{J\,mol^{-1}\,K^{-1}}$: universal gas constant), and a *power law*,

$$f(|\tau|) = |\tau|^{n-1} \tag{4.5}$$

(n: stress exponent), respectively.

Let us now generalise the non-linearly viscous flow law (4.2) for secondary creep to arbitrary deformations and stresses. To a good approximation, ice can be described as incompressible, so that the pressure p will be a free field, and the three-dimensional flow law will relate the strain-rate tensor D and the stress deviator $\mathrm{t^D}$ (compare Sect. 3.4.3). If we define a Cartesian coordinate system such that the plane of Fig. 4.4 (left panel) falls on the x-z plane (where x is the horizontal and z the vertical coordinate), then we can identify $\dot{\gamma} = 2D_{xz}$ [see Eq. (3.35)] and $\tau = t_{xz}$ (see Fig. 3.8), so that (4.2) becomes

$$D_{xz} = \frac{1}{2\eta(T, p, |t_{xz}|)}\,t_{xz}\,. \tag{4.6}$$

Since D_{xz} is the x-z component of D and t_{xz} the x-z component of $\mathrm{t^D}$, this suggests that the general flow law reads

$$\mathsf{D} = \frac{1}{2\eta(T,p,\sigma_\mathrm{e})}\,\mathsf{t}^\mathrm{D}\,. \tag{4.7}$$

The only non-straightforward point is the question how the $|t_{xz}|$ in Eq. (4.6) translates to the newly introduced scalar σ_e (*effective stress*). As a relation between two tensors, Eq. (4.7) must be independent of any particular basis (coordinate system). Therefore, the effective stress σ_e cannot be equal to a single element like $|t_{xz}|$, but it must be a *scalar invariant* of t^D. An order 2 tensor in three-dimensional space has only three independent invariants [see Eq. (2.42)], which are for t^D

$$\begin{aligned} I_{\mathsf{t}^\mathrm{D}} &= \operatorname{tr}\mathsf{t}^\mathrm{D} = 0\,, \\ I\!I_{\mathsf{t}^\mathrm{D}} &= \tfrac{1}{2}\left[\operatorname{tr}(\mathsf{t}^\mathrm{D})^2 - (\operatorname{tr}\mathsf{t}^\mathrm{D})^2\right] = \tfrac{1}{2}\operatorname{tr}(\mathsf{t}^\mathrm{D})^2 \\ &= \tfrac{1}{2}\left[(t^\mathrm{D}_{xx})^2 + (t^\mathrm{D}_{yy})^2 + (t^\mathrm{D}_{zz})^2\right] + t^2_{xy} + t^2_{xz} + t^2_{yz}\,, \\ I\!I\!I_{\mathsf{t}^\mathrm{D}} &= \det\mathsf{t}^\mathrm{D} \end{aligned} \tag{4.8}$$

(note that $t_{ij} = t^\mathrm{D}_{ij}$ for $i \neq j$). If we choose

$$\sigma_\mathrm{e} = \sqrt{I\!I_{\mathsf{t}^\mathrm{D}}} = \sqrt{\tfrac{1}{2}\operatorname{tr}(\mathsf{t}^\mathrm{D})^2}\,, \tag{4.9}$$

then we have found an invariant quantity which simplifies to $|t_{xz}|$ for the simple-shear conditions of (4.6). It is therefore reasonable to assume that (4.9) is the correct expression for the effective stress in the flow law (4.7).

As for the fluidity $1/\eta$ in Eq. (4.7), we can directly infer its functional dependence on T, p and σ_e from Eqs. (4.3), (4.4) and (4.5):

$$\frac{1}{\eta(T,p,\sigma_\mathrm{e})} = 2A(T,p)\,f(\sigma_\mathrm{e}) \tag{4.10}$$

[rate factor $A(T,p)$, creep function $f(\sigma_\mathrm{e})$], with the Arrhenius law

$$A(T,p) = A_0\,e^{-(Q+pV)/RT} \tag{4.11}$$

and the power law

$$f(\sigma_\mathrm{e}) = \sigma_\mathrm{e}^{n-1}\,. \tag{4.12}$$

The optimum value for the stress exponent n has been a matter of continuous debate, but most frequently $n = 3$ is used (Paterson 1994, van der Veen 1999, and references therein).

The melting temperature of ice, T_m, is pressure-dependent. For low pressures ($p \lesssim 100\,\mathrm{kPa}$), $T_\mathrm{m} = T_0 = 273.15\,\mathrm{K}$, and for pressures which occur typically in ice sheets and glaciers ($p \lesssim 50\,\mathrm{MPa}$) the linear relation

$$T_\mathrm{m} = T_0 - \beta\,p \tag{4.13}$$

holds. For pure ice, the *Clausius-Clapeyron constant* β has the value $\beta = 7.42 \times 10^{-8}\,\mathrm{K\,Pa^{-1}}$, but under realistic conditions the value for air-saturated ice, $\beta = 9.8 \times 10^{-8}\,\mathrm{K\,Pa^{-1}}$, is preferable (Hooke 2005). Under hydrostatic conditions, this leads to a melting-point lowering of 0.87 K per kilometre of ice thickness. With (4.13), the *temperature relative to the pressure melting point* is defined as

$$T' = T - T_\mathrm{m} + T_0 = T + \beta p\,, \tag{4.14}$$

so that the pressure melting point always corresponds to $T' = T_0 = 273.15\,\mathrm{K}$ (or 0°C). Measurements have shown that the pressure dependence in the Arrhenius law (4.11) is accounted for satisfactorily if the absolute temperature is replaced by the temperature relative to the pressure melting point, that is

$$A(T,p) = A(T') = A_0\, e^{-Q/RT'}\,. \tag{4.15}$$

Recommended values for the pre-exponential constant and the activation energy are listed in Table 4.1. The larger activation energy for $T' > 263.15\,\mathrm{K}$ is probably due to grain boundary sliding and the presence of liquid water at grain boundaries which contribute to creep in this temperature range [see the discussion by Paterson (1994), and references therein]. The two values of the pre-exponential constant yield $A(T' = 263.15\ \mathrm{K}) = 4.9 \times 10^{-25}\,\mathrm{s^{-1}\,Pa^{-3}}$ for both regimes, so that the function is continuous (Fig. 4.5). Note that these values are only reasonable for $n = 3$.

Parameter	Value	
Stress exponent, n	3	
Pre-exponential constant, A_0	$3.985 \times 10^{-13}\,\mathrm{s^{-1}\,Pa^{-3}}$	(for $T' \le 263.15\,\mathrm{K}$)
	$1.916 \times 10^{3}\,\mathrm{s^{-1}\,Pa^{-3}}$	(for $T' > 263.15\,\mathrm{K}$)
Activation energy, Q	$60\,\mathrm{kJ\,mol^{-1}}$	(for $T' \le 263.15\,\mathrm{K}$)
	$139\,\mathrm{kJ\,mol^{-1}}$	(for $T' > 263.15\,\mathrm{K}$)

Table 4.1. Stress exponent and parameters for the Arrhenius law (4.15) (Paterson 1994).

Equation (4.7) together with (4.10), (4.12) and (4.15) reads

$$\mathsf{D} = A(T')\,\sigma_\mathrm{e}^{n-1}\,\mathsf{t}^\mathrm{D}\,, \tag{4.16}$$

which is called *Nye's generalisation of Glen's flow law*, or *Glen's flow law* for short (Glen 1955, Nye 1957). Figure 4.6 shows the corresponding viscosity

$$\eta(T',\sigma_\mathrm{e}) = \frac{1}{2A(T')\,\sigma_\mathrm{e}^{n-1}} \tag{4.17}$$

for different stresses and temperatures. Evidently, the viscosity of polycrystalline ice is much larger than that of viscous fluids of everyday life. For instance, the viscosity of motor oil is of the order of 0.1 Pa s, compared to $\sim 10^{13}$ Pa s for ice at $T' = 0°\mathrm{C}$ and $\sigma_\mathrm{e} = 100\,\mathrm{kPa}$ (1 bar). On the other hand,

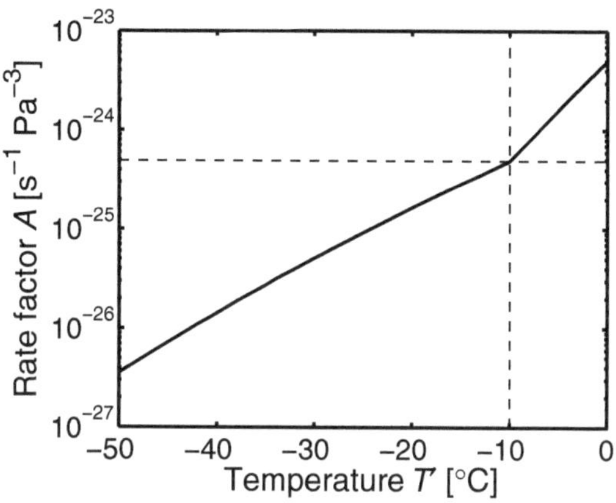

Fig. 4.5. Rate factor $A(T')$ for the temperature range from $-50°\mathrm{C}$ to $0°\mathrm{C}$ (relative to the pressure melting point) according to the Arrhenius law (4.15). The kink at $-10°\mathrm{C}$ is due to the piecewise definition of the pre-exponential constant A_0 and the activation energy Q.

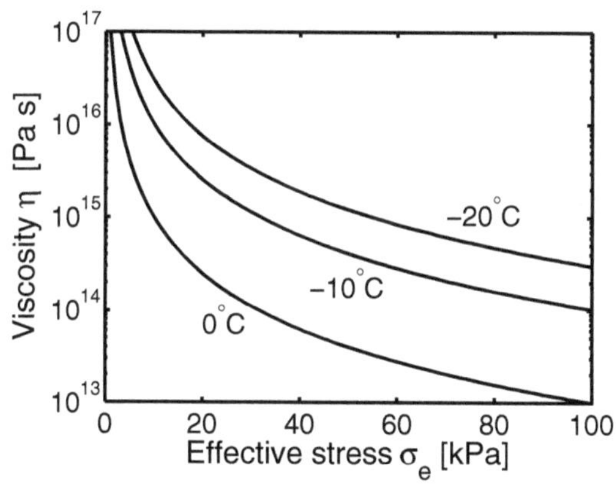

Fig. 4.6. Viscosity (4.17) for a stress exponent $n = 3$, effective stresses up to 100 kPa (1 bar) and temperatures between $-20°\mathrm{C}$ and $0°\mathrm{C}$ (relative to the pressure melting point).

the upper mantle of the Earth has a viscosity of the order of 10^{21} Pa s, which is further eight orders of magnitude stiffer, but still considered to be a fluid on geological time-scales.

In order to derive the inverse form of Glen's flow law, we define the *effective strain rate*

$$d_\mathrm{e} = \sqrt{\tfrac{1}{2}\,\mathrm{tr}\,\mathsf{D}^2}\,, \tag{4.18}$$

[square root of the second invariant of the strain-rate tensor, compare to (4.9)], for which we obtain, by inserting (4.16)

$$d_e = A(T')\,\sigma_e^{n-1}\,\sigma_e = A(T')\,\sigma_e^n \quad \Leftrightarrow \quad \sigma_e = A(T')^{-1/n}\,d_e^{1/n}\,. \tag{4.19}$$

Solving (4.16) for t^D and using (4.19) yields

$$\begin{aligned}\mathsf{t}^\mathrm{D} &= A(T')^{-1}\,\sigma_e^{-(n-1)}\mathsf{D}\\ &= A(T')^{-1}\,A(T')^{(n-1)/n}\,d_e^{-(n-1)/n}\mathsf{D}\\ &= A(T')^{-1/n}\,d_e^{-(1-1/n)}\mathsf{D}\end{aligned}$$

$$\Rightarrow \mathsf{t}^\mathrm{D} = B(T')\,d_e^{-(1-1/n)}\mathsf{D}\,, \tag{4.20}$$

where the *associated rate factor* $B(T') = A(T')^{-1/n}$ has been introduced. We may write this with the shear viscosity η as

$$\mathsf{t}^\mathrm{D} = 2\eta(T', d_e)\,\mathsf{D} \tag{4.21}$$

[see (4.7)], where

$$\eta(T', d_e) = \frac{1}{2}B(T')\,d_e^{-(1-1/n)}\,. \tag{4.22}$$

Evidently, the flow law for polycrystalline ice in the form of (4.7) or (4.21) is very similar to that of the incompressible Newtonian fluid which was discussed in Sect. 3.4.3 [see Eq. (3.135)]. The difference is that here we deal with a *non-linear* flow law, in that the viscosity depends on the effective stress or the effective strain rate.

4.3.2 Regularised Glen's Flow Law

As a consequence of Eqs. (4.10) and (4.12), Glen's flow law (4.16) with $n > 1$ yields an infinite viscosity η if the effective stress σ_e approaches zero. Physically, this is not a problem, because if the effective stress is small, the strain rate is small, too, and does not contribute significantly to the overall flow field. However, depending on the mathematical solution procedure, the infinite viscosity limit may introduce a singularity in the equations for the velocity field if the effective stress is very small, which occurs at ice divides and ice margins. In order to avoid this problem, a regularisation has been proposed, in which the power law, Eq. (4.12), is replaced by a polynomial relation,

$$f(\sigma_e) = \sigma_e^{n-1} + \sigma_0^{n-1} \tag{4.23}$$

("regularised Glen's flow law"), where the *residual stress* σ_0 is a small positive constant. The viscosity is then [see Eq. (4.17)]

$$\eta(T', \sigma_e) = \frac{1}{2A(T')\,[\sigma_e^{n-1} + \sigma_0^{n-1}]}\,, \tag{4.24}$$

which has the finite limit

$$\eta_{\sigma_e \to 0} = \frac{1}{2A(T')\sigma_0^{n-1}} \,. \tag{4.25}$$

This means that for small effective stresses the flow law of a Newtonian fluid (with temperature-dependent viscosity) results. There is some, however inconclusive, experimental evidence that polycrystalline ice may actually show this behaviour, which could serve as a physical justification for the polynomial relation (4.23). Generally, if this relation is used, the residual stress σ_0 is chosen as small as possible in order not to influence the overall solution significantly, but large enough to prevent the problems with the singularity.

Note that, for $n > 1$, an analytical inversion according to Eqs. (4.19) – (4.22) is not possible for the regularised Glen's flow law. We can still state

$$\mathsf{t}^{\mathrm{D}} = 2\eta(T', d_{\mathrm{e}})\,\mathsf{D}\,, \tag{4.26}$$

but do not obtain an explicit expression for the viscosity $\eta(T', d_{\mathrm{e}})$. Instead, by using the definitions (4.9) and (4.18), we infer from (4.26) the relation

$$\sigma_{\mathrm{e}} = 2\eta(T', d_{\mathrm{e}})\, d_{\mathrm{e}}\,. \tag{4.27}$$

Inserting this in Eq. (4.24) yields

$$\eta(T', d_{\mathrm{e}}) = \frac{1}{2A(T')\left[(2\eta(T', d_{\mathrm{e}})\, d_{\mathrm{e}})^{n-1} + \sigma_0^{n-1}\right]}$$

$$\Rightarrow\; 2^n A(T') d_{\mathrm{e}}^{n-1} \eta^n(T', d_{\mathrm{e}}) + 2A(T')\sigma_0^{n-1}\eta(T', d_{\mathrm{e}}) - 1 = 0\,, \tag{4.28}$$

which is an implicit representation of the viscosity $\eta(T', d_{\mathrm{e}})$ as a polynomial equation.

4.3.3 Smith-Morland Flow Law

An alternative flow law with finite viscosity in the limit $\sigma_{\mathrm{e}} \to 0$ was proposed by Smith and Morland (1981). It has the form

$$\mathsf{D} = A(T)\, f(\sigma_{\mathrm{e}})\, \mathsf{t}^{\mathrm{D}}\,, \tag{4.29}$$

where the dimensionless rate factor is given by the two-exponential term expression

$$A(T) = 0.7242\, \mathrm{e}^{11.9567\,\bar{T}} + 0.3438\, \mathrm{e}^{2.9494\,\bar{T}}\,, \quad \bar{T} = \frac{T - T_0}{[\Delta T]}\,, \tag{4.30}$$

and the creep function is represented by

$$f(\sigma_{\mathrm{e}}) = \frac{D_0}{\sigma_0}\left[0.3336 + 0.3200\left(\frac{\sigma_{\mathrm{e}}}{\sigma_0}\right)^2 + 0.02963\left(\frac{\sigma_{\mathrm{e}}}{\sigma_0}\right)^4\right]. \tag{4.31}$$

The constants in Eqs. (4.30) and (4.31) have the values $T_0 = 273.15\,\mathrm{K}$, $[\Delta T] = 20\,\mathrm{K}$, $D_0 = 1\,\mathrm{a}^{-1} = 3.169 \times 10^{-8}\,\mathrm{s}^{-1}$ and $\sigma_0 = 10^5\,\mathrm{Pa}$. Note that the rate factor (4.30) is a function of the absolute temperature T instead of the temperature relative to the pressure melting point T', so that it contains no pressure dependence.

The flow law (4.29) with the rate factor (4.30) and creep function (4.31) is referred to as the *Smith-Morland flow law.* Its viscosity is given by

$$\eta(T, \sigma_\mathrm{e}) = \frac{1}{2A(T)\, f(\sigma_\mathrm{e})} \,. \tag{4.32}$$

As for the case of the regularised Glen flow law (Sect. 4.3.2), the Smith-Morland flow law can formally be inverted,

$$\mathsf{t}^\mathrm{D} = 2\eta(T, d_\mathrm{e})\,\mathsf{D}\,, \tag{4.33}$$

but an explicit representation of the viscosity $\eta(T, d_\mathrm{e})$ cannot be obtained.

The Smith-Morland flow law has not been widely used, even though the authors claim that it matches laboratory data on ice deformation better than Glen's flow law. It is therefore worth being explored in more detail in future studies.

4.3.4 Flow Enhancement Factor

All flow laws of Sects. 4.3.1–4.3.3 are valid for secondary creep of isotropic polycrystalline ice. However, as we have discussed in Sect. 4.2, in regions of flowing ice sheets and glaciers with relatively high temperatures and/or stresses, tertiary creep may prevail, which goes along with the formation of an anisotropic fabric (non-uniform orientation distribution of the *c*-axes) favourable for the deformation regime at hand.

A crude, but very common way of including this effect in the flow law is by multiplying the isotropic ice fluidity for secondary creep by a *flow enhancement factor* $E > 1$ (Hooke 2005). This can be conveniently achieved by replacing the rate factor $A(T')$ for the Glen and regularised Glen flow law [or $A(T)$ for the Smith-Morland flow law] by

$$A(T') \rightarrow EA(T')\,. \tag{4.34}$$

Suggested values for the flow enhancement factor vary and depend on the deformation regime; however, in practice often an overall constant value somewhere between 1 and 10 for the considered ice sheet or glacier is chosen.

In case of Glen's flow law, we have seen that an analytical inversion is possible. Equation (4.34) yields for the associated rate factor introduced in Eq. (4.20)

$$B(T') = A(T')^{-1/n} \;\rightarrow\; [EA(T')]^{-1/n} = E_\mathrm{s}B(T')\,, \tag{4.35}$$

where $E_\mathrm{s} = E^{-1/n}$ is the *stress enhancement factor.*

4.4 Heat Flux and Internal Energy

The heat flux $\mathbf{q}$ in polycrystalline ice can be described well by Fourier's law of heat conduction [see (3.139)],

$$\mathbf{q} = -\kappa(T)\,\mathrm{grad}\,T\,, \tag{4.36}$$

with the temperature-dependent heat conductivity

$$\kappa(T) = 9.828\,e^{-0.0057\,T[\mathrm{K}]}\;\mathrm{W\,m^{-1}K^{-1}} \tag{4.37}$$

(Ritz 1987). For $T = T_0 = 273.15\,\mathrm{K}$ this yields a value of $2.07\,\mathrm{W\,m^{-1}K^{-1}}$, and it increases with decreasing temperature (Fig. 4.7, top panel).

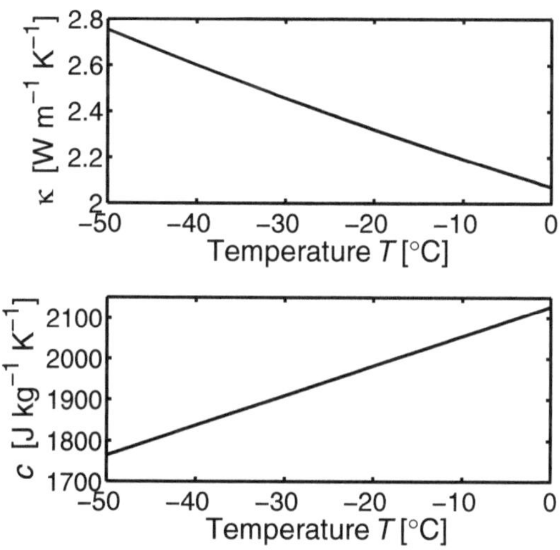

Fig. 4.7. Heat conductivity κ and specific heat c for the temperature range from −50°C until 0°C.

The caloric equation of state (constitutive equation for the internal energy) is given by

$$u = \int_{T_0}^{T} c(\bar{T})\,\mathrm{d}\bar{T}\,, \tag{4.38}$$

which is a generalisation of Eq. (3.140) with the temperature-dependent specific heat

$$c(T) = (146.3 + 7.253\,T[\mathrm{K}])\;\mathrm{J\,kg^{-1}K^{-1}} \tag{4.39}$$

(Ritz 1987). According to this formula, at $T = T_0 = 273.15\,\mathrm{K}$ one obtains $2127.5\,\mathrm{J\,kg^{-1}K^{-1}}$. Contrary to the heat conductivity, the specific heat decreases with decreasing temperature (Fig. 4.7, bottom panel).

4.5 Elasticity

For processes on short time-scales (days or less), such as the response of an ice shelf to oceanic tides, the elastic deformation of polycrystalline ice will be dominant compared to the deformation by viscous creep. In this situation, the material behaviour of ice can be described well by Hooke's law in the form (3.104), (3.106) or (3.116).

Parameter	Value
Young's modulus, E	9.33×10^9 Pa
Poisson's ratio, ν	0.325
1st Lamé parameter, λ	6.54×10^9 Pa
2nd Lamé parameter (shear modulus), μ	3.52×10^9 Pa

Table 4.2. Elastic parameters for isotropic polycrystalline ice at $T = -16°\text{C}$ (Petrenko and Whitworth 1999).

Suitable values for the material parameters at $T = -16°\text{C}$ are listed in Table 4.2. The temperature dependence is rather small; Petrenko and Whitworth (1999) give the formula

$$\begin{pmatrix} E(T) \\ \lambda(T) \\ \mu(T) \end{pmatrix} = \begin{pmatrix} E \\ \lambda \\ \mu \end{pmatrix}\Bigg|_{T=-16°\text{C}} \times \left[1 - 1.42 \times 10^{-3}\,(T[°\text{C}] + 16)\right], \qquad (4.40)$$

while the temperature dependence of Poisson's ratio ν can be neglected.

5

Large-Scale Dynamics of Ice Sheets

5.1 Full Stokes Flow Problem

5.1.1 Field Equations

With the constitutive equations given in Sects. 4.3 and 4.4, we are now able to formulate the mechanical and thermodynamical field equations for the flow of ice in an ice sheet. Figure 5.1 shows the typical geometry (cross section) of a grounded ice sheet with attached floating ice shelf (the latter will be treated in Chap. 6), as well as its interactions with the atmosphere (snowfall, melting), the lithosphere (geothermal heat flux, isostasy) and the ocean (melting, calving). Also, a Cartesian coordinate system is introduced, where x and y lie in the horizontal plane, and z is positive upward. These coordinates are naturally associated with the set of basis vectors $\{\mathbf{e}_x, \mathbf{e}_y, \mathbf{e}_z\}$. The free surface (ice-atmosphere interface) is given by the function $z = h(x, y, t)$, the ice base by $z = b(x, y, t)$ and the lithosphere surface by $z = z_\mathrm{l}(x, y, t)$. Note that for the grounded ice sheet the ice base and the lithosphere surface fall together ($b = z_\mathrm{l}$) and form the ice-lithosphere interface.

By introducing the Cartesian coordinates x, y, z, we have tacitly assumed a flat Earth. For the vertical direction, this simplification is justified because the vertical extent of ice sheets (as well as ice shelves and glaciers) is always much smaller than the mean radius of the Earth ($R_\mathrm{e} = 6371\,\mathrm{km}$), so that curvature effects are negligible. In the horizontal, the flattening can be achieved by a suitable map projection. For ice sheets, often the polar stereographic projection is used, which is illustrated in Fig. 5.2. It preserves angles, but not distances and areas. The distortions are negligible for most practical applications, though. Even for the entire Antarctic Ice Sheet (situated between $\sim 63°$S and 90°S), the distortion of the length scale nowhere exceeds 3% if the standard parallel is chosen as $\varphi_0 = 71°$S.

Since we have assumed ice to be an incompressible material, the mass balance (3.60) applies,

$$\operatorname{div} \mathbf{v} = 0 \,. \tag{5.1}$$

R. Greve, H. Blatter, *Dynamics of Ice Sheets and Glaciers*, Advances in Geophysical and Environmental Mechanics and Mathematics,
DOI 10.1007/978-3-642-03415-2_5, © Springer-Verlag Berlin Heidelberg 2009

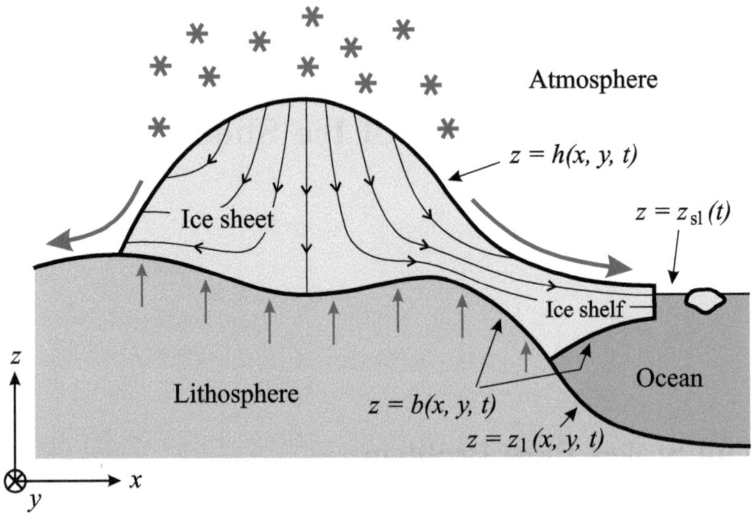

Fig. 5.1. Ice sheet geometry (with attached ice shelf) and Cartesian coordinate system. x and y span the horizontal plane, z is positive upward. $z = h(x, y, t)$ denotes the free surface, $z = b(x, y, t)$ the ice base, $z = z_l(x, y, t)$ the lithosphere surface and $z = z_{sl}(t)$ the mean sea level. Interactions with the atmosphere, the lithosphere and the ocean are indicated. Vertical exaggeration factor $\sim$ 200-500.

The flow law in the form (4.21) yields for the divergence of the stress deviator [note that, contrary to (3.137), η is not constant]

$$\begin{aligned}
(\operatorname{div} \mathbf{t}^{\mathrm{D}})_i &= 2(\eta\, D_{ij})_{,j} = 2\eta\, D_{ij,j} + 2D_{ij}\, \eta_{,j} \\
&= \eta\, (v_{i,jj} + v_{j,ij}) + (v_{i,j} + v_{j,i})\, \eta_{,j} \\
&= \eta\, [(\Delta \mathbf{v})_i + (\operatorname{div} \mathbf{v})_{,i}] + \left[\left(\operatorname{grad} \mathbf{v} + (\operatorname{grad} \mathbf{v})^{\mathrm{T}}\right) \cdot \operatorname{grad} \eta\right]_i \\
&= \eta\, (\Delta \mathbf{v})_i + \left[\left(\operatorname{grad} \mathbf{v} + (\operatorname{grad} \mathbf{v})^{\mathrm{T}}\right) \cdot \operatorname{grad} \eta\right]_i .
\end{aligned} \tag{5.2}$$

The volume force $\mathbf{f}$ acting on an ice sheet on the rotating Earth consists of the force of gravity, the centrifugal force and the Coriolis force (the latter two are inertial forces). Since the centrifugal force depends only on position, it is usually combined with the actual force of gravity to form the effective force of gravity $\rho\mathbf{g}$, where $\rho = 910\,\mathrm{kg\,m^{-3}}$ is the density of ice, and $\mathbf{g}$ is the gravitational acceleration. On the surface of the Earth, the gravitational acceleration takes values between $\sim$ 9.78 and $9.83\,\mathrm{m\,s^{-2}}$ depending on latitude. Since this variability is negligible for our purposes, we adopt the constant standard value $g = |\mathbf{g}| = 9.81\,\mathrm{m\,s^{-2}}$ instead. The vector $\mathbf{g}$ is directed downward, so that $\mathbf{g} = -g\,\mathbf{e}_z$. The Coriolis force depends on the flow velocity $\mathbf{v}$ and the angular velocity $\boldsymbol{\Omega}$ of the Earth. The vector $\boldsymbol{\Omega}$ points northward parallel to the rotational axis of the Earth, and its value is $\Omega = |\boldsymbol{\Omega}| = 2\pi/(23.9345\,\mathrm{h}) = 7.2921 \times 10^{-5}\,\mathrm{s^{-1}}$. Hence, the volume force reads

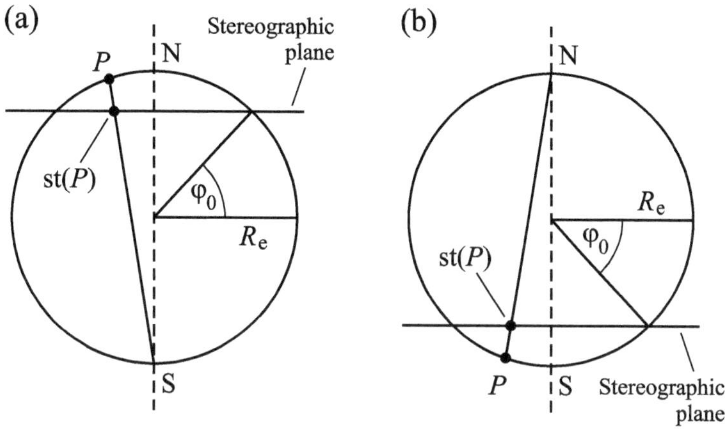

Fig. 5.2. Polar stereographic projection for (a) the northern and (b) the southern hemisphere. The stereographic plane is parallel to the equatorial plane and defined by the standard parallel φ_0 (often chosen as 71°N or 71°S). A point P on the surface of the Earth is projected on the point st(P) by intersecting the line $\overline{PS}$ (case a) or $\overline{PN}$ (case b) with the stereographic plane.

$$\mathbf{f} = \rho\mathbf{g} - 2\rho\boldsymbol{\Omega} \times \mathbf{v}\,. \tag{5.3}$$

By inserting Eqs. (5.2) and (5.3) in the momentum balance (3.72) we obtain the equation of motion

$$\begin{aligned} \rho\frac{\mathrm{d}\mathbf{v}}{\mathrm{d}t} &= -\operatorname{grad} p + \eta\,\Delta\mathbf{v} + \big(\operatorname{grad}\mathbf{v} + (\operatorname{grad}\mathbf{v})^{\mathrm{T}}\big)\cdot\operatorname{grad}\eta \\ &\quad + \rho\mathbf{g} - 2\rho\boldsymbol{\Omega}\times\mathbf{v}\,. \end{aligned} \tag{5.4}$$

In this equation, let us compare the acceleration term on the left-hand side with the pressure-gradient term on the right-hand side. To this end, we introduce *typical values* for the horizontal and vertical extent of an ice sheet, the horizontal and vertical flow velocities, the pressure and the time as follows,

$$\begin{aligned} \text{typical horizontal extent } [L] &= 1000\,\mathrm{km}\,, \\ \text{typical vertical extent } [H] &= 1\,\mathrm{km}\,, \\ \text{typical horizontal velocity } [U] &= 100\,\mathrm{m\,a^{-1}}\,, \\ \text{typical vertical velocity } [W] &= 0.1\,\mathrm{m\,a^{-1}}\,, \\ \text{typical pressure } [P] &= \rho g[H] \approx 10\,\mathrm{MPa}\,, \\ \text{typical time-scale } [t] &= [L]/[U] = [H]/[W] = 10^4\,\mathrm{a}\,. \end{aligned} \tag{5.5}$$

Further, the *aspect ratio* ε is defined as the ratio of vertical to horizontal extents and velocities, respectively:

$$\varepsilon = \frac{[H]}{[L]} = \frac{[W]}{[U]} = 10^{-3}\,. \tag{5.6}$$

For the horizontal direction, the ratio of acceleration and pressure gradient, called the *Froude number* Fr, is then

$$Fr = \frac{\rho[U]/[t]}{[P]/[L]} = \frac{\rho[U]^2/[L]}{\rho g[H]/[L]} = \frac{[U]^2}{g[H]} \approx 10^{-15} \tag{5.7}$$

(note that $1\,\mathrm{a} = 31556926\,\mathrm{s} \approx \sqrt{10^{15}}\,\mathrm{s}$), and for the vertical direction we obtain the ratio

$$\frac{\rho[W]/[t]}{[P]/[H]} = \frac{\rho[W]^2/[H]}{\rho g[H]/[H]} = \frac{[W]^2}{g[H]} = \varepsilon^2 Fr \approx 10^{-21}\,. \tag{5.8}$$

Consequently, for the flow of ice sheets, the acceleration term in the equation of motion (5.4) is negligible.

In a similar way, we estimate the ratio between the Coriolis and pressure-gradient terms in Eq. (5.4). Since the cross product in the Coriolis term mixes horizontal and vertical contributions, we apply the common scales $2\rho\Omega[U]$ and $[P]/[L]$, respectively, for both the horizontal and vertical direction. By introducing the *Rossby number* Ro as

$$Ro = \frac{[U]}{2\Omega[L]} \approx 2 \times 10^{-8}\,, \tag{5.9}$$

the Coriolis-force-to-pressure-gradient ratio yields

$$\frac{2\rho\Omega[U]}{[P]/[L]} = \frac{2\Omega[U][L]}{g[H]} = \frac{[U]^2}{g[H]}\,\frac{2\Omega[L]}{[U]} = \frac{Fr}{Ro} \approx 5 \times 10^{-8}\,, \tag{5.10}$$

which is seven orders of magnitude larger than the Froude number, but still very small. Hence, the Coriolis term in the equation of motion (5.4) is also negligible, and it can be simplified to

$$-\operatorname{grad} p + \eta\,\Delta\mathbf{v} + \left(\operatorname{grad}\mathbf{v} + (\operatorname{grad}\mathbf{v})^{\mathrm{T}}\right)\cdot\operatorname{grad}\eta + \rho\mathbf{g} = \mathbf{0}\,. \tag{5.11}$$

This is the *Stokes equation*, and the resulting type of flow is called *Stokes flow*.

Since the Stokes equation is a differential equation for the velocity field, it is favourable to employ the form of the viscosity η which depends via the effective strain rate d_e on the velocity gradient. Therefore, $\eta = \eta(T', d_\mathrm{e})$, and, depending on whether the normal or the regularised Glen's flow law shall be used, it is determined either by Eq. (4.22) or by Eq. (4.28).

Owing to the temperature dependence of the viscosity, a thermo-mechanically coupled problem applies, and its complete formulation requires an evolution equation for the temperature field. As it was demonstrated in Sect. 3.4.3, this equation can be derived by inserting the constitutive equations for the stress deviator (4.21), the heat flux (4.36) and the internal energy (4.38) in the internal-energy balance (3.92). We obtain

$$\frac{\mathrm{d}u}{\mathrm{d}t} = c\frac{\mathrm{d}T}{\mathrm{d}t}\,, \qquad \operatorname{div}\mathbf{q} = -\operatorname{div}(\kappa\operatorname{grad}T) \tag{5.12}$$

and

$$\mathrm{tr}\,(\mathsf{t}\cdot\mathsf{D}) = \mathrm{tr}\,[(-p\,\mathsf{I} + 2\eta\mathsf{D})\cdot\mathsf{D}] = 2\eta\,\mathrm{tr}\,\mathsf{D}^2 = 4\eta\,d_{\mathrm{e}}^2 \qquad (5.13)$$

[see (4.18)]. Further, except for the very uppermost few centimetres of ice exposed to sunlight, the radiation r is negligible in an ice sheet, so that we obtain the temperature evolution equation in the form

$$\rho c\frac{\mathrm{d}T}{\mathrm{d}t} = \mathrm{div}\,(\kappa\,\mathrm{grad}\,T) + 4\eta\,d_{\mathrm{e}}^2\,. \qquad (5.14)$$

Since the ice temperature must not exceed the pressure melting point, the solution of (5.14) is subject to the secondary condition $T \le T_{\mathrm{m}}$. With the continuity equation (5.1), the equation of motion (5.11), the expressions (4.22) or (4.28) for the viscosity and the temperature evolution equation (5.14), we have found a closed system of six equations for the six unknown fields v_x, v_y, v_z, η, p and T of the thermo-mechanical Stokes flow problem.

5.1.2 Boundary Conditions

In order to provide a solvable problem, the above system of equations needs to be completed by appropriate *boundary conditions* at the free surface and the ice base (see Fig. 5.1). The possible presence of attached ice shelves will be ignored for now.

Free Surface

Like any boundary, the free surface of an ice sheet can be regarded as a singular surface in the sense of Sect. 3.3.3. If we denote it in implicit form by the equation

$$F_{\mathrm{s}}(\mathbf{x},t) = z - h(x,y,t) = 0\,, \qquad (5.15)$$

then it can be interpreted as a zero-equipotential surface of the function $F_{\mathrm{s}}(\mathbf{x},t)$, where the unit normal vector is the normalised gradient

$$\mathbf{n} = \frac{\mathrm{grad}\,F_{\mathrm{s}}}{|\mathrm{grad}\,F_{\mathrm{s}}|} = \frac{1}{N_{\mathrm{s}}}\begin{pmatrix} -\dfrac{\partial h}{\partial x} \\ -\dfrac{\partial h}{\partial y} \\ 1 \end{pmatrix}, \qquad (5.16)$$

which points into the atmosphere (Fig. 5.3). Note that the abbreviation N_{s} stands for the gradient norm,

$$N_{\mathrm{s}} = |\mathrm{grad}\,F_{\mathrm{s}}| = \left(1 + \left(\frac{\partial h}{\partial x}\right)^2 + \left(\frac{\partial h}{\partial y}\right)^2\right)^{1/2}. \qquad (5.17)$$

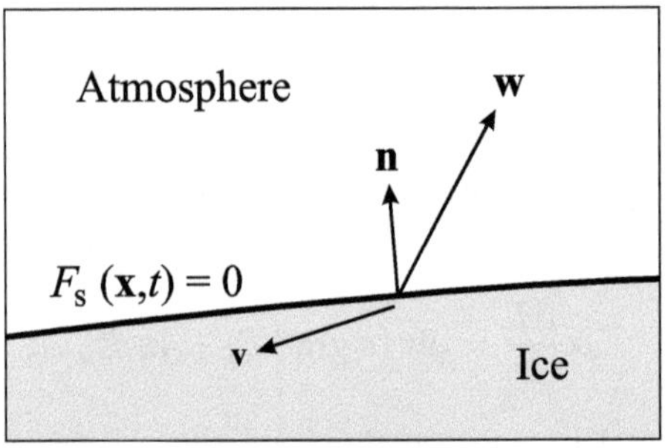

Fig. 5.3. Geometry of the free surface $F_s(\mathbf{x}, t) = 0$. $\mathbf{n}$ is the unit normal vector, $\mathbf{v}$ the ice velocity and $\mathbf{w}$ the velocity of the free surface.

As a direct consequence of Eq. (5.15), the time derivative of F_s *following the motion of the free surface* with velocity $\mathbf{w}$ must vanish,

$$\frac{\mathrm{d}_w F_s}{\mathrm{d}t} = \frac{\partial F_s}{\partial t} + (\operatorname{grad} F_s) \cdot \mathbf{w} = 0 \tag{5.18}$$

[compare Eq. (3.23)]. Let $\mathbf{v}$ be the ice surface velocity, then we can introduce the ice volume flux through the free surface,

$$a_s^\perp = (\mathbf{w} - \mathbf{v}) \cdot \mathbf{n}\,, \tag{5.19}$$

which is also known as the *accumulation-ablation function* or *surface mass balance* (perpendicular to the free surface). The sign is chosen such that a supply (accumulation) is counted as positive and a loss (ablation) as negative. With this definition and (5.16), Eq. (5.18) can be rewritten as

$$\frac{\partial F_s}{\partial t} + (\operatorname{grad} F_s) \cdot \mathbf{v} = -N_s a_s^\perp\,, \tag{5.20}$$

or, by inserting $F_s = z - h$ [see (5.15)],

$$\frac{\partial h}{\partial t} + v_x \frac{\partial h}{\partial x} + v_y \frac{\partial h}{\partial y} - v_z = N_s a_s^\perp\,. \tag{5.21}$$

Since this condition has been derived by geometrical considerations only, it is called the *kinematic boundary condition.* Provided that the accumulation-ablation function $a_s^\perp$ is known, it evidently governs the evolution of the free surface.

If we identify the positive side of the free surface with the atmosphere and the negative side with the ice, then the momentum jump condition (3.73) (note that the free surface is a non-material surface if $a_s^\perp \neq 0$) yields

$$\begin{aligned} &\mathsf{t}_{\text{atm}} \cdot \mathbf{n} - \mathsf{t} \cdot \mathbf{n} - \rho((\mathbf{v} - \mathbf{w}) \cdot \mathbf{n})\, [\![\mathbf{v}]\!] \\ &= \mathsf{t}_{\text{atm}} \cdot \mathbf{n} - \mathsf{t} \cdot \mathbf{n} + \rho a_s^\perp\, [\![\mathbf{v}]\!] = \mathbf{0} \end{aligned} \tag{5.22}$$

[in the first line, the term $\rho((\mathbf{v}-\mathbf{w})\cdot\mathbf{n})$ can be factored out of the jump brackets because of the mass balance (3.61)]. The advective term $\rho a_\mathrm{s}^\perp\,[\![\mathbf{v}]\!]$ is very small and thus negligible due to the small ice flow velocities, and the atmospheric stress $\mathbf{t}_\mathrm{atm}\cdot\mathbf{n}$ (atmospheric pressure plus wind stress) is also small compared to the typical stresses in an ice sheet. Thus, we can neglect both terms, and obtain the *stress-free condition*

$$\mathbf{t}\cdot\mathbf{n}=\mathbf{0}\,. \tag{5.23}$$

This is the *dynamic boundary condition* for the free surface.

For the temperature evolution equation (5.14), it is further required to provide a *thermodynamic boundary condition.* This can be simply done by prescribing the surface temperature T_s,

$$T=T_\mathrm{s}\,. \tag{5.24}$$

Measurements have shown that T_s can be well approximated by the mean-annual surface air temperature, as long as the latter is $\le 0°\mathrm{C}$.

Ice Base

In a similar manner to the free surface, a *kinematic boundary condition* for the ice base can be derived. Let

$$F_\mathrm{b}(\mathbf{x},t)=b(x,y,t)-z=0 \tag{5.25}$$

be its implicit representation, then the unit normal vector is

$$\mathbf{n}=\frac{\operatorname{grad}F_\mathrm{b}}{|\operatorname{grad}F_\mathrm{b}|}=\frac{1}{N_\mathrm{b}}\begin{pmatrix}\dfrac{\partial b}{\partial x}\\ \dfrac{\partial b}{\partial y}\\ -1\end{pmatrix}, \tag{5.26}$$

which points into the bedrock (Fig. 5.4). The abbreviation N_b denotes the gradient norm,

$$N_\mathrm{b}=|\operatorname{grad}F_\mathrm{b}|=\left(1+\left(\frac{\partial b}{\partial x}\right)^2+\left(\frac{\partial b}{\partial y}\right)^2\right)^{1/2}. \tag{5.27}$$

Analogous to Eq. (5.18), the time derivative of F_b following the motion of the ice base vanishes,

$$\frac{\mathrm{d}_w F_\mathrm{b}}{\mathrm{d}t}=\frac{\partial F_\mathrm{b}}{\partial t}+(\operatorname{grad}F_\mathrm{b})\cdot\mathbf{w}=0\,, \tag{5.28}$$

where $\mathbf{w}$ is the velocity of the ice base. With the ice volume flux through the base,

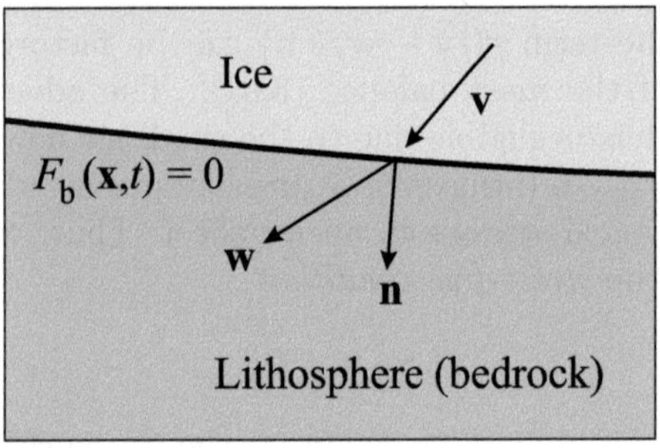

Fig. 5.4. Geometry of the ice base $F_b(\mathbf{x}, t) = 0$. $\mathbf{n}$ is the unit normal vector, $\mathbf{v}$ the ice velocity and $\mathbf{w}$ the velocity of the ice base.

$$a_b^{\perp} = (\mathbf{v} - \mathbf{w}) \cdot \mathbf{n}\,, \tag{5.29}$$

where the sign has been chosen such that a mass loss due to basal melting (and subsequent penetration of the meltwater into the ground) is counted as positive, and which is therefore called the *basal melting rate* (perpendicular to the ice base), we obtain

$$\frac{\partial F_b}{\partial t} + (\mathrm{grad}\, F_b) \cdot \mathbf{v} = N_b a_b^{\perp}\,, \tag{5.30}$$

and, by inserting $F_b = b - z$ [Eq. (5.25)],

$$\frac{\partial b}{\partial t} + v_x \frac{\partial b}{\partial x} + v_y \frac{\partial b}{\partial y} - v_z = N_b a_b^{\perp}\,. \tag{5.31}$$

We identify the positive side of the ice base with the lithosphere and the negative side with the ice. Then, corresponding to Eq. (5.22), the momentum jump condition reads

$$\mathsf{t}_{\mathrm{lith}} \cdot \mathbf{n} - \mathsf{t} \cdot \mathbf{n} - \rho a_b^{\perp}\, [\![\mathbf{v}]\!] = \mathbf{0}\,. \tag{5.32}$$

Again, the advective term $\rho a_b^{\perp}\, [\![\mathbf{v}]\!]$ is very small and therefore negligible, so that

$$\mathsf{t} \cdot \mathbf{n} = \mathsf{t}_{\mathrm{lith}} \cdot \mathbf{n}\,. \tag{5.33}$$

This is to say that the stress vector is continuous across the interface. However, since we do not have any information about the stress conditions in the bedrock, this finding does not provide a boundary condition for the basal stress in the ice. Instead, an empirical *sliding law* will serve as the required dynamic boundary condition. It is reasonable to assume that the ice is frozen to the ground if the basal temperature T_b is below the pressure melting point T_m, so that *no-slip conditions* prevail. By contrast, if the basal temperature is at the pressure melting point, basal sliding can be expected, and its amount can be related to the basal drag $\boldsymbol{\tau}_b$ and the basal normal stress N_b in the form of a power law (*Weertman-type sliding law*). To this end, we split up the basal stress vector $\mathbf{t_n}|_{z=b} = \mathsf{t}|_{z=b} \cdot \mathbf{n}$ into its normal component, $\mathbf{N}_\mathrm{b} = -N_\mathrm{b}\mathbf{n}$ (the minus sign causes N_b to be positive for compression, which is the expected situation), and its shear component, $\boldsymbol{\tau}_\mathrm{b} = \tau_\mathrm{b}\mathbf{e}_\mathrm{t}$,

$$\mathbf{t_n}|_{z=b} = \mathbf{N}_\mathrm{b} + \boldsymbol{\tau}_\mathrm{b} = -N_\mathrm{b}\mathbf{n} + \tau_\mathrm{b}\mathbf{e}_\mathrm{t}\,, \tag{5.34}$$

where $\mathbf{e}_\mathrm{t}$ denotes the direction of the basal shear stress in the plane tangential to the ice base ($\mathbf{e}_\mathrm{t} \perp \mathbf{n}$). The basal sliding velocity $\mathbf{v}_\mathrm{b}$ is then expressed as

$$\mathbf{v}_\mathrm{b} = \begin{cases} \mathbf{0}\,, & \text{if } T_\mathrm{b} < T_\mathrm{m}\,, \\ -C_\mathrm{b}\dfrac{\tau_\mathrm{b}^p}{N_\mathrm{b}^q}\,\mathbf{e}_\mathrm{t}\,, & \text{if } T_\mathrm{b} = T_\mathrm{m}\,, \end{cases} \tag{5.35}$$

where p and q are the basal sliding exponents. As for the stress exponent n in the creep function (4.12) of Glen's flow law, the best choice for their values are a matter of debate, but commonly used values are $(p, q) = (3, 1)$ or $(3, 2)$ for sliding on hard rock, and $(p, q) = (1, 0)$ for sliding on soft, deformable sediment.

As for the *thermodynamic boundary condition*, sufficient information on the spatio-temporal distribution of the basal temperature is not available, so that it cannot be prescribed directly. Instead, we will have to formulate the energy jump condition (3.94) for the ice base. We obtain

$$\mathbf{q}_\mathrm{lith} \cdot \mathbf{n} - \mathbf{q} \cdot \mathbf{n} - [\![\mathbf{v}]\!] \cdot \mathsf{t} \cdot \mathbf{n} + \rho((\mathbf{v} - \mathbf{w}) \cdot \mathbf{n}) \left[\!\!\left[u + \frac{v^2}{2} \right]\!\!\right] = 0\,. \tag{5.36}$$

Note that the term $\rho((\mathbf{v} - \mathbf{w}) \cdot \mathbf{n})$ has been factored out of the jump brackets because of the mass balance (3.61), and the stress vector $\mathsf{t}\cdot\mathbf{n}$ has been factored out due to (5.33). The term $-\mathbf{q}_\mathrm{lith} \cdot \mathbf{n}$ can be identified with the *geothermal heat flux* $q_\mathrm{geo}^\perp$, that is, the heat flow which enters the ice body from below due to the warmer Earth's interior. Further, we insert Eqs. (4.36) and (5.29), and neglect the very small kinetic energy $v^2/2$ in comparison with the internal energy u, so that

$$\kappa\,(\mathrm{grad}\,T \cdot \mathbf{n}) - q_\mathrm{geo}^\perp - [\![\mathbf{v}]\!] \cdot \mathsf{t} \cdot \mathbf{n} + \rho a_\mathrm{b}^\perp\,[\![u]\!] = 0\,. \tag{5.37}$$

Two cases are to be distinguished. For a *cold base*, that is, a basal temperature below the pressure melting point, there cannot be any basal melting ($a_\mathrm{b}^\perp = 0$), and no-slip conditions prevail [see $(5.35)_1$], so that $[\![\mathbf{v}]\!] = \mathbf{0}$. Thus, Eq. (5.37) simplifies to

$$\kappa\,(\mathrm{grad}\,T \cdot \mathbf{n}) = q_\mathrm{geo}^\perp\,, \tag{5.38}$$

which is a Neumann-type boundary condition for the basal temperature. By contrast, in case of a *temperate base* (basal temperature at the pressure melting point), the basal temperature itself is known, namely

$$T = T_\mathrm{m} \tag{5.39}$$

(Dirichlet-type condition). As a consequence, the energy jump condition (5.37) is no longer needed as a boundary condition for the temperature field, but it can be used to compute the unknown basal melting rate $a_\mathrm{b}^\perp$. Since there will be meltwater on the positive (lithosphere) side and ice on the negative (ice) side of the base, the jump of the internal energy $[\![u]\!]$ is equal to the *latent heat* L of ice melt, that is, $[\![u]\!] = L$. Furthermore, the velocity jump is given by the sliding velocity $(5.35)_2$, so that $[\![\mathbf{v}]\!] = -\mathbf{v}_\mathrm{b}$. With these settings, the energy jump condition (5.37), solved for $a_\mathrm{b}^\perp$, yields

$$a_\mathrm{b}^\perp = \frac{q_\mathrm{geo}^\perp - \kappa\,(\operatorname{grad} T \cdot \mathbf{n}) - \mathbf{v}_\mathrm{b} \cdot \mathbf{t} \cdot \mathbf{n}}{\rho L}\,. \tag{5.40}$$

Evidently, the situation is different from that of the free surface, where the accumulation-ablation function $a_\mathrm{s}^\perp$ must be prescribed as climatic input (along with the surface temperature T_s), whereas the basal melting rate $a_\mathrm{b}^\perp$ can be computed. Instead, at the ice base the geothermal heat flux $q_\mathrm{geo}^\perp$ must be prescribed as an input quantity.

5.1.3 Ice Thickness Equation

By combining the continuity equation (5.1) with the kinematic boundary conditions (5.21) and (5.31), we can now derive an evolution equation for the ice thickness $H(x, y, t) = h(x, y, t) - b(x, y, t)$. To this end, we write (5.1) in component form,

$$\frac{\partial v_x}{\partial x} + \frac{\partial v_y}{\partial y} + \frac{\partial v_z}{\partial z} = 0\,, \tag{5.41}$$

and integrate it from the ice base to the free surface:

$$\int_b^h \frac{\partial v_x}{\partial x}\,\mathrm{d}z + \int_b^h \frac{\partial v_y}{\partial y}\,\mathrm{d}z + \int_b^h \frac{\partial v_z}{\partial z}\,\mathrm{d}z = 0\,. \tag{5.42}$$

The first two terms can be modified using Leibniz's rule,

$$\frac{\partial}{\partial x}\int_b^h v_x\,\mathrm{d}z = \int_b^h \frac{\partial v_x}{\partial x}\,\mathrm{d}z + v_x|_{z=h}\,\frac{\partial h}{\partial x} - v_x|_{z=b}\,\frac{\partial b}{\partial x} \tag{5.43}$$

(and accordingly for the y-derivative), and the third term is simply

$$\int_b^h \frac{\partial v_z}{\partial z}\,\mathrm{d}z = v_z|_{z=h} - v_z|_{z=b}\,, \tag{5.44}$$

so that

$$\frac{\partial}{\partial x}\int_b^h v_x\,\mathrm{d}z + \frac{\partial}{\partial y}\int_b^h v_y\,\mathrm{d}z - v_x\big|_{z=h}\,\frac{\partial h}{\partial x} - v_y\big|_{z=h}\,\frac{\partial h}{\partial y} + v_z\big|_{z=h}$$
$$+\, v_x\big|_{z=b}\,\frac{\partial b}{\partial x} + v_y\big|_{z=b}\,\frac{\partial b}{\partial y} - v_z\big|_{z=b} = 0\,. \tag{5.45}$$

With the kinematic conditions (5.21) and (5.31), this yields

$$\frac{\partial}{\partial x}\int_b^h v_x\,\mathrm{d}z + \frac{\partial}{\partial y}\int_b^h v_y\,\mathrm{d}z + \frac{\partial h}{\partial t} - N_\mathrm{s}a_\mathrm{s}^\perp - \frac{\partial b}{\partial t} + N_\mathrm{b}a_\mathrm{b}^\perp = 0\,. \tag{5.46}$$

By introducing the *volume flux* $\mathbf{Q}$ as the vertically integrated horizontal velocity, that is,

$$\mathbf{Q} = \begin{pmatrix} Q_x \\ Q_y \end{pmatrix} = \begin{pmatrix} \displaystyle\int_b^h v_x\,\mathrm{d}z \\ \displaystyle\int_b^h v_y\,\mathrm{d}z \end{pmatrix}, \tag{5.47}$$

rearranging (5.46) and introducing the ice thickness $H = h - b$, we obtain

$$\frac{\partial H}{\partial t} = -\mathrm{div}\,\mathbf{Q} + N_\mathrm{s}a_\mathrm{s}^\perp - N_\mathrm{b}a_\mathrm{b}^\perp \tag{5.48}$$

(note that $\mathrm{div}\,\mathbf{Q} = \partial Q_x/\partial x + \partial Q_y/\partial y$). This result is known as the *ice thickness equation.*

Recall that the accumulation-ablation function (surface mass balance) $a_\mathrm{s}^\perp$ and the basal melting rate $a_\mathrm{b}^\perp$ are fluxes perpendicular to the free surface and the ice base, respectively. However, since the term $\partial H/\partial t$ in Eq. (5.48) refers to the vertical direction, it is desirable to introduce new quantities a_s and a_b, which are also taken in the vertical direction.

Let $\Delta V^\perp$ be the ice volume which is accumulated on the area $\Delta A^\perp$ on the surface of the ice sheet during the time Δt (Fig. 5.5). The accumulation-ablation function perpendicular to the free surface is then

$$a_\mathrm{s}^\perp = \lim_{\Delta t\to 0}\frac{\Delta V^\perp}{\Delta A^\perp\,\Delta t} \tag{5.49}$$

(where $\Delta V^\perp$ is positive in case of positive $a_\mathrm{s}^\perp$ and negative in case of negative $a_\mathrm{s}^\perp$). Similarly, the accumulation-ablation function in the vertical direction is

$$a_\mathrm{s} = \lim_{\Delta t\to 0}\frac{\Delta V}{\Delta A\,\Delta t}\,. \tag{5.50}$$

Since $\Delta V = \Delta V^\perp$ and $\Delta A = \Delta A^\perp\cos\alpha$ (where α is the surface inclination angle, see Fig. 5.5), the two accumulation-ablation functions are related by

$$a_\mathrm{s} = \frac{a_\mathrm{s}^\perp}{\cos\alpha}\,. \tag{5.51}$$

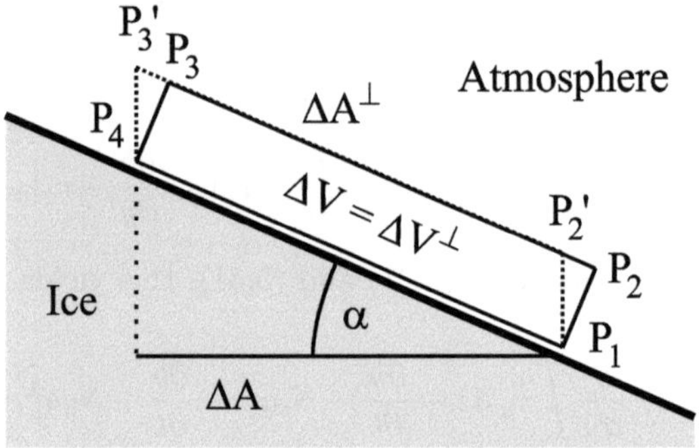

Fig. 5.5. On the accumulation-ablation functions in the vertical direction (a_s) and perpendicular to the free surface ($a_s^\perp$). The volumes $\Delta V^\perp = P_1P_2P_3P_4$ and $\Delta V = P_1P_2'P_3'P_4$ are equal. The areas $\Delta A^\perp = P_1P_4 = P_2P_3$ and ΔA (projection of $\Delta A^\perp$ into the horizontal plane) are related by $\Delta A = \Delta A^\perp \cos\alpha$.

Without loss of generality, let us assume for the moment that the coordinates x and y are oriented such that x is parallel to the line of steepest descent, so that $\partial h/\partial x = \tan\alpha$ and $\partial h/\partial y = 0$. Then, due to Eq. (5.17),

$$\begin{aligned} N_s &= \left(1 + \left(\frac{\partial h}{\partial x}\right)^2\right)^{1/2} = \left(1+\tan^2\alpha\right)^{1/2} = \left(\frac{\cos^2\alpha + \sin^2\alpha}{\cos^2\alpha}\right)^{1/2} \\ &= \frac{1}{\cos\alpha}\,. \end{aligned} \tag{5.52}$$

By inserting (5.52) in (5.51), we find

$$a_s = N_s\, a_s^\perp\,. \tag{5.53}$$

With the same arguments, an analogous relation can be established for the basal melting rates,

$$a_b = N_b\, a_b^\perp\,. \tag{5.54}$$

We can now insert Eqs. (5.53) and (5.54) in the ice thickness equation (5.48) in order to obtain the simplified form

$$\frac{\partial H}{\partial t} = -\operatorname{div}\mathbf{Q} + a_s - a_b\,. \tag{5.55}$$

The ice thickness equation is usually presented in this form. It is the central evolution equation in ice sheet dynamics.

5.2 Hydrostatic Approximation

In order to derive a simplified, approximated system of equations for the large-scale dynamics of ice sheets, we go back to the momentum balance (3.72), and write it in component form. By neglecting again the acceleration and Coriolis terms [see (5.7)–(5.10)] and setting $\mathbf{f} = \rho\mathbf{g} = -\rho g\,\mathbf{e}_z$, we obtain

$$\begin{aligned} \frac{\partial t_{xx}}{\partial x} + \frac{\partial t_{xy}}{\partial y} + \frac{\partial t_{xz}}{\partial z} &= 0\,, \\ \frac{\partial t_{xy}}{\partial x} + \frac{\partial t_{yy}}{\partial y} + \frac{\partial t_{yz}}{\partial z} &= 0\,, \\ \frac{\partial t_{xz}}{\partial x} + \frac{\partial t_{yz}}{\partial y} + \frac{\partial t_{zz}}{\partial z} &= \rho g\,. \end{aligned} \tag{5.56}$$

In all parts of an ice sheet, the shear stresses t_{xz} and t_{yz} ($\lesssim 100\,\mathrm{kPa}$) are small compared to the vertical normal stress t_{zz}, which is approximately equal to the pressure p, so that $[t_{zz}] \approx [P] = \rho g[H] \approx 10\,\mathrm{MPa}$ [see (5.5)]. Consequently, the vertical momentum balance $(5.56)_3$ can be reduced to a balance between the vertical gradient of t_{zz} and the gravity force,

$$\frac{\partial t_{zz}}{\partial z} = \rho g\,. \tag{5.57}$$

The same approximation in the vertical component of the stress-free condition at the free surface (5.23) yields

$$t_{zz}|_{z=h} = 0\,, \tag{5.58}$$

so that Eq. (5.57) can readily be integrated,

$$t_{zz} = -\rho g(h - z)\,. \tag{5.59}$$

Evidently, the vertical normal stress t_{zz} is hydrostatic. With this result, the pressure p reads

$$\begin{aligned} p &= p - t^{\mathrm{D}}_{xx} - t^{\mathrm{D}}_{yy} - t^{\mathrm{D}}_{zz} = -t^{\mathrm{D}}_{xx} - t^{\mathrm{D}}_{yy} - t_{zz} \\ &= \rho g(h - z) - t^{\mathrm{D}}_{xx} - t^{\mathrm{D}}_{yy}\,. \end{aligned} \tag{5.60}$$

Thus, the horizontal normal stresses t_{xx}, t_{yy} can be expressed as

$$\begin{aligned} t_{xx} &= -p + t^{\mathrm{D}}_{xx} = 2t^{\mathrm{D}}_{xx} + t^{\mathrm{D}}_{yy} - \rho g(h - z)\,, \\ t_{yy} &= -p + t^{\mathrm{D}}_{yy} = 2t^{\mathrm{D}}_{yy} + t^{\mathrm{D}}_{xx} - \rho g(h - z)\,. \end{aligned} \tag{5.61}$$

Inserting these in the horizontal (x, y) components of the momentum balance $(5.56)_{1,2}$ yields

$$\begin{aligned} 2\frac{\partial t^{\mathrm{D}}_{xx}}{\partial x} + \frac{\partial t^{\mathrm{D}}_{yy}}{\partial x} + \frac{\partial t_{xy}}{\partial y} + \frac{\partial t_{xz}}{\partial z} &= \rho g \frac{\partial h}{\partial x}\,, \\ 2\frac{\partial t^{\mathrm{D}}_{yy}}{\partial y} + \frac{\partial t^{\mathrm{D}}_{xx}}{\partial y} + \frac{\partial t_{xy}}{\partial x} + \frac{\partial t_{yz}}{\partial z} &= \rho g \frac{\partial h}{\partial y}\,, \end{aligned} \tag{5.62}$$

and the viscous rheology of ice [Eqs. (4.7), (4.21)] in Cartesian coordinates is

$$\begin{aligned} t^{\mathrm{D}}_{xx} &= 2\eta\,\frac{\partial v_x}{\partial x}\,, \\ t^{\mathrm{D}}_{yy} &= 2\eta\,\frac{\partial v_y}{\partial y}\,, \\ t^{\mathrm{D}}_{zz} &= 2\eta\,\frac{\partial v_z}{\partial z}\,, \\ t_{xz} &= \eta\left(\frac{\partial v_x}{\partial z}+\frac{\partial v_z}{\partial x}\right), \\ t_{yz} &= \eta\left(\frac{\partial v_y}{\partial z}+\frac{\partial v_z}{\partial y}\right), \\ t_{xy} &= \eta\left(\frac{\partial v_x}{\partial y}+\frac{\partial v_y}{\partial x}\right). \end{aligned} \tag{5.63}$$

Since the trace of the deviatoric stress tensor vanishes ($t^{\mathrm{D}}_{xx}+t^{\mathrm{D}}_{yy}+t^{\mathrm{D}}_{zz}=0$), only two out of the first three equations of (5.63) are independent. Inserting $(5.63)_{1,2,4,5,6}$ in (5.62) yields

$$\begin{aligned} 4\frac{\partial}{\partial x}\left(\eta\frac{\partial v_x}{\partial x}\right)+2\frac{\partial}{\partial x}\left(\eta\frac{\partial v_y}{\partial y}\right)+\frac{\partial}{\partial y}\left(\eta\Big(\frac{\partial v_x}{\partial y}+\frac{\partial v_y}{\partial x}\Big)\right) \\ +\frac{\partial}{\partial z}\left(\eta\Big(\frac{\partial v_x}{\partial z}+\frac{\partial v_z}{\partial x}\Big)\right) = \rho g\frac{\partial h}{\partial x}\,, \\ 4\frac{\partial}{\partial y}\left(\eta\frac{\partial v_y}{\partial y}\right)+2\frac{\partial}{\partial y}\left(\eta\frac{\partial v_x}{\partial x}\right)+\frac{\partial}{\partial x}\left(\eta\Big(\frac{\partial v_x}{\partial y}+\frac{\partial v_y}{\partial x}\Big)\right) \\ +\frac{\partial}{\partial z}\left(\eta\Big(\frac{\partial v_y}{\partial z}+\frac{\partial v_z}{\partial y}\Big)\right) = \rho g\frac{\partial h}{\partial y}\,. \end{aligned} \tag{5.64}$$

As for the full Stokes flow problem discussed above, we complement these differential equations for the velocity field by the functional form $\eta(T',d_{\mathrm{e}})$ of the viscosity, which is either given by Eq. (4.22) (for the normal Glen flow law) or by Eq. (4.28) (for the regularised Glen flow law). The component form of the effective strain rate d_{e} reads

$$\begin{aligned} d_{\mathrm{e}} &= \sqrt{\tfrac{1}{2}\,\mathrm{tr}\,\mathsf{D}^2} = \sqrt{\tfrac{1}{2}\,D_{ij}D_{ij}} \\ &= \sqrt{\tfrac{1}{2}\,(D_{11}^2+D_{22}^2+D_{33}^2+2D_{12}^2+2D_{13}^2+2D_{23}^2)}\,, \end{aligned} \tag{5.65}$$

and, due to the continuity equation (5.1), we have $\mathrm{div}\,\mathbf{v}=\mathrm{tr}\,\mathsf{D}=D_{11}+D_{22}+D_{33}=0$, so that

$$d_{\mathrm{e}} = \sqrt{\tfrac{1}{2}\,[D_{11}^2+D_{22}^2+(-D_{11}-D_{22})^2+2D_{12}^2+2D_{13}^2+2D_{23}^2]}$$

$$= \sqrt{D_{11}^2 + D_{22}^2 + D_{11}D_{22} + D_{12}^2 + D_{13}^2 + D_{23}^2}$$

$$= \left\{ \left(\frac{\partial v_x}{\partial x}\right)^2 + \left(\frac{\partial v_y}{\partial y}\right)^2 + \frac{\partial v_x}{\partial x}\frac{\partial v_y}{\partial y} + \frac{1}{4}\left(\frac{\partial v_x}{\partial y} + \frac{\partial v_y}{\partial x}\right)^2 \right.$$

$$\left. + \frac{1}{4}\left(\frac{\partial v_x}{\partial z} + \frac{\partial v_z}{\partial x}\right)^2 + \frac{1}{4}\left(\frac{\partial v_y}{\partial z} + \frac{\partial v_z}{\partial y}\right)^2 \right\}^{1/2} . \qquad (5.66)$$

Equations (5.64) and (4.22) or (4.28), respectively, together with the continuity equation (5.41) and the temperature evolution equation (5.14), which is

$$\rho c\left(\frac{\partial T}{\partial t} + v_x\frac{\partial T}{\partial x} + v_y\frac{\partial T}{\partial y} + v_z\frac{\partial T}{\partial z}\right)$$
$$= \frac{\partial}{\partial x}\left(\kappa\frac{\partial T}{\partial x}\right) + \frac{\partial}{\partial y}\left(\kappa\frac{\partial T}{\partial y}\right) + \frac{\partial}{\partial z}\left(\kappa\frac{\partial T}{\partial z}\right) + 4\eta\, d_e^2 , \qquad (5.67)$$

are five equations for the five unknown fields v_x, v_y, v_z, η and T. This set of field equations is called the *hydrostatic approximation*. Compared to the full Stokes flow problem formulated in Sect. 5.1.1, the pressure has been eliminated, which is a substantial simplification of the problem. The field equations are completed by the kinematic boundary condition (5.21), the stress-free condition (5.23) and the temperature condition (5.24) at the free surface, the kinematic condition (5.31), the sliding law (5.35) and the thermodynamic conditions (5.38), (5.39), (5.40) at the ice base, and the ice thickness equation (5.55).

5.3 First Order Approximation

With the typical values (5.5) and the aspect ratio (5.6), we find for the ratio of components of the velocity gradient

$$\frac{\partial v_z}{\partial x} \Big/ \frac{\partial v_x}{\partial z} , \quad \frac{\partial v_z}{\partial y} \Big/ \frac{\partial v_y}{\partial z} \sim \frac{[W]}{[L]} \Big/ \frac{[U]}{[H]} = \frac{[W]}{[U]}\frac{[H]}{[L]} = \varepsilon^2 \sim 10^{-6} , \qquad (5.68)$$

so that horizontal derivatives of the vertical velocity are negligible compared to vertical derivatives of the horizontal velocity. This allows us to neglect the terms containing horizontal derivatives of the vertical velocity in the viscous rheology $(5.63)_{1,2,4,5,6}$, which yields

$$t_{xx}^{\mathrm{D}} = 2\eta\,\frac{\partial v_x}{\partial x} ,$$
$$t_{yy}^{\mathrm{D}} = 2\eta\,\frac{\partial v_y}{\partial y} ,$$
$$t_{xz} = \eta\,\frac{\partial v_x}{\partial z} , \qquad (5.69)$$

$$t_{yz} = \eta\,\frac{\partial v_y}{\partial z}\,,$$

$$t_{xy} = \eta\left(\frac{\partial v_x}{\partial y} + \frac{\partial v_y}{\partial x}\right).$$

Consequently, the momentum balance (5.64) simplifies to

$$\begin{aligned}
&4\frac{\partial}{\partial x}\left(\eta\frac{\partial v_x}{\partial x}\right) + 2\frac{\partial}{\partial x}\left(\eta\frac{\partial v_y}{\partial y}\right) + \frac{\partial}{\partial y}\left(\eta\Big(\frac{\partial v_x}{\partial y} + \frac{\partial v_y}{\partial x}\Big)\right)\\
&\qquad + \frac{\partial}{\partial z}\left(\eta\,\frac{\partial v_x}{\partial z}\right) = \rho g\frac{\partial h}{\partial x}\,,\\
&4\frac{\partial}{\partial y}\left(\eta\frac{\partial v_y}{\partial y}\right) + 2\frac{\partial}{\partial y}\left(\eta\frac{\partial v_x}{\partial x}\right) + \frac{\partial}{\partial x}\left(\eta\Big(\frac{\partial v_x}{\partial y} + \frac{\partial v_y}{\partial x}\Big)\right)\\
&\qquad + \frac{\partial}{\partial z}\left(\eta\,\frac{\partial v_y}{\partial z}\right) = \rho g\frac{\partial h}{\partial y}\,.
\end{aligned} \tag{5.70}$$

Again, the viscosity is taken as $\eta(T', d_\mathrm{e})$ and given by either Eq. (4.22) or Eq. (4.28), with

$$\begin{aligned}
d_\mathrm{e} &= \left\{\left(\frac{\partial v_x}{\partial x}\right)^2 + \left(\frac{\partial v_y}{\partial y}\right)^2 + \frac{\partial v_x}{\partial x}\frac{\partial v_y}{\partial y}\right.\\
&\qquad\left. + \frac{1}{4}\left(\frac{\partial v_x}{\partial y} + \frac{\partial v_y}{\partial x}\right)^2 + \frac{1}{4}\left(\frac{\partial v_x}{\partial z}\right)^2 + \frac{1}{4}\left(\frac{\partial v_y}{\partial z}\right)^2\right\}^{1/2}\\
&= \left\{\left(\frac{\partial v_x}{\partial x}\right)^2 + \left(\frac{\partial v_y}{\partial y}\right)^2 + \frac{\partial v_x}{\partial x}\frac{\partial v_y}{\partial y} + \frac{1}{2}\frac{\partial v_x}{\partial y}\frac{\partial v_y}{\partial x}\right.\\
&\qquad\left. + \frac{1}{4}\left(\frac{\partial v_x}{\partial y}\right)^2 + \frac{1}{4}\left(\frac{\partial v_y}{\partial x}\right)^2 + \frac{1}{4}\left(\frac{\partial v_x}{\partial z}\right)^2 + \frac{1}{4}\left(\frac{\partial v_y}{\partial z}\right)^2\right\}^{1/2}.
\end{aligned} \tag{5.71}$$

In this simplified hydrostatic approximation, generally called the *first order approximation*, Eqs. (5.70) and (5.71) [and therefore the viscosity $\eta(T', d_\mathrm{e})$] contain only the horizontal components of the velocity, v_x and v_y. Thus, the solution of these equations is fully decoupled from the determination of the vertical velocity v_z via the continuity equation (5.41), in contrast to the hydrostatic approximation and the full Stokes flow problem. Once the horizontal velocity has been computed, the vertical velocity can be obtained by integrating Eq. (5.41) from $z = b$ to z,

$$v_z = v_z|_{z=b} - \int_b^z \left(\frac{\partial v_x}{\partial x} + \frac{\partial v_y}{\partial y}\right) \mathrm{d}\bar{z}\,. \tag{5.72}$$

The vertical velocity at the ice base $v_z|_{z=b}$ is obtained from the kinematic condition (5.31).

5.4 Shallow Ice Approximation

Everywhere in an ice sheet, except the immediate vicinity of ice domes (up to and within $\sim$ 10 km in the horizontal direction) and ice margins, the flow regime is essentially simple, bed-parallel shear, and the slopes of the free surface and the ice base are small (Fig. 5.6). Under these conditions, the relevant components of the stress deviator t^{D} are the shear stresses in the horizontal plane, t_{xz} and t_{yz}, which are supported by the basal drag. The normal stress deviators t^{D}_{xx}, t^{D}_{yy} and t^{D}_{zz} as well as the shear stress in the vertical planes, t_{xy}, are consequently negligible.

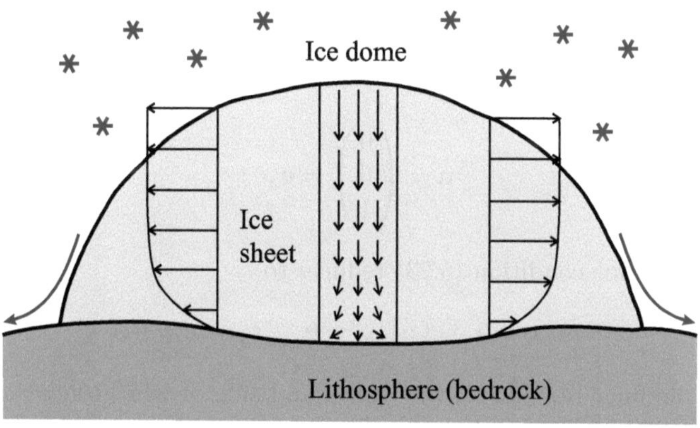

Fig. 5.6. Flow regimes in an ice sheet. In most regions, simple, bed-parallel shear flow prevails. By contrast, in the vicinity of an ice dome, the flow direction is essentially downward, which leads to vertical compression and horizontal extension. Close to the ice margin, the slope of the free surface can be large. Vertical exaggeration factor $\sim$ 200-500.

This allows further simplifications of the hydrostatic approximation to be made, which go beyond the first order approximation, and are known as the *shallow ice approximation* (SIA) (Hutter 1983, Morland 1984). All normal stresses are equal to the negative pressure,

$$t_{xx} = t_{yy} = t_{zz} = -p\,, \tag{5.73}$$

so that the vertical momentum balance (5.57) reads

$$\frac{\partial p}{\partial z} = -\rho g\,, \tag{5.74}$$

and its integrated form [see (5.59)] gives the hydrostatic pressure distribution

$$p = p_{\mathrm{hyd}} = \rho g(h - z)\,. \tag{5.75}$$

The horizontal components of the momentum balance, $(5.56)_{1,2}$ or $(5.62)_{1,2}$, simplify to

$$\begin{aligned} \frac{\partial t_{xz}}{\partial z} &= \frac{\partial p}{\partial x} = \rho g \frac{\partial h}{\partial x}\,, \\ \frac{\partial t_{yz}}{\partial z} &= \frac{\partial p}{\partial y} = \rho g \frac{\partial h}{\partial y}\,. \end{aligned} \tag{5.76}$$

Due to the small surface slope, the derivatives $\partial h/\partial x$ and $\partial h/\partial y$ are small, typically of the order of the aspect ratio ε:

$$\frac{\partial h}{\partial x}, \frac{\partial h}{\partial y} \sim \frac{[H]}{[L]} = \varepsilon \tag{5.77}$$

[see Eqs. (5.5) and (5.6)]. Thus, the unit normal vector of the free surface (5.16) is approximately vertical,

$$\mathbf{n} = \begin{pmatrix} 0 \\ 0 \\ 1 \end{pmatrix} = \mathbf{e}_z\,, \tag{5.78}$$

and the stress-free condition (5.23) reduces to

$$p|_{z=h} = 0\,, \quad t_{xz}|_{z=h} = 0\,, \quad t_{yz}|_{z=h} = 0\,. \tag{5.79}$$

With this finding, Eq. (5.76), the right-hand side of which does not depend on z, can readily be integrated and yields

$$\begin{aligned} t_{xz} &= -\rho g (h-z) \frac{\partial h}{\partial x}\,, \\ t_{yz} &= -\rho g (h-z) \frac{\partial h}{\partial y}\,. \end{aligned} \tag{5.80}$$

Equations (5.75) and (5.80) tell us that in the SIA the stress field, the only non-negligible components of which are p, t_{xz} and t_{yz}, is fully determined if the geometry of the ice sheet is known. The effective stress σ_e [see $(4.8)_2$ and (4.9)] is then

$$\begin{aligned} \sigma_\mathrm{e} &= \sqrt{t_{xz}^2 + t_{yz}^2} \\ &= \rho g (h-z) \left(\left(\frac{\partial h}{\partial x}\right)^2 + \left(\frac{\partial h}{\partial y}\right)^2 \right)^{1/2} = \rho g (h-z)\, |\mathrm{grad}\, h|\,. \end{aligned} \tag{5.81}$$

Since in the SIA, the infinite viscosity limit of Glen's flow law for small stresses does not cause any mathematical problems, we do not consider its regularised version of Sect. 4.3.2 here. The above results are therefore inserted in the x-z and y-z components of Glen's flow law in the form (4.16):

$$
\begin{aligned}
\frac{1}{2}\left(\frac{\partial v_x}{\partial z}+\frac{\partial v_z}{\partial x}\right) &= A(T')\,\sigma_{\rm e}^{n-1}\,t_{xz} \\
&= -A(T')\,[\rho g(h-z)]^n|\mathrm{grad}\,h|^{n-1}\frac{\partial h}{\partial x}\,, \\
\frac{1}{2}\left(\frac{\partial v_y}{\partial z}+\frac{\partial v_z}{\partial y}\right) &= A(T')\,\sigma_{\rm e}^{n-1}\,t_{yz} \\
&= -A(T')\,[\rho g(h-z)]^n|\mathrm{grad}\,h|^{n-1}\frac{\partial h}{\partial y}\,.
\end{aligned}
\tag{5.82}
$$

Based on the order of magnitudes of the spatial derivatives of velocity components, Eq. (5.68), the horizontal derivatives of the vertical velocity are negligible. This yields

$$
\begin{aligned}
\frac{\partial v_x}{\partial z} &= -2A(T')\,[\rho g(h-z)]^n|\mathrm{grad}\,h|^{n-1}\frac{\partial h}{\partial x}\,, \\
\frac{\partial v_y}{\partial z} &= -2A(T')\,[\rho g(h-z)]^n|\mathrm{grad}\,h|^{n-1}\frac{\partial h}{\partial y}\,,
\end{aligned}
\tag{5.83}
$$

which can be integrated from the ice base $z = b$ to an arbitrary position z in the ice sheet in order to compute the horizontal velocities,

$$
\begin{aligned}
v_x &= v_{\mathrm{b}x} - 2(\rho g)^n|\mathrm{grad}\,h|^{n-1}\frac{\partial h}{\partial x}\int_b^z A(T')\,(h-\bar{z})^n\,\mathrm{d}\bar{z}\,, \\
v_y &= v_{\mathrm{b}y} - 2(\rho g)^n|\mathrm{grad}\,h|^{n-1}\frac{\partial h}{\partial y}\int_b^z A(T')\,(h-\bar{z})^n\,\mathrm{d}\bar{z}\,,
\end{aligned}
\tag{5.84}
$$

where $v_{\mathrm{b}x}$ and $v_{\mathrm{b}y}$ are the respective velocities at the ice base. Since the bed slopes are of the same order of magnitude as the surface slopes [see (5.77)],

$$
\frac{\partial b}{\partial x},\ \frac{\partial b}{\partial y} \sim \frac{[H]}{[L]} = \varepsilon\,,
\tag{5.85}
$$

and the unit normal vector of the ice base (5.26) is approximately vertical,

$$
\mathbf{n} = \begin{pmatrix} 0 \\ 0 \\ -1 \end{pmatrix} = -\mathbf{e}_z\,,
\tag{5.86}
$$

the tangential plane to the ice base is approximately equal to the horizontal plane. Therefore, $v_{\mathrm{b}x}$ and $v_{\mathrm{b}y}$ are the two components of the basal sliding velocity $\mathbf{v}_\mathrm{b}$ given by the Weertman-type sliding law (5.35). The basal drag $\boldsymbol{\tau}_\mathrm{b} = \tau_\mathrm{b}\mathbf{e}_\mathrm{t}$ consists of the x- and y-components of the stress vector $\mathbf{t}_\mathbf{n}|_{z=b} = \mathsf{t}|_{z=b}\cdot\mathbf{n} = -\mathsf{t}|_{z=b}\cdot\mathbf{e}_z$, that is,

$$
\boldsymbol{\tau}_\mathrm{b} = -\begin{pmatrix} t_{xz}|_{z=b} \\ t_{yz}|_{z=b} \end{pmatrix} = \rho g H \begin{pmatrix} \dfrac{\partial h}{\partial x} \\[2mm] \dfrac{\partial h}{\partial y} \end{pmatrix}\,.
\tag{5.87}
$$

Thus,

$$\tau_\mathrm{b} = \left| -\begin{pmatrix} t_{xz}|_{z=b} \\ t_{yz}|_{z=b} \end{pmatrix} \right| = \sqrt{t_{xz}^2|_{z=b} + t_{yz}^2|_{z=b}} = \rho g H \, |\mathrm{grad}\, h| \tag{5.88}$$

and

$$\mathbf{e}_\mathrm{t} = -\frac{1}{\tau_\mathrm{b}} \begin{pmatrix} t_{xz}|_{z=b} \\ t_{yz}|_{z=b} \end{pmatrix} = \frac{1}{|\mathrm{grad}\, h|} \begin{pmatrix} \dfrac{\partial h}{\partial x} \\ \dfrac{\partial h}{\partial y} \end{pmatrix} . \tag{5.89}$$

The basal normal stress $\mathbf{N}_\mathrm{b} = -N_\mathrm{b}\mathbf{n} = N_\mathrm{b}\mathbf{e}_z$ is equal to the z-component of the stress vector,

$$\mathbf{N}_\mathrm{b} = -t_{zz}|_{z=b}\, \mathbf{e}_z \quad \Rightarrow \quad N_\mathrm{b} = -t_{zz}|_{z=b} = \rho g H . \tag{5.90}$$

Thus we obtain

$$\begin{aligned} v_{\mathrm{b}x} &= \begin{cases} 0\,, & \text{if } T_\mathrm{b} < T_\mathrm{m}\,, \\ -C_\mathrm{b}(\rho g H)^{p-q} |\mathrm{grad}\, h|^{p-1} \dfrac{\partial h}{\partial x}\,, & \text{if } T_\mathrm{b} = T_\mathrm{m}\,, \end{cases} \\ v_{\mathrm{b}y} &= \begin{cases} 0\,, & \text{if } T_\mathrm{b} < T_\mathrm{m}\,, \\ -C_\mathrm{b}(\rho g H)^{p-q} |\mathrm{grad}\, h|^{p-1} \dfrac{\partial h}{\partial y}\,, & \text{if } T_\mathrm{b} = T_\mathrm{m}\,. \end{cases} \end{aligned} \tag{5.91}$$

By introducing the horizontal velocity vector

$$\mathbf{v}_\mathrm{h} = \begin{pmatrix} v_x \\ v_y \end{pmatrix} \tag{5.92}$$

and the scalar function

$$C = \begin{cases} 2(\rho g)^n |\mathrm{grad}\, h|^{n-1} \displaystyle\int_b^z A(T')\,(h-\bar{z})^n \,\mathrm{d}\bar{z}\,, & \text{if } T_\mathrm{b} < T_\mathrm{m}\,, \\ C_\mathrm{b}(\rho g H)^{p-q} |\mathrm{grad}\, h|^{p-1} & \\ \quad +2(\rho g)^n |\mathrm{grad}\, h|^{n-1} \displaystyle\int_b^z A(T')\,(h-\bar{z})^n \,\mathrm{d}\bar{z}\,, & \text{if } T_\mathrm{b} = T_\mathrm{m}\,, \end{cases} \tag{5.93}$$

we can express the horizontal velocity (5.84) as

$$\begin{pmatrix} v_x \\ v_y \end{pmatrix} = -C \begin{pmatrix} \dfrac{\partial h}{\partial x} \\ \dfrac{\partial h}{\partial y} \end{pmatrix} , \quad \text{or} \quad \mathbf{v}_\mathrm{h} = -C \,\mathrm{grad}\, h . \tag{5.94}$$

That is, in the shallow ice approximation, the direction of the horizontal velocity is anti-parallel to the gradient of the free surface. In other words, the

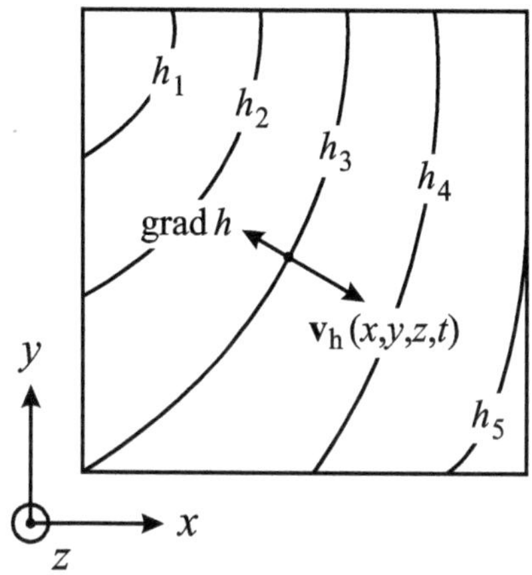

Fig. 5.7. Surface topography and horizontal velocity in the shallow ice approximation. The horizontal velocity $\mathbf{v}_\mathrm{h}$ is anti-parallel to the direction of the surface gradient grad h. Surface-topography contours: $h_1 > h_2 > h_3 > h_4 > h_5$.

ice always flows down the steepest surface slope, irrespective of the bedrock topography (see Fig. 5.7). This holds for any particle of the ice sheet, even the near-basal ice. However, note that this result is only valid as long as the bed slopes are sufficiently small, as described by Eq. (5.85).

As for the first order approximation, the vertical velocity can now be computed by integrating the continuity equation (5.41) from $z = b$ to z,

$$v_z = v_z|_{z=b} - \int_b^z \left(\frac{\partial v_x}{\partial x} + \frac{\partial v_y}{\partial y} \right) \mathrm{d}\bar{z} \,. \tag{5.95}$$

In this equation, v_x and v_y are given by (5.94), and the vertical velocity at the ice base $v_z|_{z=b}$ is determined by the kinematic condition (5.31).

In order to formulate the ice thickness equation in the shallow ice approximation, we compute the volume flux $\mathbf{Q}$ [see Eq. (5.47)] with the horizontal velocities (5.94). This yields

$$\begin{pmatrix} Q_x \\ Q_y \end{pmatrix} = -D \begin{pmatrix} \dfrac{\partial h}{\partial x} \\ \dfrac{\partial h}{\partial y} \end{pmatrix} , \quad \text{with} \quad D = \int_b^h C \,\mathrm{d}z \,. \tag{5.96}$$

The function C depends on z only via the integral term in (5.93), for which we find by integration by parts

$$\begin{aligned}
\int_b^h \int_b^z A(T')\,(h-\bar{z})^n\,\mathrm{d}\bar{z}\,\mathrm{d}z &= \int_b^h 1 \int_b^z A(T')\,(h-\bar{z})^n\,\mathrm{d}\bar{z}\,\mathrm{d}z \\
&= \left[z \int_b^z A(T')\,(h-\bar{z})^n\,\mathrm{d}\bar{z} \right]_b^h - \int_b^h z A(T')\,(h-z)^n\,\mathrm{d}z \\
&= h \int_b^h A(T')\,(h-\bar{z})^n\,\mathrm{d}\bar{z} - \int_b^h z A(T')\,(h-z)^n\,\mathrm{d}z \\
&= \int_b^h A(T')\,(h-z)^{n+1}\,\mathrm{d}z\,.
\end{aligned} \tag{5.97}$$

Thus, the function D is

$$D = \begin{cases} 2(\rho g)^n |\mathrm{grad}\,h|^{n-1} \displaystyle\int_b^h A(T')\,(h-z)^{n+1}\,\mathrm{d}z\,, & \text{if } T_\mathrm{b} < T_\mathrm{m}\,, \\ C_\mathrm{b} H (\rho g H)^{p-q} |\mathrm{grad}\,h|^{p-1} & \\ \quad + 2(\rho g)^n |\mathrm{grad}\,h|^{n-1} \displaystyle\int_b^h A(T')\,(h-z)^{n+1}\,\mathrm{d}z\,, & \text{if } T_\mathrm{b} = T_\mathrm{m}\,. \end{cases} \tag{5.98}$$

By inserting the volume flux (5.96) with the function (5.98) into the ice thickness equation (5.55), we obtain

$$\frac{\partial H}{\partial t} = \frac{\partial}{\partial x}\left(D \frac{\partial h}{\partial x}\right) + \frac{\partial}{\partial y}\left(D \frac{\partial h}{\partial y}\right) + a_\mathrm{s} - a_\mathrm{b}\,, \tag{5.99}$$

or, alternatively, expressed as an evolution equation for the surface topography h,

$$\frac{\partial h}{\partial t} = \frac{\partial}{\partial x}\left(D \frac{\partial h}{\partial x}\right) + \frac{\partial}{\partial y}\left(D \frac{\partial h}{\partial y}\right) + a_\mathrm{s} - a_\mathrm{b} + \frac{\partial b}{\partial t}\,. \tag{5.100}$$

Note also that, due to Eqs. (5.77) and (5.85), we have $N_\mathrm{s} \approx 1$ and $N_\mathrm{b} \approx 1$, so that the accumulation-ablation functions and basal melting rates in the vertical direction and perpendicular to the respective interfaces are approximately equal,

$$a_\mathrm{s} \approx a_\mathrm{s}^\perp\,, \quad a_\mathrm{b} \approx a_\mathrm{b}^\perp\,. \tag{5.101}$$

Mathematically, Eq. (5.100) is a non-linear diffusion equation (the function D depends itself on h) with additional source terms. As described at the end of Sect. 5.1.2, the accumulation-ablation function a_s is a climatic input quantity, and the basal melting rate a_b is determined by Eq. (5.40). The unknown variation $\partial b/\partial t$ of the ice-base/bedrock topography can be obtained from a model of glacial isostasy, which will be treated later (see Chap. 8).

As for the temperature evolution equation (5.67), we complement the typical values (5.5) by the typical temperature variation magnitude

$$[\Delta T] = 20\,\mathrm{K}\,. \tag{5.102}$$

Then, we find for the ratio of horizontal and vertical heat conduction

$$\frac{\partial}{\partial x}\left(\kappa\frac{\partial T}{\partial x}\right)\Big/\frac{\partial}{\partial z}\left(\kappa\frac{\partial T}{\partial z}\right),\quad \frac{\partial}{\partial y}\left(\kappa\frac{\partial T}{\partial y}\right)\Big/\frac{\partial}{\partial z}\left(\kappa\frac{\partial T}{\partial z}\right)$$
$$\sim \kappa\frac{[\Delta T]}{[L]^2}\Big/\kappa\frac{[\Delta T]}{[H]^2} = \frac{[H]^2}{[L]^2} = \varepsilon^2 \sim 10^{-6}\,, \quad (5.103)$$

so that horizontal heat conduction is negligible. Furthermore, in the dissipation term we express the effective strain rate d_e in terms of the effective stress σ_e by using $(4.19)_1$. This yields

$$\rho c\left(\frac{\partial T}{\partial t} + v_x\frac{\partial T}{\partial x} + v_y\frac{\partial T}{\partial y} + v_z\frac{\partial T}{\partial z}\right) = \frac{\partial}{\partial z}\left(\kappa\frac{\partial T}{\partial z}\right) + 4\eta\, A^2(T')\,\sigma_\mathrm{e}^{2n}\,, \quad (5.104)$$

and, by inserting (4.10) and (4.12), we obtain

$$\rho c\left(\frac{\partial T}{\partial t} + v_x\frac{\partial T}{\partial x} + v_y\frac{\partial T}{\partial y} + v_z\frac{\partial T}{\partial z}\right) = \frac{\partial}{\partial z}\left(\kappa\frac{\partial T}{\partial z}\right) + 2A(T')\,\sigma_\mathrm{e}^{n+1}\,, \quad (5.105)$$

where σ_e is given by Eq. (5.81). The boundary conditions for this equation are the prescribed surface temperature (5.24), the temperature gradient (5.38) for a cold base and the melting temperature (5.39) for a temperate base.

The shallow ice approximation simplifies the problem of large-scale ice sheet flow drastically. The stress field is given by the simple, analytic expressions (5.75), (5.80) and (5.81), and the velocity field depends only on the *local* ice sheet geometry and temperature via (5.94) and (5.95), whereas in the full Stokes flow problem, the hydrostatic approximation and the first order approximation systems of non-linear differential equations [Eqs. (5.11), (5.64) and (5.70), respectively] must be solved. The remaining "hard work" is the solution of the surface evolution equation (5.100) and of the temperature evolution equation (5.105).

5.5 Driving Stress

Equation (5.80) represents the bed-parallel shear stress in the shallow ice approximation. Evaluating this equation at the bed ($z = b$) yields the vector

$$\boldsymbol{\tau}_\mathrm{d} = -\rho g H \begin{pmatrix} \dfrac{\partial h}{\partial x} \\[2mm] \dfrac{\partial h}{\partial y} \end{pmatrix}, \quad (5.106)$$

which is often called the *driving stress*.

By construction, the driving stress corresponds to the basal shear stress in the shallow ice approximation. However, its definition is not limited to the shallow ice approximation, and it can be interpreted in general as the action which drives the flow of the ice sheet. In the shallow ice approximation, the driving stress $\boldsymbol{\tau}_\mathrm{d}$ is equal to the negative of the basal drag $\boldsymbol{\tau}_\mathrm{b}$ [see Eq. (5.87)], which means that the driving forces and the resistive forces are balanced locally at the bed of the ice sheet. In the full Stokes problem, the hydrostatic approximation and the first order approximation such a local balance does not hold; however, the driving stress and the basal drag still balance *on average* for the entire domain of the ice sheet (van der Veen 1999). Local imbalances are compensated by the deviatoric normal stresses t^D_{xx} and t^D_{yy} as well as the shear stress in the vertical plane t_{xy}.

5.6 Analytical Solutions

5.6.1 Simplified Problem

For very simple, idealised cases, the equations of the shallow ice approximation derived above can be solved analytically. Let us consider the following situation:

- Plane strain approximation: two-dimensional flow in the x-z plane, any lateral effects neglected (see Sect. 3.4.3).
- Steady-state conditions: $\partial(\cdot)/\partial t = 0$ for all field quantities.
- Flat, rigid bed: $b(x) = 0$.
- Ice sheet extent between $x = -L$ and $x = L$, symmetric around the ice divide at $x = 0$.
- No basal melting ($a_\mathrm{b} = 0$), no basal sliding ($C_\mathrm{b} = 0$).
- Constant rate factor: $A(T') = A = \text{const.}$

The last assumption decouples the mechanical from the thermodynamical problem, and therefore we do not have to deal with the temperature evolution equation (5.105).

With the assumptions made above, the ice thickness equation (5.55) becomes

$$\operatorname{div}\mathbf{Q} = a_\mathrm{s}\,, \tag{5.107}$$

where, according to Eqs. (5.47) and (5.98),

$$\begin{aligned}\mathbf{Q} &= -2A(\rho g)^n\,|\operatorname{grad} h|^{n-1}\operatorname{grad} h\int_b^h (h-z)^{n+1}\,\mathrm{d}z\\ &= -2A(\rho g)^n\,|\operatorname{grad} h|^{n-1}\operatorname{grad} h\left[-\frac{(h-z)^{n+2}}{n+2}\right]_b^h\\ &= -\frac{2A(\rho g)^n}{n+2}\,H^{n+2}\,|\operatorname{grad} h|^{n-1}\operatorname{grad} h\\ &= -A_0\,H^{n+2}\,|\operatorname{grad} h|^{n-1}\operatorname{grad} h\,.\end{aligned} \tag{5.108}$$

In the last step, the abbreviation

$$A_0 = \frac{2A(\rho g)^n}{n+2} \tag{5.109}$$

has been introduced.

Due to the flat bed, the ice thickness H is equal to the surface topography h, and due to the plane strain approximation, the volume flux $\mathbf{Q} = Q\,\mathbf{e}_x$ and the surface gradient $\operatorname{grad} h = (\mathrm{d}h/\mathrm{d}x)\,\mathbf{e}_x$. Hence,

$$Q = -A_0\, h^{n+2} \left|\frac{\mathrm{d}h}{\mathrm{d}x}\right|^{n-1} \frac{\mathrm{d}h}{\mathrm{d}x}\,, \tag{5.110}$$

and

$$\frac{\mathrm{d}Q}{\mathrm{d}x} = -\frac{\mathrm{d}}{\mathrm{d}x}\left(A_0\, h^{n+2} \left|\frac{\mathrm{d}h}{\mathrm{d}x}\right|^{n-1} \frac{\mathrm{d}h}{\mathrm{d}x}\right) = a_\mathrm{s}\,. \tag{5.111}$$

In order to find analytical solutions of the steady-state ice thickness equation (5.111), we recognise that it is a separable ordinary differential equation, provided that the surface mass balance a_s is a function of x only. The assumption of symmetry with respect to $x = 0$ implies $Q(0) = 0$ and $(\mathrm{d}h/\mathrm{d}x)_{x=0} = 0$. Thus, a first integral of Eq. (5.111) yields

$$A_0\, h^{n+2} \left|\frac{\mathrm{d}h}{\mathrm{d}x}\right|^{n-1} \frac{\mathrm{d}h}{\mathrm{d}x} = -\int_0^x a_\mathrm{s}(x')\,\mathrm{d}x' = -Q(x)\,. \tag{5.112}$$

For the half-domain $0 < x < L$, the surface topography h decreases monotonically from the ice divide at $x = 0$ to the margin at $x = L$, so that $Q > 0$ and $\mathrm{d}h/\mathrm{d}x < 0$. This allows to take the nth root of Eq. (5.112),

$$h^{(n+2)/n}\,\frac{\mathrm{d}h}{\mathrm{d}x} = -\left(\frac{Q(x)}{A_0}\right)^{1/n}, \tag{5.113}$$

and, by separation of variables, compute the second integral

$$h(x)^{(2n+2)/n} = h_0^{(2n+2)/n} - \frac{2n+2}{n}\int_0^x \left(\frac{Q(x')}{A_0}\right)^{1/n} \mathrm{d}x'\,, \tag{5.114}$$

where h_0 is the surface elevation at the ice divide. The ice thickness equation (5.111) has thus been reduced to a quadrature, which can be solved analytically depending on the mass balance function $a_\mathrm{s}(x)$.

5.6.2 Vialov Profile

In order to simplify the problem further, we assume that the surface mass balance a_s is a positive constant over the entire domain. From Eq. (5.112), we obtain the volume flux

$$Q(x) = a_\mathrm{s} x\,, \tag{5.115}$$

so that Eq. (5.114) reads

$$h(x)^{(2n+2)/n} = h_0^{(2n+2)/n} - 2\left(\frac{a_s}{A_0}\right)^{1/n} x^{(n+1)/n}. \tag{5.116}$$

With the boundary condition $h(L) = 0$, we obtain for the surface elevation at the ice divide

$$h_0 = 2^{n/(2n+2)} \left(\frac{a_s}{A_0}\right)^{1/(2n+2)} L^{1/2}. \tag{5.117}$$

Inserting this result into Eq. (5.116) yields

$$\begin{aligned} h^{(2n+2)/n} &= 2\left(\frac{a_s}{A_0}\right)^{1/n} \left(L^{(n+1)/n} - x^{(n+1)/n}\right) \\ &= 2\left(\frac{a_s}{A_0}\right)^{1/n} L^{(n+1)/n} \left[1 - \left(\frac{x}{L}\right)^{(n+1)/n}\right], \end{aligned} \tag{5.118}$$

which can be written in simpler form as

$$h = h_0 \left[1 - \left(\frac{x}{L}\right)^{(n+1)/n}\right]^{n/(2n+2)}. \tag{5.119}$$

This solution is called the *Vialov profile* (Vialov 1958). Note that for negative values of x, that is, for the half-domain $-L < x < 0$, the variable x must be replaced by $|x|$ in order to maintain the symmetry of the profile.

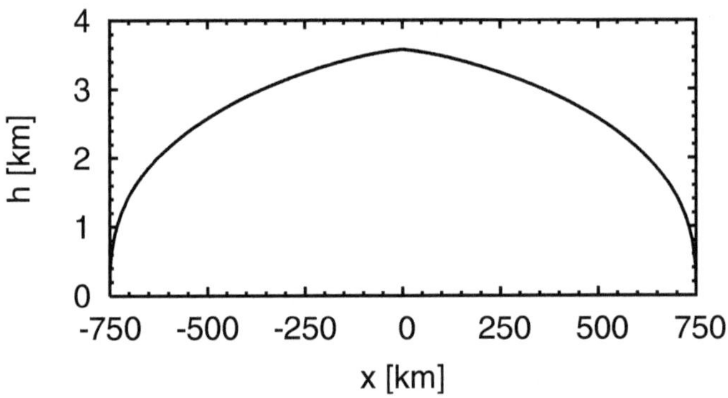

Fig. 5.8. Vialov profile (5.119) for $L = 750\,\mathrm{km}$, $n = 3$, $a_s = 0.3\,\mathrm{m\,a^{-1}}$, $A = 10^{-16}\,\mathrm{a^{-1}\,Pa^{-3}}$, $\rho = 910\,\mathrm{kg\,m^{-3}}$ and $g = 9.81\,\mathrm{m\,s^{-2}}$.

An example is shown in Fig. 5.8. The parameters are those of the EISMINT model intercomparison exercise described by Huybrechts et al. (1996), which resemble the conditions of the Greenland Ice Sheet. For a half-span of $L =$

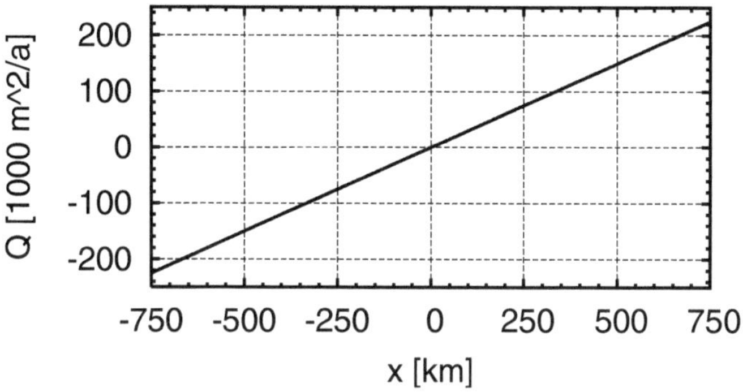

Fig. 5.9. Volume flux for the Vialov profile shown in Fig. 5.8.

750 km, the maximum elevation resulting from (5.117) is $h_0 = 3575.1\,\mathrm{m}$. It is interesting to note that the profile appears pointed at the divide. This is the case since at $x = 0$ the curvature (second derivative) of the function $h(x)$ defined by (5.119) is infinite, a problem which is related to the fact that the shallow ice approximation is not valid there. Further, the slope (first derivative) of the profile is infinite at the margins $x = \pm L$, which violates the assumption of small surface slopes.

The volume flux Q follows from Eq. (5.115). For the above EISMINT parameters, it is shown in Fig. 5.9. The volume flux vanishes at the ice divide, and it increases linearly away from the divide with a gradient of $\mathrm{d}Q/\mathrm{d}x = a_\mathrm{s} = 0.3\,\mathrm{m}^2\,\mathrm{a}^{-1}/\mathrm{m}$ in order to balance the ice accumulation. At the margin, the volume flux reaches a value of $Q(L) = 2.25 \times 10^5\,\mathrm{m}^2\,\mathrm{a}^{-1}$, which can be interpreted as the *calving rate* into a surrounding ocean.

An unrealistic feature of the Vialov profile is the behaviour of the shear stress near the margin. According to Eq. (5.106), the basal shear stress in the shallow ice approximation (driving stress) is proportional to the product of the ice thickness and the inclination of the ice surface, which yields for the Vialov profile (5.119)

$$\tau_\mathrm{d} = -\rho g h \frac{\mathrm{d}h}{\mathrm{d}x} = \frac{\rho g h_0^2}{2L} \left[1 - \left(\frac{x}{L} \right)^{(n+1)/n} \right]^{-1/(n+1)} \left(\frac{x}{L} \right)^{1/n} . \qquad (5.120)$$

In the limit of $x \to L$ this expression diverges, thus the basal shear stress is unbounded at the ice margin (see also below, Fig. 5.12).

5.6.3 Bueler Profile

Analytical solutions of Eq. (5.114) for variable mass balance functions $a_\mathrm{s}(x)$ exist under the condition that the integral of $Q^{1/n}(x)$ can be computed in

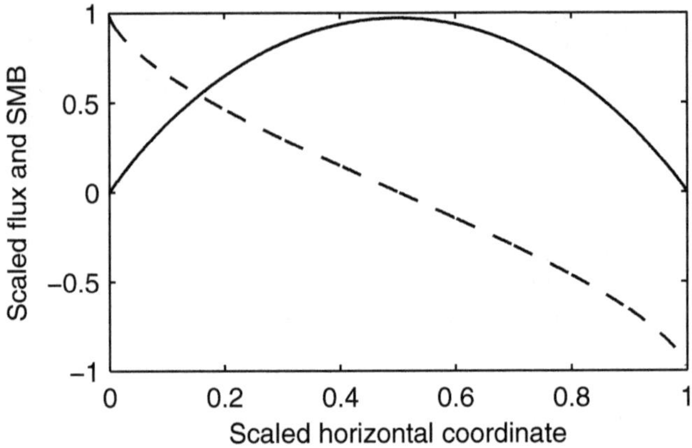

Fig. 5.10. Volume flux (solid line) and surface mass balance (dashed line) of the Bueler profile for the half-domain $0 < x < L$ in scaled units.

closed form. This allows to match physically reasonable boundary conditions to the ice flux, such as a prescribed calving rate at the margin, $Q(L) = Q_L$.

An example is the combination of power functions of x (Bueler 2003, Bueler et al. 2005),

$$Q(x) = \alpha \left[\left(\frac{x}{L}\right)^{1/n} + \left(1 - \frac{x}{L}\right)^{1/n} - 1 \right]^n , \tag{5.121}$$

where $\alpha > 0$ is an adjustable parameter. This volume flux fulfills the symmetry condition $Q(0) = 0$ at the ice divide and the no-flux condition $Q(L) = 0$ at the margin. The corresponding mass balance function results from Eq. (5.111),

$$\begin{aligned} a_\mathrm{s}(x) = \frac{\mathrm{d}Q}{\mathrm{d}x} = \frac{\alpha}{L} &\left[\left(\frac{x}{L}\right)^{1/n} + \left(1 - \frac{x}{L}\right)^{1/n} - 1 \right]^{n-1} \\ &\times \left[\left(\frac{x}{L}\right)^{(1-n)/n} - \left(1 - \frac{x}{L}\right)^{(1-n)/n} \right] . \end{aligned} \tag{5.122}$$

Figure 5.10 shows the volume flux (5.121) and the mass balance function (5.122). The surface mass balance is positive (accumulation) in the interior, high-elevation part of the ice sheet and negative (ablation) in the low-elevation part near the margin. However, an unrealistic feature is the steep increase towards the ice divide.

The solution of Eq. (5.114) with the volume flux (5.121) is

$$\begin{aligned} h(x)^{(2n+2)/n} = h_0^{(2n+2)/n} &- 2L \left(\frac{\alpha}{A_0}\right)^{1/n} \\ \times &\left[1 - \frac{n+1}{n}\frac{x}{L} + \left(\frac{x}{L}\right)^{(n+1)/n} - \left(1 - \frac{x}{L}\right)^{(n+1)/n} \right] , \end{aligned} \tag{5.123}$$

and the surface elevation at the ice divide follows from the boundary condition $h(L) = 0$ as

$$h_0^{(2n+2)/n} = 2L \left(\frac{\alpha}{A_0}\right)^{1/n} \frac{n-1}{n} . \qquad (5.124)$$

Insertion of Eq. (5.124) into Eq. (5.123) yields the *Bueler profile*

$$\begin{aligned} h(x) = {} & \frac{h_0}{(n-1)^{n/(2n+2)}} \\ & \times \left[(n+1)\frac{x}{L} - n\left(\frac{x}{L}\right)^{(n+1)/n} \right. \\ & \left. \qquad + n\left(1-\frac{x}{L}\right)^{(n+1)/n} - 1 \right]^{n/(2n+2)} . \end{aligned} \qquad (5.125)$$

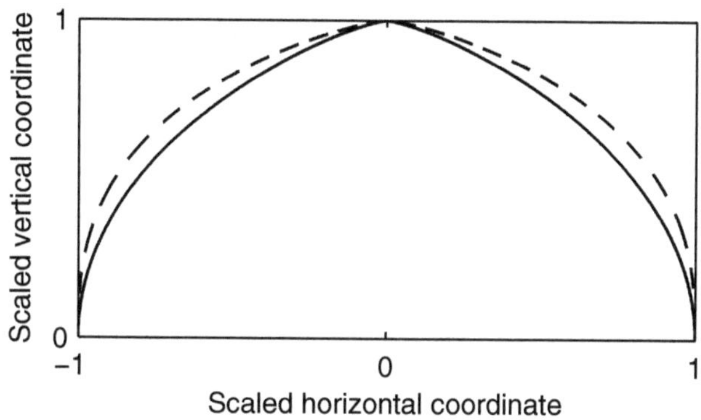

Fig. 5.11. Bueler profile (solid line) and Vialov profile (dashed line) in scaled coordinates.

A comparison of the Bueler and Vialov profiles is shown in Fig. 5.11. The Bueler profile shows similar features as the Vialov profile. The inclination of the surface is zero at the ice divide, but the curvature is infinite there, and the inclination at the margin is unbounded. However, an important difference is the behaviour of the basal shear stress (driving stress), which is for the Bueler solution

$$\tau_\mathrm{d} = -\rho g h \frac{\mathrm{d}h}{\mathrm{d}x} \qquad [\text{with } h(x) \text{ from Eq. (5.125)}] . \qquad (5.126)$$

A substitution $x/L = 1-\xi$ and a subsequent first order expansion in ξ (which shall not be detailed here) shows that τ_d remains finite for $\xi \to 0$ $(x \to L)$ in the Bueler solution, whereas it is unbounded in the Vialov solution. This is illustrated in Fig. 5.12.

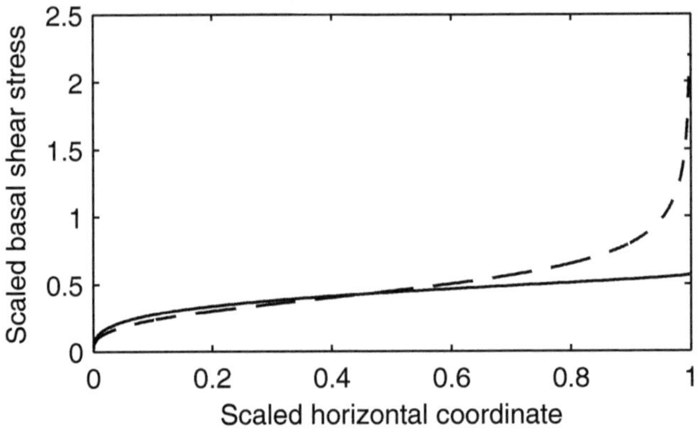

Fig. 5.12. Basal shear stress of the Bueler (solid line) and Vialov (dashed line) solutions for the half-domain $0 < x < L$ in scaled units.

If for verification purposes more realistic solutions are required, Eq. (5.114) may be used, even if the integral is not analytically solvable. For smooth mass balance functions, numerical quadrature is unproblematic and can be carried out with high accuracy. This extends the range of possible benchmark solutions of the ice thickness equation considerably.

5.7 Numerical Methods

Apart from idealised cases like those of Sect. 5.6, the field equations and boundary conditions of the shallow ice approximation for the flow and temperature fields in ice sheets are too complicated to be solved analytically. In general situations, it is therefore required to solve the equations by means of numerical techniques. Most existing ice sheet models do this by employing the *finite difference method*, which will now be described.

The spatial domain under consideration is covered by a regular, three-dimensional grid. The equations are then re-written for each grid point by replacing the differentials by differences of the field variables between the neighbouring grid points. This yields a set of algebraic equations, which can then be solved by various methods, such as explicit forward integration, numerical quadrature or solving a set of linear or non-linear equations for the unknown field variables at the given grid points. Of course, there are many different ways of realizing this. Here, we shall describe a simplified, yet fully operational version of the numerical scheme employed by the well-established ice sheet model SICOPOLIS ("SImulation COde for POLythermal Ice Sheets"; see http://sicopolis.greveweb.net/).

5.7.1 Terrain-Following Coordinate Transformation

The most straightforward choice for a numerical grid is that of a regular, rectangular grid with constant grid spacings Δx, Δy and Δz in the respective directions. However, such a grid cannot match an irregular domain (like a real ice sheet) exactly. The surfaces of the domain generally fall in between the grid points, so that the values of field variables at the true surfaces must be interpolated. This makes the book-keeping of the values at the surfaces awkward and introduces additional inaccuracies into the computation.

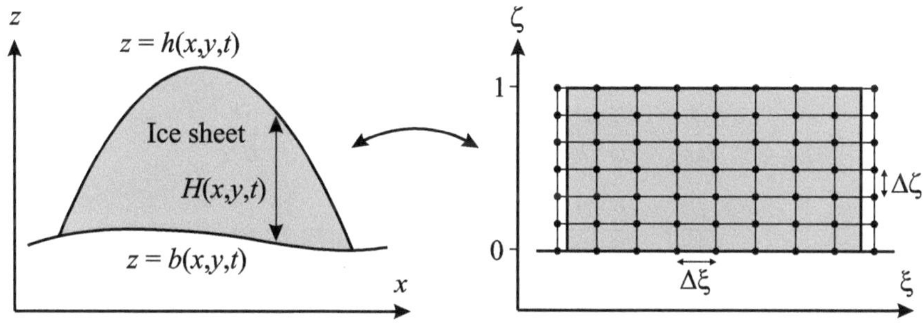

Fig. 5.13. Terrain-following sigma transformation (y and φ directions not shown). In the transformed domain (right), a regular, rectangular grid with spacings $\Delta\xi$ and $\Delta\zeta$ is shown.

In order to avoid these difficulties, it is suitable to introduce a terrain-following coordinate transformation that maps the local ice thickness onto unity (Fig. 5.13),

$$\xi = x\,, \quad \varphi = y\,, \quad \zeta = \frac{z - b(x,y,t)}{H(x,y,t)}\,, \quad \tau = t\,, \tag{5.127}$$

where (x, y, z) and (ξ, φ, ζ) are the natural Cartesian coordinates and the curvilinear, transformed coordinates, respectively, and t and τ are the time. This transformation, which is often referred to as the *sigma transformation*, maps the ice surface $h = h(x, y, t)$ to $\zeta = 1$ and the ice base $b = b(x, y, t)$ to $\zeta = 0$. In the transformed domain, a regular, rectangular grid with spacings $\Delta\xi$, $\Delta\varphi$ and $\Delta\zeta$ can easily be defined such that the uppermost layer of grid points matches the ice surface and the lowermost layer the ice base. However, since the transformation leaves the coordinates in the horizontal plane unchanged, the ice margin does not necessarily coincide with the grid points (Fig. 5.13). This may affect the accuracy of the computed position of the ice margin, especially during advance or retreat stages of the modelled ice sheet when it changes rapidly over time.

A further difficulty of the sigma transformation is the introduced singularity at the ice margin and outside the ice-covered area, where a zero ice thickness is mapped onto the unity interval [note the $1/H$ term in Eq. $(5.127)_3$]. This requires special consideration in the numerical scheme in order to avoid division-by-zero errors.

Equation (5.127) is a special case of the general, spatio-temporal coordinate transformation

$$\begin{aligned} \xi &= \xi(x,y,z,t)\,, \\ \varphi &= \varphi(x,y,z,t)\,, \\ \zeta &= \zeta(x,y,z,t)\,, \\ \tau &= \tau(x,y,z,t)\,. \end{aligned} \tag{5.128}$$

The differentials of any scalar field are transformed according to the chain rule,

$$\begin{aligned} \frac{\partial}{\partial x} &= \frac{\partial \xi}{\partial x}\frac{\partial}{\partial \xi} + \frac{\partial \varphi}{\partial x}\frac{\partial}{\partial \varphi} + \frac{\partial \zeta}{\partial x}\frac{\partial}{\partial \zeta} + \frac{\partial \tau}{\partial x}\frac{\partial}{\partial \tau}\,, \\ \frac{\partial}{\partial y} &= \frac{\partial \xi}{\partial y}\frac{\partial}{\partial \xi} + \frac{\partial \varphi}{\partial y}\frac{\partial}{\partial \varphi} + \frac{\partial \zeta}{\partial y}\frac{\partial}{\partial \zeta} + \frac{\partial \tau}{\partial y}\frac{\partial}{\partial \tau}\,, \\ \frac{\partial}{\partial z} &= \frac{\partial \xi}{\partial z}\frac{\partial}{\partial \xi} + \frac{\partial \varphi}{\partial z}\frac{\partial}{\partial \varphi} + \frac{\partial \zeta}{\partial z}\frac{\partial}{\partial \zeta} + \frac{\partial \tau}{\partial z}\frac{\partial}{\partial \tau}\,, \\ \frac{\partial}{\partial t} &= \frac{\partial \xi}{\partial t}\frac{\partial}{\partial \xi} + \frac{\partial \varphi}{\partial t}\frac{\partial}{\partial \varphi} + \frac{\partial \zeta}{\partial t}\frac{\partial}{\partial \zeta} + \frac{\partial \tau}{\partial t}\frac{\partial}{\partial \tau}\,. \end{aligned} \tag{5.129}$$

It is evident that for the concrete realisation (5.127) many of the coordinate differentials are equal to zero. Thus, Eq. (5.129) simplifies to

$$\begin{aligned} \frac{\partial}{\partial x} &= \frac{\partial}{\partial \xi} + \frac{\partial \zeta}{\partial x}\frac{\partial}{\partial \zeta}\,, \\ \frac{\partial}{\partial y} &= \frac{\partial}{\partial \varphi} + \frac{\partial \zeta}{\partial y}\frac{\partial}{\partial \zeta}\,, \\ \frac{\partial}{\partial z} &= \frac{\partial \zeta}{\partial z}\frac{\partial}{\partial \zeta}\,, \\ \frac{\partial}{\partial t} &= \frac{\partial}{\partial \tau} + \frac{\partial \zeta}{\partial t}\frac{\partial}{\partial \zeta}\,. \end{aligned} \tag{5.130}$$

Note that, although $\xi = x$, $\varphi = y$ and $\tau = t$ [Eq. $(5.127)_{1,2,4}$], the respective derivatives are not the same. It is therefore imperative to consider the entire, spatio-temporal transformation (5.127), even though only the vertical coordinate is changed. The relations for the second derivatives are

$$\frac{\partial^2}{\partial x^2} = \frac{\partial}{\partial x}\left(\frac{\partial}{\partial x}\right) = \frac{\partial^2}{\partial \xi^2} + \left(\frac{\partial \zeta}{\partial x}\right)^2 \frac{\partial^2}{\partial \zeta^2} + 2\frac{\partial \zeta}{\partial x}\frac{\partial^2}{\partial \xi\,\partial \zeta}$$

$$+\left[\left(\frac{\partial}{\partial \xi}\frac{\partial \zeta}{\partial x}\right)+\frac{\partial \zeta}{\partial x}\left(\frac{\partial}{\partial \zeta}\frac{\partial \zeta}{\partial x}\right)\right]\frac{\partial}{\partial \zeta},$$

$$\frac{\partial^2}{\partial y^2}=\frac{\partial}{\partial y}\left(\frac{\partial}{\partial y}\right)=\frac{\partial^2}{\partial \varphi^2}+\left(\frac{\partial \zeta}{\partial y}\right)^2\frac{\partial^2}{\partial \zeta^2}+2\frac{\partial \zeta}{\partial y}\frac{\partial^2}{\partial \varphi\,\partial \zeta}$$
$$+\left[\left(\frac{\partial}{\partial \varphi}\frac{\partial \zeta}{\partial y}\right)+\frac{\partial \zeta}{\partial y}\left(\frac{\partial}{\partial \zeta}\frac{\partial \zeta}{\partial y}\right)\right]\frac{\partial}{\partial \zeta}, \qquad (5.131)$$
$$\frac{\partial^2}{\partial z^2}=\left(\frac{\partial \zeta}{\partial z}\right)^2\frac{\partial^2}{\partial \zeta^2},$$
$$\frac{\partial^2}{\partial t^2}=\frac{\partial}{\partial t}\left(\frac{\partial}{\partial t}\right)=\frac{\partial^2}{\partial \tau^2}+\left(\frac{\partial \zeta}{\partial t}\right)^2\frac{\partial^2}{\partial \zeta^2}+2\frac{\partial \zeta}{\partial t}\frac{\partial^2}{\partial \tau\,\partial \zeta}$$
$$+\left[\left(\frac{\partial}{\partial \tau}\frac{\partial \zeta}{\partial t}\right)+\frac{\partial \zeta}{\partial t}\left(\frac{\partial}{\partial \zeta}\frac{\partial \zeta}{\partial t}\right)\right]\frac{\partial}{\partial \zeta},$$

and the coordinate differentials in Eqs. (5.130) and (5.131) take the forms

$$\frac{\partial \zeta}{\partial z}=\frac{1}{H},$$
$$\frac{\partial \zeta}{\partial x}=-\frac{(1-\zeta)b_{,\xi}+\zeta h_{,\xi}}{H},$$
$$\frac{\partial}{\partial \zeta}\frac{\partial \zeta}{\partial x}=-\frac{h_{,\xi}-b_{,\xi}}{H},$$
$$\frac{\partial}{\partial \xi}\frac{\partial \zeta}{\partial x}=-\frac{\partial \zeta}{\partial x}\frac{h_{,\xi}-b_{,\xi}}{H}-\frac{(1-\zeta)b_{,\xi\xi}+\zeta h_{,\xi\xi}}{H},$$
$$\frac{\partial \zeta}{\partial y}=-\frac{(1-\zeta)b_{,\eta}+\zeta h_{,\eta}}{H},$$
$$\frac{\partial}{\partial \zeta}\frac{\partial \zeta}{\partial y}=-\frac{h_{,\eta}-b_{,\eta}}{H}, \qquad (5.132)$$
$$\frac{\partial}{\partial \varphi}\frac{\partial \zeta}{\partial y}=-\frac{\partial \zeta}{\partial y}\frac{h_{,\eta}-b_{,\eta}}{H}-\frac{(1-\zeta)b_{,\eta\eta}+\zeta h_{,\eta\eta}}{H},$$
$$\frac{\partial \zeta}{\partial t}=-\frac{(1-\zeta)b_{,\tau}+\zeta h_{,\tau}}{H},$$
$$\frac{\partial}{\partial \zeta}\frac{\partial \zeta}{\partial t}=-\frac{h_{,\tau}-b_{,\tau}}{H},$$
$$\frac{\partial}{\partial \tau}\frac{\partial \zeta}{\partial t}=-\frac{\partial \zeta}{\partial x}\frac{h_{,\tau}-b_{,\tau}}{H}-\frac{(1-\zeta)b_{,\tau\tau}+\zeta h_{,\tau\tau}}{H}.$$

Recall that the compact comma notation denotes partial derivatives, that is, $(\cdot)_{,\xi} = \partial(\cdot)/\partial\xi$, etc. Further, due to Eq. (5.130)$_{1,2,4}$, the derivatives with respect to x, y, t and ξ, φ, τ are identical for field quantities which do not depend on z or ζ, such as h, b and H.

When dealing with the sigma transformation for ice sheet dynamics, the transformation for vertical integrals is also required. According to the general substitution rule for integrals, it is

$$\int_{z_1}^{z_2} (\cdot)\,\mathrm{d}z = \int_{\zeta_1}^{\zeta_2} (\cdot)\,\frac{\partial z}{\partial \zeta}\,\mathrm{d}\zeta = \int_{\zeta_1}^{\zeta_2} (\cdot)\,H\,\mathrm{d}\zeta\,. \tag{5.133}$$

5.7.2 Plane Strain Shallow Ice Equations

For simplicity, in the following we will employ the plane strain approximation, which was already used above for the analytical solutions (Sect. 5.6). That is, we will only consider a two-dimensional problem in the vertical x-z plane, and ignore any dependencies of the transverse y-direction. In other words, we assume that ice flow occurs only parallel to the x-z plane, and that the conditions are homogeneous in the y-direction, so that $\partial/\partial y = 0$ and $v_y = 0$. From Glen's flow law in the form (4.20), it follows readily that $t^{\mathrm{D}}_{yy} = t_{yz} = t_{xy} = 0$. Furthermore, the unit normal vectors $\mathbf{n}$ of the free surface and the ice base are parallel to the x-z plane, thus $n_y = 0$.

By subjecting Eqs. (5.94), (5.95), (5.100) and (5.105) to these assumptions, we obtain the following reduced set of equations:

Horizontal velocity:

$$v_x = -C\frac{\partial h}{\partial x}\,, \tag{5.134}$$

with the scalar function

$$C = \begin{cases} 2(\rho g)^n \left|\dfrac{\partial h}{\partial x}\right|^{n-1} \displaystyle\int_b^z A(T')\,(h-\bar{z})^n\,\mathrm{d}\bar{z}\,, & \text{if } T_\mathrm{b} < T_\mathrm{m}\,, \\[2ex] C_\mathrm{b}(\rho g H)^{p-q} \left|\dfrac{\partial h}{\partial x}\right|^{p-1} & \\[2ex] \quad +2(\rho g)^n \left|\dfrac{\partial h}{\partial x}\right|^{n-1} \displaystyle\int_b^z A(T')\,(h-\bar{z})^n\,\mathrm{d}\bar{z}\,, & \text{if } T_\mathrm{b} = T_\mathrm{m}\,. \end{cases} \tag{5.135}$$

Vertical velocity:

$$\begin{aligned} v_z &= v_z\big|_{z=b} - \int_b^z \frac{\partial v_x}{\partial x}\,\mathrm{d}z' \\ &= v_x\big|_{z=b}\frac{\partial b}{\partial x} - \int_b^z \frac{\partial v_x}{\partial x}\,\mathrm{d}z'\,. \end{aligned} \tag{5.136}$$

In order to eliminate the unknown vertical velocity at the base ($v_z|_{z=b}$) and obtain the second line, the kinematic condition (5.31) has been inserted in Eq. (5.95). Further, it has been assumed that the ice base is rigid ($\partial b/\partial t = 0$), and that the basal melting rate is negligible ($a_\mathrm{b} = 0$). These assumptions are not crucial though; they have only been made for reasons of simplicity.

Evolution of the ice surface:

$$\frac{\partial h}{\partial t} = \frac{\partial}{\partial x}\left(D\frac{\partial h}{\partial x}\right) + a_\mathrm{s}\,, \tag{5.137}$$

with the diffusivity

$$D = \int_b^h C\,\mathrm{d}z\,. \tag{5.138}$$

Again, note that the vertical movement of the ice base ($\partial b/\partial t$) and the basal melting rate (a_b) have been neglected.

Evolution of the ice temperature:

$$\begin{aligned}\rho c\left(\frac{\partial T}{\partial t} + v_x\frac{\partial T}{\partial x} + v_z\frac{\partial T}{\partial z}\right) &= \frac{\partial}{\partial z}\left(\kappa\frac{\partial T}{\partial z}\right) + 2A(T')\,\sigma_\mathrm{e}^{n+1}\\ &= \frac{\partial}{\partial z}\left(\kappa\frac{\partial T}{\partial z}\right) + 2A(T')\,[\rho g(h-z)]^{n+1}\left|\frac{\partial h}{\partial x}\right|^{n+1}\,.\end{aligned} \tag{5.139}$$

In order to derive the second line, the effective stress (5.81) has been inserted in (5.105). The boundary conditions result from Eqs. (5.24), (5.38) and (5.39),

$$T|_{z=h} = T_\mathrm{s} \tag{5.140}$$

and

$$\begin{aligned}\kappa\frac{\partial T}{\partial z}\bigg|_{z=b} &= -q_\mathrm{geo}^\perp\,, &&\text{if } T_\mathrm{b} < T_\mathrm{m}\,,\\ T|_{z=b} &= T_\mathrm{m}\,, &&\text{otherwise}\,.\end{aligned} \tag{5.141}$$

Transformed Equations

The shallow ice equations in the plane strain approximation, Eqs. (5.134) to (5.141), shall now be subjected to the two-dimensional form of the sigma transformation (5.127),

$$\xi = x\,,\quad \zeta = \frac{z - b(x,t)}{H(x,t)}\,,\quad \tau = t\,. \tag{5.142}$$

Applying this transformation together with Eqs. (5.130) and (5.132), and using the notation $u = v_x$, $w = v_z$, yields:

Horizontal velocity:

$$u = -C\frac{\partial h}{\partial \xi}, \tag{5.143}$$

with the scalar function

$$C = \begin{cases} 2(\rho g)^n \left|\dfrac{\partial h}{\partial \xi}\right|^{n-1} H \displaystyle\int_0^\zeta A(T')\,(1-\zeta')^n\,\mathrm{d}\zeta', & \text{if } T_\mathrm{b} < T_\mathrm{m}, \\[2ex] C_\mathrm{b}(\rho g H)^{p-q} \left|\dfrac{\partial h}{\partial \xi}\right|^{p-1} & \\[2ex] +2(\rho g)^n \left|\dfrac{\partial h}{\partial \xi}\right|^{n-1} H \displaystyle\int_0^\zeta A(T')\,(1-\zeta')^n\,\mathrm{d}\zeta', & \text{if } T_\mathrm{b} = T_\mathrm{m}. \end{cases} \tag{5.144}$$

Vertical velocity:

$$w = u|_{\zeta=0}\frac{\partial b}{\partial \xi} - H\int_0^\zeta \left(\frac{\partial u}{\partial \xi} - \frac{(1-\zeta')b_{,\xi} + \zeta' h_{,\xi}}{H}\frac{\partial u}{\partial \zeta'}\right)\mathrm{d}\zeta'. \tag{5.145}$$

Evolution of the ice surface:

$$\frac{\partial h}{\partial \tau} = \frac{\partial}{\partial \xi}\left(D\frac{\partial h}{\partial \xi}\right) + a_\mathrm{s}, \tag{5.146}$$

with the diffusivity

$$D = H\int_0^1 C\,\mathrm{d}\zeta. \tag{5.147}$$

Evolution of the ice temperature:

$$\begin{aligned} \rho c &\left[\frac{\partial T}{\partial \tau} + u\frac{\partial T}{\partial \xi}\right. \\ &\quad \left. + \frac{-[(1-\zeta)b_{,\tau} + \zeta h_{,\tau}] - u\,[(1-\zeta)b_{,\xi} + \zeta h_{,\xi}] + w}{H}\frac{\partial T}{\partial \zeta}\right] \\ &= \frac{1}{H^2}\frac{\partial}{\partial \zeta}\left(\kappa\frac{\partial T}{\partial \zeta}\right) + 2A(T')\,[\rho g H(1-\zeta)]^{n+1}\left|\frac{\partial h}{\partial \xi}\right|^{n+1}, \end{aligned} \tag{5.148}$$

with the boundary conditions

$$T|_{\zeta=1} = T_\mathrm{s} \tag{5.149}$$

and

$$\begin{aligned} \frac{\kappa}{H}\frac{\partial T}{\partial \zeta}\bigg|_{\zeta=0} &= -q_\mathrm{geo}^\perp, & \text{if } T_\mathrm{b} < T_\mathrm{m}, \\ T|_{\zeta=0} &= T_\mathrm{m}, & \text{otherwise}. \end{aligned} \tag{5.150}$$

Note that, while applying the sigma transformation, we have not attempted to write the velocity vector **v** in components of the curvilinear, transformed coordinates. Instead, the Cartesian components $u = v_x$ and $w = v_z$ are treated like scalar fields which are not affected by the transformation. This is common practise in connection with the sigma transformation; otherwise, the transformed equations would become very complicated due to the non-orthogonality of the transformed coordinates.

5.7.3 Discretised Ice Sheet Equations

A regular, rectangular grid as sketched in Fig. 5.13 is now defined in the (ξ, ζ) space of the transformed coordinates. The grid consists of $I + 1$ and $K + 1$ grid points in the ξ and ζ directions, numbered by $i = 0, \dots, I$ and $k = 0, \dots, K$, respectively. It is presupposed that the grid covers the entire area of the ice sheet at all times, so that the ice thickness and all velocity and stress components are zero at the end points $i = 0$ and $i = I$. The grid points indexed by $k = 0$ correspond to the ice base ($\zeta = 0$), and those indexed by $k = K$ match the free surface ($\zeta = 1$).

For reasons of stability of the numerical scheme, it is not practical to employ a single grid for all unknowns. Velocity and flux quantities are preferably defined on a secondary grid, often called a *staggered grid*, with grid lines in between the main grid lines. The main grid lines are then numbered by integers, $i = 0, 1, 2, \dots, I-1, I$ for the ξ-direction and $k = 0, 1, 2, \dots, K-1, K$ for the ζ-direction, and the secondary grid lines by half-numbers, $\frac{1}{2}, \frac{3}{2}, \frac{5}{2}, \dots, I-\frac{3}{2}, I-\frac{1}{2}$ and $\frac{1}{2}, \frac{3}{2}, \frac{5}{2}, \dots, K-\frac{3}{2}, K-\frac{1}{2}$ (Fig. 5.14).

Let ξ_0 be the origin of the model domain, then the positions ξ_i of the main grid points are situated at

$$\xi_i = \xi_0 + i\Delta\xi \,. \tag{5.151}$$

Similarly, the positions ζ_k of the main grid points in the vertical direction are

$$\zeta_k = k\Delta\zeta = \frac{k}{K} \,. \tag{5.152}$$

Time is discretised by the time step $\Delta\tau$, such that the time after n steps is

$$\tau_n = \tau_0 + n\Delta\tau \,, \tag{5.153}$$

where τ_0 is the initial time of the numerical simulation.

The staggered grid can be realised in different ways. It turns out that a very suitable choice for the shallow ice equations is the *Arakawa C grid*, for which the velocity components u and w are defined on secondary grid points as follows,

$$u_{i\pm\frac{1}{2},k,n} \,, \quad w_{i,k\pm\frac{1}{2},n} \,. \tag{5.154}$$

The secondary grid points which result from this choice are indicated in Fig. 5.14. The positions $(i \pm \frac{1}{2}, k \pm \frac{1}{2})$ are not used as secondary grid points in the Arakawa C grid.

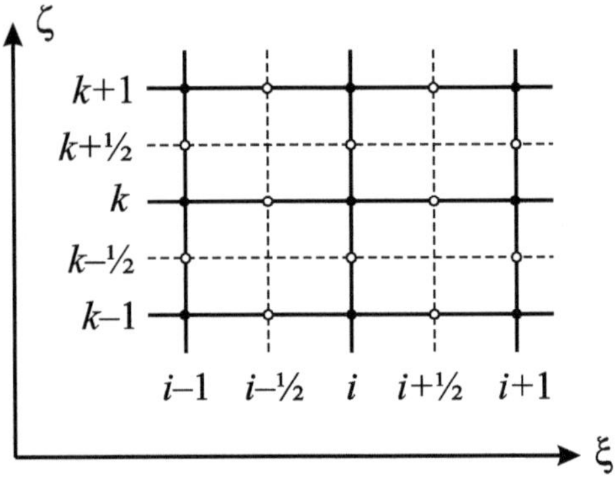

Fig. 5.14. Arakawa C grid for the plane strain approximation (two-dimensional problem). Full circles and solid lines denote main grid points and grid lines with integer indices, open circles and dashed lines secondary grid points and grid lines with half-numbered indices.

Furthermore, the volume flux $Q = Q_x$ (which is independent of z or ζ) is defined on the staggered grid,

$$Q_{i\pm\frac{1}{2},n}\,, \tag{5.155}$$

which is a natural consequence of its definition as the vertically integrated horizontal velocity. All other quantities are defined on the main grid points,

$$T_{i,k,n}\,, \tag{5.156}$$

$$h_{i,n}\,,\quad b_{i,n}\,,\quad H_{i,n}\,. \tag{5.157}$$

If one of the quantities which is defined on the secondary grid is required on the main grid, it is interpolated by the arithmetic mean of the values on the neighbouring secondary grid points,

$$\bar{u}_{i,k,n} = \frac{1}{2}\left(u_{i+\frac{1}{2},k,n} + u_{i-\frac{1}{2},k,n}\right), \tag{5.158}$$

$$\bar{w}_{i,k,n} = \frac{1}{2}\left(w_{i,k+\frac{1}{2},n} + w_{i,k-\frac{1}{2},n}\right), \tag{5.159}$$

$$\bar{Q}_{i,n} = \frac{1}{2}\left(Q_{i+\frac{1}{2},n} + Q_{i-\frac{1}{2},n}\right). \tag{5.160}$$

Conversely, if a quantity which is defined on the main grid is required on the secondary grid, the possible interpolations are

$$\bar{T}_{i+\frac{1}{2},k,n} = \frac{1}{2}\left(T_{i,k,n} + T_{i+1,k,n}\right), \tag{5.161}$$

$$\bar{T}_{i,k+\frac{1}{2},n} = \frac{1}{2}\left(T_{i,k,n} + T_{i,k+1,n}\right), \tag{5.162}$$

$$\bar{h}_{i+\frac{1}{2},n} = \frac{1}{2}\left(h_{i,n} + h_{i+1,n}\right). \tag{5.163}$$

In all above cases, the interpolation is indicated by the bar symbol over the variable.

In the following, we are going to discretise the sigma-transformed plane strain shallow ice equations derived above [Eqs. (5.143) to (5.150)]. They take the following forms:

Horizontal velocity:

First, the scalar function $C_{i,k,n}$ is computed from Eq. (5.144). The derivative $\partial h/\partial\xi$ is approximated by central differences,

$$\left.\frac{\partial h}{\partial \xi}\right|_{i,n} \sim \frac{h_{i+1,n} - h_{i-1,n}}{2\Delta\xi}, \tag{5.164}$$

and the integral is approximated by the trapezoidal rule,

$$\begin{aligned}
&\left.\int_0^\zeta A(T')\,(1-\zeta')^n\,\mathrm{d}\zeta'\right|_{i,k,n} \\
&\quad\sim \left[\frac{1}{2}A(T'_{i,0,n}) + \sum_{k'=1}^{k-1} A(T'_{i,k',n})\,(1-k'\Delta\zeta)^n \right.\\
&\qquad\qquad \left. + \frac{1}{2}A(T'_{i,k,n})\,(1-k\Delta\zeta)^n\right]\Delta\zeta\,.
\end{aligned} \tag{5.165}$$

The horizontal velocity itself follows from Eq. (5.143),

$$u_{i+\frac{1}{2},k,n} = -\bar{C}_{i+\frac{1}{2},k,n}\frac{h_{i+1,n} - h_{i,n}}{\Delta\xi}, \tag{5.166}$$

where the scalar function C needs to be interpolated onto the secondary grid points, and central differences have been applied for the derivative $\partial h/\partial\xi$.

Vertical velocity:

Let us abbreviate the integrand of Eq. (5.145) as follows,

$$U = \frac{\partial u}{\partial \xi} - \frac{(1-\zeta)b_{,\xi} + \zeta h_{,\xi}}{H}\frac{\partial u}{\partial \zeta}. \tag{5.167}$$

It is discretised by central differences,

$$\begin{aligned}
U_{i,k,n} = {} & \frac{u_{i+\frac{1}{2},k,n} - u_{i-\frac{1}{2},k,n}}{\Delta\xi} \\
& - \frac{(1-k\Delta\zeta)\,(b_{i+1,n} - b_{i-1,n}) + k\Delta\zeta\,(h_{i+1,n} - h_{i-1,n})}{2\Delta\xi\, H_{i,n}} \\
& \times \frac{\bar{u}_{i,k+1,n} - \bar{u}_{i,k-1,n}}{2\Delta\zeta}\,.
\end{aligned} \tag{5.168}$$

For the numerical integration of Eq. (5.145), due to the geometry of the staggered grid, it is more convenient to employ Gaussian quadrature instead of the trapezoidal rule,

$$w_{i,k+\frac{1}{2},n} = \bar{u}_{i,0,n}\,\frac{b_{i+1,n} - b_{i-1,n}}{2\Delta\xi} - H_{i,n}\left[\frac{1}{2}U_{i,0,n} + \sum_{k'=1}^{k} U_{i,k',n}\right]\Delta\zeta\,. \tag{5.169}$$

Evolution of the ice surface:

For the time derivative $\partial h/\partial\tau$ in Eq. (5.146), Euler forward stepping is employed,

$$\left.\frac{\partial h}{\partial\tau}\right|_{i,n} \sim \frac{h_{i,n+1} - h_{i,n}}{\Delta\tau}\,. \tag{5.170}$$

The diffusivity $D_{i,n}$ is computed from Eq. (5.147) with the trapezoidal rule,

$$D_{i,n} = H_{i,n}\left[\frac{1}{2}C_{i,0,n} + \sum_{k=1}^{K-1} C_{i,k,n} + \frac{1}{2}C_{i,K,n}\right]\Delta\zeta\,, \tag{5.171}$$

which allows as a side result the computation of the volume flux Q,

$$Q_{i+\frac{1}{2},n} = -\bar{D}_{i+\frac{1}{2},n}\,\frac{h_{i+1,n} - h_{i,n}}{\Delta\xi}\,. \tag{5.172}$$

The non-linear diffusion term in Eq. (5.146) is approximated by central differences,

$$\left.\frac{\partial}{\partial\xi}\left(D\frac{\partial h}{\partial\xi}\right)\right|_{i,n} \sim \frac{1}{\Delta\xi}\left[\left.\left(D\frac{\partial h}{\partial\xi}\right)\right|_{i+\frac{1}{2},n} - \left.\left(D\frac{\partial h}{\partial\xi}\right)\right|_{i-\frac{1}{2},n}\right], \tag{5.173}$$

where

$$\begin{aligned}\left.\left(D\frac{\partial h}{\partial\xi}\right)\right|_{i+\frac{1}{2},n} &\sim \bar{D}_{i+\frac{1}{2},n}\,\frac{h_{i+1,n} - h_{i,n}}{\Delta\xi}\,,\\ \left.\left(D\frac{\partial h}{\partial\xi}\right)\right|_{i-\frac{1}{2},n} &\sim \bar{D}_{i-\frac{1}{2},n}\,\frac{h_{i,n} - h_{i-1,n}}{\Delta\xi}\,.\end{aligned} \tag{5.174}$$

This yields an explicit scheme. Provided that all quantities are known at the old time τ_n, then Eqs. (5.171) – (5.174) do not contain any unknowns, and with the prescribed accumulation-ablation function $(a_\mathrm{s})_{i,n}$, the discretised ice surface equation can readily be solved for the unknown surface elevation $h_{i,n+1}$ at the new time τ_{n+1} [from Eq. (5.170)].

In order to allow larger time steps $\Delta\tau$ while keeping the scheme numerically stable, it is also possible to introduce some implicitness. Replacing the surface elevations $h_{i+1,n}$, $h_{i,n}$ and $h_{i-1,n}$ in Eq. (5.174) by their counterparts at the new time, $h_{i+1,n+1}$, $h_{i,n+1}$ and $h_{i-1,n+1}$, yields a semi-implicit scheme. At each time step, it requires solving a system of linear equations for all new surface elevations $h_{i,n+1}$ $(i = 0 \ldots I)$ simultaneously. In addition to that, the diffusivities in Eq. (5.174), which depend non-linearly on the surface elevations, can also be taken at the new time, $\bar{D}_{i+\frac{1}{2},n+1}$ and $\bar{D}_{i-\frac{1}{2},n+1}$, which renders the scheme fully implicit. The price to pay is that a system of non-linear equations must be solved at each time-step in order to obtain the new surface elevations.

Evolution of the ice temperature:

As for the evolution equation of the ice surface, Euler forward stepping is used for the approximation of the time derivative $\partial T/\partial \tau$ in Eq. (5.148),

$$\left.\frac{\partial T}{\partial \tau}\right|_{i,k,n} \sim \frac{T_{i,k,n+1} - T_{i,k,n}}{\Delta \tau}. \tag{5.175}$$

For the horizontal advection term $u\,\partial T/\partial \xi$, central differences are not suitable, because this leads to numerical instabilities. Instead, an asymmetric, "upstream" scheme is employed,

$$u\left.\frac{\partial T}{\partial \xi}\right|_{i,k,n} \sim \begin{cases} u_{i-\frac{1}{2},k,n}\,\dfrac{T_{i,k,n} - T_{i-1,k,n}}{\Delta \xi}, & \text{if } \bar{u}_{i,k,n} > 0\,, \\[2ex] u_{i+\frac{1}{2},k,n}\,\dfrac{T_{i+1,k,n} - T_{i,k,n}}{\Delta \xi}, & \text{if } \bar{u}_{i,k,n} < 0\,. \end{cases} \tag{5.176}$$

This scheme accounts for the fact that the flow of information is from the upstream direction by shifting the discretisation anti-parallel to the flow (sign of $\bar{u}_{i,k,n}$) by half a grid point.

A similar method is chosen for the vertical advection term. Let us abbreviate the pre-factor of $\partial T/\partial \zeta$ by

$$W = \frac{-[(1-\zeta)b_{,\tau} + \zeta h_{,\tau}] - u\,[(1-\zeta)b_{,\xi} + \zeta h_{,\xi}] + w}{H}. \tag{5.177}$$

Owing to our assumption of a rigid ice base [see discussion of Eq. (5.136)], $b_{,\tau} = 0$. The further derivatives can be discretised by

$$\begin{aligned} (h_{,\tau})_{i,n+1} &\sim \frac{h_{i,n+1} - h_{i,n}}{\Delta \tau}, \\ (b_{,\xi})_{i,n+1} &\sim \frac{b_{i+1,n+1} - b_{i-1,n+1}}{2\Delta \xi}, \\ (h_{,\xi})_{i,n+1} &\sim \frac{h_{i+1,n+1} - h_{i-1,n+1}}{2\Delta \xi}, \end{aligned} \tag{5.178}$$

where the results of the computation of the ice surface elevation at the new time τ_{n+1} are used. We now set

$$
\begin{aligned}
W_{i,k+\frac{1}{2},n+1} = &-\frac{(k+\frac{1}{2})\Delta\zeta\,(h_{,\tau})_{i,n+1}}{H_{i,n+1}} \\
&-\bar{u}_{i,k+\frac{1}{2},n} \\
&\quad\times\frac{[1-(k+\frac{1}{2})\Delta\zeta]\,(b_{,\xi})_{i,n+1}+(k+\frac{1}{2})\Delta\zeta\,(h_{,\xi})_{i,n+1}}{H_{i,n+1}} \\
&+\frac{w_{i,k+\frac{1}{2},n}}{H_{i,n+1}},
\end{aligned} \tag{5.179}
$$

and approximate the vertical advection term by

$$
W\frac{\partial T}{\partial\zeta}\bigg|_{i,k,n+1} \sim \begin{cases} W_{i,k-\frac{1}{2},n+1}\,\dfrac{T_{i,k,n+1}-T_{i,k-1,n+1}}{\Delta\zeta}, & \\ & \text{if } \bar{W}_{i,k,n}>0, \\ W_{i,k+\frac{1}{2},n+1}\,\dfrac{T_{i,k+1,n+1}-T_{i,k,n+1}}{\Delta\zeta}, & \\ & \text{if } \bar{W}_{i,k,n}<0. \end{cases} \tag{5.180}
$$

Note that, in addition to the upstreaming, this is an implicit scheme in which the temperatures are taken at the new time τ_{n+1}, that is, they are unknown at the time of computation. The necessity to use an implicit discretisation for the vertical derivatives arises from the sigma transformation. Where the ice thickness is small (for instance, close to the margin), the spacing $\Delta\zeta$ in the transformed coordinates corresponds to a very small spacing Δz in the physical space. Using an explicit scheme for the vertical derivatives would therefore lead to prohibitively small time steps in order to keep the integration numerically stable.

For the same reason, an implicit scheme, now with central differences, is applied for the diffusion term,

$$
\begin{aligned}
&\frac{\partial}{\partial\zeta}\left(\kappa\frac{\partial T}{\partial\zeta}\right)\bigg|_{i,k,n+1} \\
&\qquad\sim\frac{1}{\Delta\zeta}\left[\left(\kappa\frac{\partial T}{\partial\zeta}\right)\bigg|_{i,k+\frac{1}{2},n+1}-\left(\kappa\frac{\partial T}{\partial\zeta}\right)\bigg|_{i,k-\frac{1}{2},n+1}\right],
\end{aligned} \tag{5.181}
$$

where

$$
\begin{aligned}
\left(\kappa\frac{\partial T}{\partial\zeta}\right)\bigg|_{i,k+\frac{1}{2},n+1} &\sim \bar{\kappa}_{i,k+\frac{1}{2},n}\,\frac{T_{i,k+1,n+1}-T_{i,k,n+1}}{\Delta\zeta}, \\
\left(\kappa\frac{\partial T}{\partial\zeta}\right)\bigg|_{i,k-\frac{1}{2},n+1} &\sim \bar{\kappa}_{i,k-\frac{1}{2},n}\,\frac{T_{i,k,n+1}-T_{i,k-1,n+1}}{\Delta\zeta}.
\end{aligned} \tag{5.182}
$$

Discretisation of the dissipation term is straightforward, with the only occurring derivative being the surface gradient, which is also taken at the new time τ_{n+1},

$$\left.\frac{\partial h}{\partial \xi}\right|_{i,k,n+1} \sim \frac{h_{i+1,k,n+1} - h_{i-1,k,n+1}}{2\Delta\xi} . \tag{5.183}$$

The boundary conditions (5.149) and (5.150) enter the discretised problem as follows. At the free surface, the temperature is set to the prescribed surface temperature,

$$T_{i,K,n+1} = (T_\mathrm{s})_{i,n+1} . \tag{5.184}$$

At the ice base, we obtain

$$\begin{aligned} \frac{\kappa_{i,0,n}}{H_{i,n+1}} \frac{T_{i,1,n+1} - T_{i,0,n+1}}{\Delta\zeta} &= -q_\mathrm{geo}^\perp , \qquad \text{if } T_{i,0,n+1} < (T_\mathrm{m})_{i,0,n+1} , \\ T_{i,0,n+1} &= (T_\mathrm{m})_{i,0,n+1} , \qquad \text{otherwise} . \end{aligned} \tag{5.185}$$

Since it is not known *a priori* which of the two cases is applicable for a certain ice column defined by the index i, a trial-and-error procedure must be applied. This can be done by first assuming case $(5.185)_1$, computing the new temperatures for the ice column, checking whether the result fulfills the condition of a basal temperature below the melting point, and, if not, discarding the result and repeating the computation for case $(5.185)_2$. The column-wise computation of the new temperatures is possible because the presented scheme is only implicit with respect to vertical derivatives, but explicit with respect to horizontal derivatives. Therefore, for each ice column indexed by i, a separate system of linear equations must be solved in order to compute $T_{i,k,n+1}$ $(k = 0 \ldots K)$ simultaneously.

Recall that the numerical scheme discussed in this section is based on the shallow ice equations in the plane strain approximation. However, this limitation serves didactic purposes only by limiting the length of the discretised equations and keeping them as clear as possible. The method itself is not limited to two spatial dimensions, and the extension of the numerical scheme to the full, three-dimensional case is straightforward.

5.7.4 Example: The EGIG Line of the Greenland Ice Sheet

As an example of a numerical ice-sheet-modelling study using the plane strain approximation, let us consider the application of a two-dimensional version of the ice sheet model SICOPOLIS to the EGIG line of the Greenland Ice Sheet. The EGIG [Expédition Glaciologique International au Grœnland; Hofmann (1974)] line is a transect across the ice sheet essentially in the west-east direction at about 70–72°N. For most of its length, the EGIG line follows a flowline of the ice sheet (along the gradient of the ice surface); however, in the easternmost part it deviates to the north. Therefore, Abe-Ouchi (1993) defined a modified profile (here referred to as “EGIG1”) in which the flowline is extended to the eastern ice margin (Fig. 5.15). The EGIG1 profile is well suited for the plane strain approximation because of small ice surface slopes in the transverse direction.

For the definition of the several input quantities, we follow Greve (1994). The present-day (interglacial) surface temperature is given by

$$T_s^0(h) = T_{sl} - \gamma h\,, \tag{5.186}$$

where $T_{sl} = -3.5°\mathrm{C}$ is the temperature at sea level and $\gamma = 8°\mathrm{C\,km^{-1}}$ the atmospheric lapse rate. The surface-temperature anomaly $\Delta T_s(t)$ at an arbitrary time t is prescribed by the cosine function

$$\Delta T_s(t) = \frac{10°\mathrm{C}}{2}\left(\cos\left(\frac{2\pi t}{100\,\mathrm{ka}}\right) - 1\right)\,, \tag{5.187}$$

which represents idealised glacial-interglacial cycles with a period of 100 ka and glacial minima 10°C colder than interglacial maxima (see below, top panel of Fig. 5.17). With Eqs. (5.186) and (5.187), the space- and time-dependent surface temperature is

$$T_s(h,t) = T_s^0(h) + \Delta T_s(t)\,. \tag{5.188}$$

The surface mass balance a_s is the difference between snowfall (accumulation) and melting (ablation). We employ a simple parameterisation with three parameters, the snowfall rate S_0, the melting gradient m_0 and the equilibrium line altitude h_{EL},

$$a_s(h,t) = \min[S_0(t),\, m_0(h - h_{EL}(t))] \tag{5.189}$$

(Fig. 5.16). For a present-day (interglacial) climate, the parameters have the values

$$\begin{aligned} S_0 &= 0.3\,\mathrm{m\,a^{-1}}\,, \\ h_{EL} &= 1100\,\mathrm{m}\,, \\ m_0 &= 0.005\,\mathrm{a^{-1}}\,, \end{aligned} \tag{5.190}$$

and a glacial climate is characterised by

$$\begin{aligned} S_0 &= 0.15\,\mathrm{m\,a^{-1}}\,, \\ h_{EL} &= 100\,\mathrm{m}\,, \\ m_0 &= 0.005\,\mathrm{a^{-1}}\,. \end{aligned} \tag{5.191}$$

For arbitrary times t, we assume that these parameters are related linearly to the surface-temperature anomaly $\Delta T_s(t)$. Thus,

$$\begin{aligned} S_0(t) &= (0.3 + 0.015\Delta T_s(t)[°\mathrm{C}])\,\mathrm{m\,a^{-1}}\,, \\ h_{EL}(t) &= (1100 + 100\Delta T_s(t)[°\mathrm{C}])\,\mathrm{m}\,, \\ m_0 &= 0.005\,\mathrm{a^{-1}}\,, \end{aligned} \tag{5.192}$$

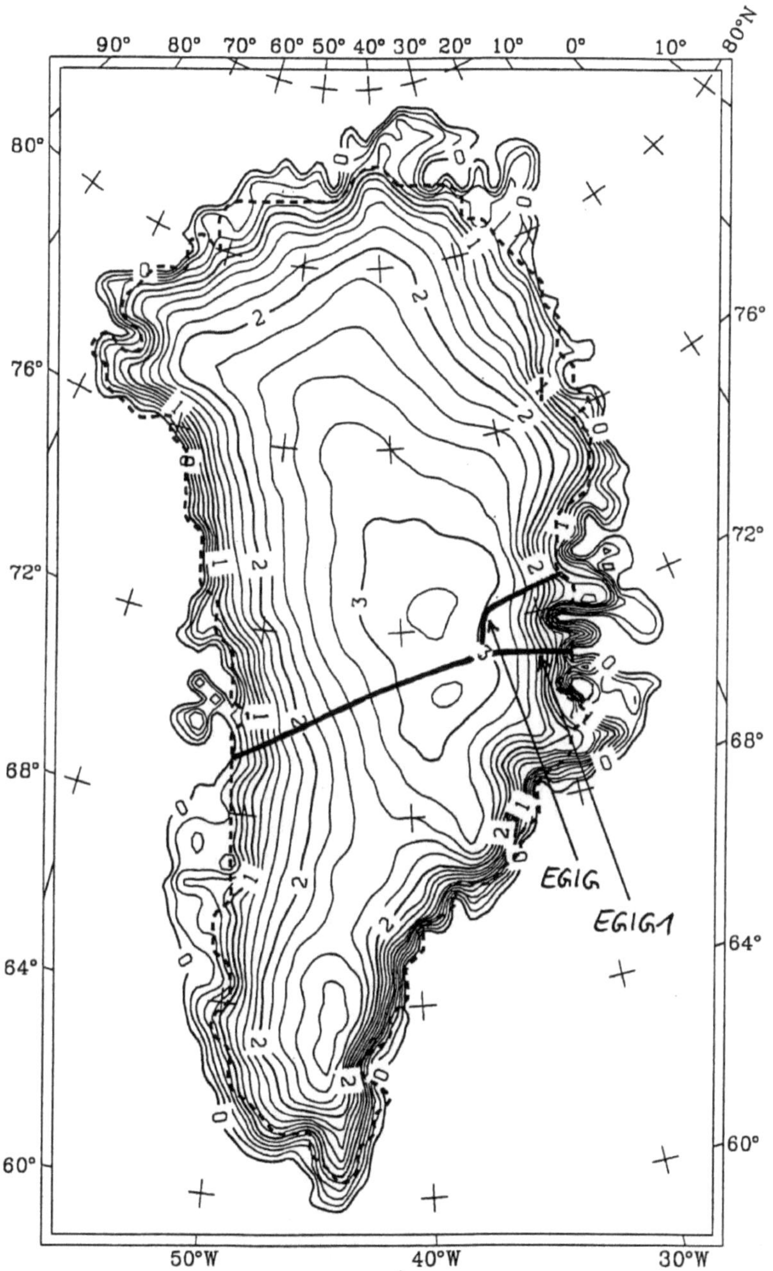

Fig. 5.15. Surface topography of the Greenland Ice Sheet (in km AMSL, contour spacing 200 m). The original EGIG profile and the modified, flowline-following EGIG1 profile are indicated as bold lines. Adapted from Calov (1994), © R. Calov.

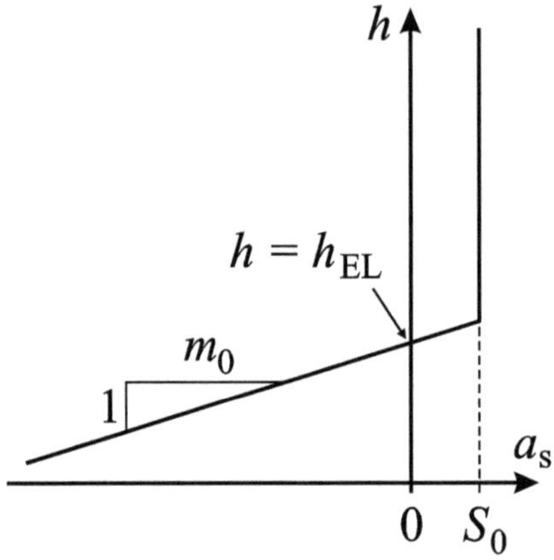

Fig. 5.16. Parameterisation of the surface mass balance a_s as a function of the surface elevation h. The three parameters are the snowfall rate S_0, the melting gradient m_0 and the equilibrium line altitude h_{EL}.

which fulfills Eq. (5.190) for $\Delta T_s = 0°C$ and Eq. (5.191) for $\Delta T_s = -10°C$.

At the base, the parameters for the Weertman-type sliding law (5.91) are chosen as $p = 3$, $q = 2$ and $C_b = 6 \times 10^4\,a^{-1}/(\rho g) = 6.72\,m\,a^{-1}\,Pa^{-1}$. The geothermal heat flux is set to the often-applied value for Precambrian rocks, $q_{geo}^{\perp} = 42\,mW\,m^{-2}$.

The spatio-temporal discretisation of the model domain is characterised by a horizontal resolution of $\Delta\xi = 20\,km$, a vertical resolution of $\Delta\zeta = 0.0125$ and a time step of $\Delta\tau = 2.5\,a$. The simulation runs from $t = 0$ until $t = 400\,ka$. The first 200 ka are the spin-up time with a temporally constant, interglacial forcing $[\Delta T_s(t) \equiv 0°C]$, and from $t = 200\,ka$ until $t = 400\,ka$ the idealised glacial-interglacial cycles according to Eq. (5.187) are applied.

Figure 5.17 shows the prescribed surface-temperature anomaly ΔT_s, the simulated maximum surface elevation above mean sea level h_{max}, cross-sectional area A_{ges} and length of the temperate base $L_{t,b}$ as functions of time for the two glacial-interglacial cycles (from $t = 200\,ka$ until $t = 400\,ka$). At $t = 250\,ka$, that is, 50 ka after the onset of the time-dependent climate forcing, the system has largely forgotten the initial state provided by the spin-up period, and from then on all shown quantities follow the surface-temperature anomaly approximately sinusoidally. The positive correlation between the surface-temperature anomaly, the maximum surface elevation and the cross-sectional area is counter-intuitive at first glance, as one would expect more ice in a colder climate and less ice in a warmer climate. However, the glacial climate is also drier [compare Eqs. $(5.190)_1$ and $(5.191)_1$], and the reduced snowfall outweighs the reduced melting and slower glacial flow, at least for the EGIG1 profile. By contrast, the positive correlation between the surface-temperature anomaly and the length of the temperate base is imme-

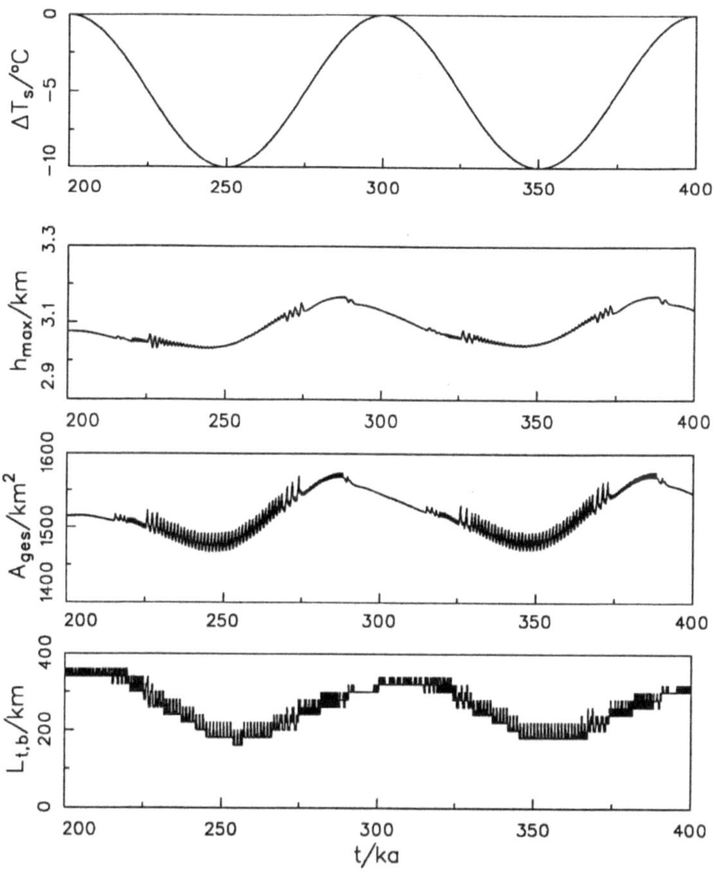

Fig. 5.17. Time series of the prescribed surface-temperature anomaly ΔT_s and the simulated quantities: maximum surface elevation above mean sea level h_{max}, cross-sectional area A_{ges} and length of the temperate base $L_{t,b}$, over two idealised glacial cycles.

diately intuitive. Note the ~ 10 ka phase shift between these two quantities, which is a result of the transfer time of changes of the surface temperature to the ice base.

The length of the temperate base, the cross-sectional area and to a lesser extent the maximum surface elevation show high-frequency oscillations with an average period of ~ 1.2 ka. These oscillations are more pronounced during the cold phases of the sinusoidal climate forcing and can be interpreted as cycles of mini-surges and subsequent recovery phases of the ice sheet. While there is observational evidence that ice sheets can undergo even large-scale surges on millennial time scales (Heinrich 1988, Bond et al. 1992, Bond and

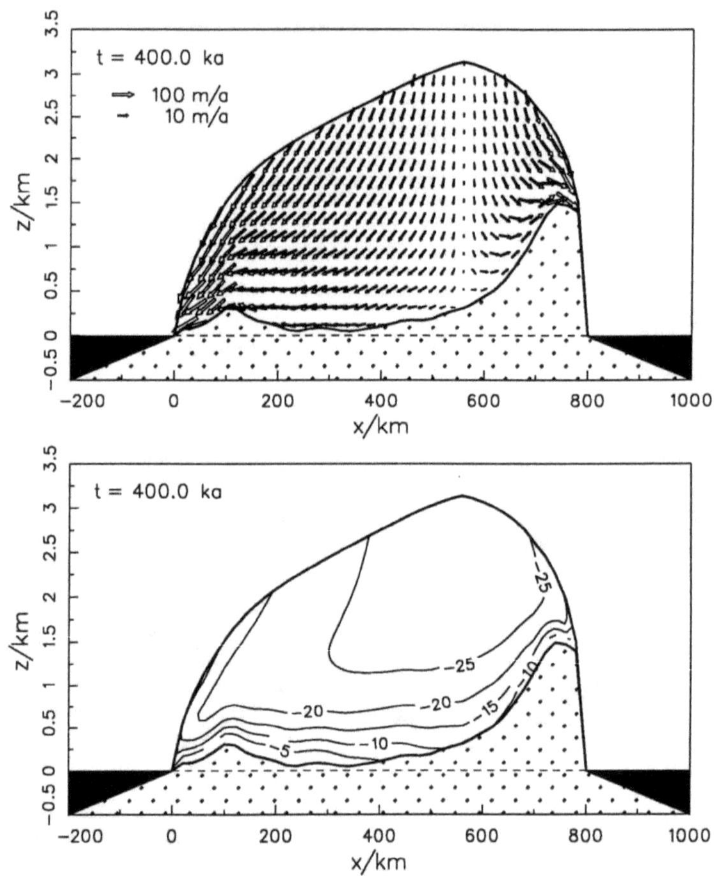

Fig. 5.18. Final state of the simulated EGIG1 profile at $t = 400\,\text{ka}$. Top: Velocity field. Bottom: Temperature field (in °C, relative to the pressure melting point).

Lotti 1995), the relevance of such simulated surges (physical process vs. numerical artifact) is a matter of current debate. The ongoing ISMIP (Ice Sheet Model Intercomparison Project) HEINO (Heinrich Event INtercOmparison) model intercomparison topic is supposed to shed more light on this problem (Calov and Greve 2006, Greve et al. 2006; final paper in preparation).

The simulated velocity and temperature fields of the EGIG1 profile at the end of the simulation ($t = 400\,\text{ka}$) are depicted in Fig. 5.18. These results can be considered as an approximation of the real, present-day state. The velocity field nicely shows the downward and outward flow of the ice sheet, and the near-margin velocities are significantly larger than their counterparts in the center. The temperature contours become denser towards the base, which is due to the advective, downward transport of cold surface ice. In the western half, and to a lesser extent close to the eastern margin, the temperature field even shows an inversion: starting from the surface, the ice temperature first decreases with depth, and only from a certain point on does the normal behaviour of an increase with depth appear. Melting conditions at the ice base prevail on 320 km of the 800-km long profile, of which 260 km are found near the western margin (low-lying ground) and 60 km near the eastern margin (high-lying ground).

Of course, numerical ice-sheet-modelling studies are not restricted to the two-dimensional plane strain approach presented here. Three-dimensional ice sheet models based on the shallow ice approximation are computationally much less expensive than Atmosphere or Ocean General Circulation Models (AGCM/OGCM) of comparable complexity and can even be run on modern PCs. Therefore, a number of ice sheet models have been developed and applied to problems of past, present and future glaciation of Greenland, Antarctica, North America, Scandinavia/Eurasia and even the polar ice caps of the planet Mars [e.g., Greve (2000) and references therein, Forsström et al. (2003), Huybrechts et al. (2004), Saito and Abe-Ouchi (2004), Lunt et al. (2008)]. Some recent efforts have aimed at devising models which go beyond the shallow ice approximation (Pattyn et al. 2008, and references therein); however, transient simulations for whole ice sheets have not been accomplished yet with such models.

[illegible]

6

Large-Scale Dynamics of Ice Shelves

6.1 Full Stokes Flow Problem

6.1.1 Field Equations, Boundary Conditions at the Free Surface

Ice shelves are floating ice masses, which are connected to and nourished by a grounded ice sheet (see Fig. 5.1). Most ice shelves, like the three major ice shelves of Antarctica (Ross Ice Shelf, Filchner-Rønne Ice Shelf, Amery Ice Shelf), are confined by large embayments. Smaller ice shelves can also be unconfined. In the latter case, stabilisation typically results from the contact with small islands or grounding on shoals.

Ice shelves are smaller and thinner than ice sheets and, on average, the ice flow is distinctly faster. Suitable typical values are

$$\begin{aligned}
\text{typical horizontal extent} \quad [L] &= 500\,\text{km}\,, \\
\text{typical vertical extent} \quad [H] &= 500\,\text{m}\,, \\
\text{typical horizontal velocity} \quad [U] &= 1000\,\text{m}\,\text{a}^{-1}\,, \\
\text{typical vertical velocity} \quad [W] &= 1\,\text{m}\,\text{a}^{-1}\,, \\
\text{typical pressure} \quad [P] &= \rho g [H] \approx 5\,\text{MPa}\,, \\
\text{typical time-scale} \quad [t] &= [L]/[U] = [H]/[W] = 500\,\text{a}\,, \\
\text{typical temperature variation} \quad [\Delta T] &= 20\,\text{K}\,.
\end{aligned} \tag{6.1}$$

[compare Eqs. (5.5) and (5.102) for ice sheets]. Note the resulting time-scale $[t]$, which is twenty times smaller than that of ice sheets. The scales (6.1) give an aspect ratio of

$$\varepsilon = \frac{[H]}{[L]} = \frac{[W]}{[U]} = 10^{-3}\,, \tag{6.2}$$

a Froude number of

$$Fr = \frac{[U]^2}{g[H]} \approx 2 \times 10^{-13}\,, \tag{6.3}$$

a Rossby number of

R. Greve, H. Blatter, *Dynamics of Ice Sheets and Glaciers*, Advances in Geophysical and Environmental Mechanics and Mathematics,
DOI 10.1007/978-3-642-03415-2_6,

$$Ro = \frac{[U]}{2\Omega[L]} \approx 4 \times 10^{-7}\,, \tag{6.4}$$

and a Coriolis-force-to-pressure-gradient ratio of

$$\frac{2\rho\Omega[U]}{[P]/[L]} = \frac{Fr}{Ro} \approx 5 \times 10^{-7}\,. \tag{6.5}$$

Evidently, the aspect ratio is the same as for ice sheets, whereas the Froude number and Coriolis-force-to-pressure-gradient ratio are larger by factors 200 and 10, respectively, but still very small. As a consequence, the flow of ice shelves is also governed by *Stokes flow*, and the equation of motion (5.11) remains valid. Also, the mass balance (5.1), the temperature evolution equation (5.14), the kinematic boundary condition (5.21), the stress-free condition (5.23) and the temperature boundary condition (5.24) for the free surface hold without any changes.

6.1.2 Boundary Conditions at the Ice Base

The situation is different at the ice base. First of all, it is an interface between ice and sea water and therefore situated *above* the lithosphere surface ($b \ge z_\mathrm{l}$, see Fig. 5.1). Both melting and freezing processes can occur, so that the quantity $a_\mathrm{b}^\perp$ defined in (5.29) can have either sign and must be interpreted as a *basal melting-freezing rate*. It is positive for melting and negative for freezing. Apart from this extended interpretation of $a_\mathrm{b}^\perp$, the kinematic boundary condition (5.31) remains valid. Consequently, the ice thickness equation (5.55), which has been derived by using the mass balance and the kinematic conditions at the free surface and the ice base, holds as well.

Analogous to Eq. (5.32), the momentum jump condition is

$$\mathbf{t}_\mathrm{sea} \cdot \mathbf{n} - \mathbf{t} \cdot \mathbf{n} - \rho a_\mathrm{b}^\perp \, [\![\mathbf{v}]\!] = \mathbf{0}\,, \tag{6.6}$$

where the lithospheric stress $\mathbf{t}_\mathrm{lith}$ has been replaced by the stress at the sea side of the interface, $\mathbf{t}_\mathrm{sea}$. Whereas the lithospheric stress under an ice sheet is unknown, the stress conditions in the sea can be described by a hydrostatic pressure p_sea plus a shear stress $\boldsymbol{\tau}_\mathrm{sea}$ induced by the circulating sea water. The former corresponds to the weight of the water column from the water surface (mean sea level $z = z_\mathrm{sl}$) to the ice base ($z = b$),

$$p_\mathrm{sea} = \rho_\mathrm{sw} g (z_\mathrm{sl} - b)\,, \tag{6.7}$$

where $\rho_\mathrm{sw} = 1028\,\mathrm{kg\,m^{-3}}$ is the density of sea water. The latter can be related to the flow velocity of the subglacial sea water, $\mathbf{v}_\mathrm{sea}$, by the empirical relation

$$\boldsymbol{\tau}_\mathrm{sea} = C_\mathrm{wi}\rho_\mathrm{sw}|\mathbf{v}_\mathrm{sea}|^2\,\mathbf{e}_\mathrm{t}\,, \tag{6.8}$$

where C_wi is the dimensionless water-ice drag coefficient ($C_\mathrm{wi} \approx 2.5 \times 10^{-3}$, depending on the roughness of the basal ice), and $\mathbf{e}_\mathrm{t}$ is the direction of $\boldsymbol{\tau}_\mathrm{sea}$

in the tangential plane to the ice base ($\mathbf{e}_\mathrm{t} \perp \mathbf{n}$), assumed parallel to $\mathbf{v}_\mathrm{sea}$ (so that $\mathbf{e}_\mathrm{t} = \mathbf{v}_\mathrm{sea}/|\mathbf{v}_\mathrm{sea}|$). Thus,

$$\mathsf{t}_\mathrm{sea} \cdot \mathbf{n} = -p_\mathrm{sea}\,\mathbf{n} + \boldsymbol{\tau}_\mathrm{sea} = -\rho_\mathrm{sw} g(z_\mathrm{sl} - b)\,\mathbf{n} + C_\mathrm{wi}\rho_\mathrm{sw}|\mathbf{v}_\mathrm{sea}|^2\,\mathbf{e}_\mathrm{t}\,. \tag{6.9}$$

Further, the advective term $\rho a_\mathrm{b}^\perp\,[\![\mathbf{v}]\!]$ in the momentum jump condition (6.6) is negligible. Therefore, we obtain the *stress condition*

$$\mathsf{t} \cdot \mathbf{n} = -\rho_\mathrm{sw} g(z_\mathrm{sl} - b)\,\mathbf{n} + C_\mathrm{wi}\rho_\mathrm{sw}|\mathbf{v}_\mathrm{sea}|^2\,\mathbf{e}_\mathrm{t}\,. \tag{6.10}$$

This relation serves as a dynamic boundary condition at the base of an ice shelf. Consequently, the additional formulation of an empirical sliding law like (5.35) is not required.

The temperature of the ice at the ice base is equal to the temperature of the sea water immediately below,

$$T = T_\mathrm{sea}\,, \tag{6.11}$$

which will be at the freezing point under the prevailing pressure and salinity conditions (typically around −2°C). The situation is therefore comparable to that of a temperate base of an ice sheet [see Eq. (5.39)]. In addition to (6.11), we can formulate an energy jump condition similar to (5.37),

$$\kappa\,(\operatorname{grad} T \cdot \mathbf{n}) - q_\mathrm{sea}^\perp - [\![\mathbf{v}]\!] \cdot \mathsf{t} \cdot \mathbf{n} + \rho a_\mathrm{b}^\perp\,[\![u]\!] = 0\,, \tag{6.12}$$

where $q_\mathrm{sea}^\perp$ is the heat flux entering the ice body from the sea water below. For the third term of (6.12), we find, using (6.8) and (6.10)

$$[\![\mathbf{v}]\!] \cdot \mathsf{t} \cdot \mathbf{n} = -\rho_\mathrm{sw} g(z_\mathrm{sl} - b)\;[\![\mathbf{v} \cdot \mathbf{n}]\!] + [\![\mathbf{v}]\!] \cdot \boldsymbol{\tau}_\mathrm{sea}\,. \tag{6.13}$$

Using the general mass jump condition (3.61) and the definition (5.29) of $a_\mathrm{b}^\perp$ yields

$$\begin{aligned} &\rho_\mathrm{sw}(\mathbf{v}_\mathrm{sea} - \mathbf{w}) \cdot \mathbf{n} = \rho(\mathbf{v} - \mathbf{w}) \cdot \mathbf{n} = \rho a_\mathrm{b}^\perp \\ &\Rightarrow\ (\mathbf{v}_\mathrm{sea} - \mathbf{w}) \cdot \mathbf{n} = \frac{\rho}{\rho_\mathrm{sw}} a_\mathrm{b}^\perp\,, \quad (\mathbf{v} - \mathbf{w}) \cdot \mathbf{n} = a_\mathrm{b}^\perp\,, \end{aligned} \tag{6.14}$$

so that the jump of the normal velocity in (6.13) is

$$[\![\mathbf{v} \cdot \mathbf{n}]\!] = (\mathbf{v}_\mathrm{sea} - \mathbf{w}) \cdot \mathbf{n} - (\mathbf{v} - \mathbf{w}) \cdot \mathbf{n} = \frac{\rho - \rho_\mathrm{sw}}{\rho_\mathrm{sw}}\,a_\mathrm{b}^\perp\,. \tag{6.15}$$

Further, we define the *frictional dissipation* of the boundary-layer current below the ice shelf as

$$\begin{aligned} \delta_\mathrm{sea} &= [\![\mathbf{v}]\!] \cdot \boldsymbol{\tau}_\mathrm{sea} \\ &\overset{(6.8)}{=} C_\mathrm{wi}\rho_\mathrm{sw}|\mathbf{v}_\mathrm{sea}|^2\;[\![\mathbf{v} \cdot \mathbf{e}_\mathrm{t}]\!] \\ &= C_\mathrm{wi}\rho_\mathrm{sw}|\mathbf{v}_\mathrm{sea}|^2\,(\mathbf{v}_\mathrm{sea} \cdot \mathbf{e}_\mathrm{t} - \mathbf{v} \cdot \mathbf{e}_\mathrm{t}) \\ &\approx C_\mathrm{wi}\rho_\mathrm{sw}|\mathbf{v}_\mathrm{sea}|^3\,. \end{aligned} \tag{6.16}$$

In the last step, the assumptions that the sea-water flow is essentially parallel to the ice base ($\mathbf{v}_{\text{sea}} \approx |\mathbf{v}_{\text{sea}}|\,\mathbf{e}_{\text{t}}$) and much larger than the flow of the ice shelf ($\mathbf{v}_{\text{sea}} \gg \mathbf{v}$) are made. Equation (6.13) then becomes

$$[\![\mathbf{v}]\!] \cdot \mathsf{t} \cdot \mathbf{n} = (\rho_{\text{sw}} - \rho) g (z_{\text{sl}} - b)\, a_{\text{b}}^{\perp} + \delta_{\text{sea}} \,. \tag{6.17}$$

For the fourth term of (6.12), we state that, as for an ice sheet with a temperate base, the jump of the internal energy is approximately equal to the latent heat of ice melt ($[\![u]\!] = L$). By inserting this and (6.17) in (6.12), we obtain

$$\kappa\,(\text{grad}\, T \cdot \mathbf{n}) - (\rho_{\text{sw}} - \rho) g (z_{\text{sl}} - b)\, a_{\text{b}}^{\perp} + \rho L a_{\text{b}}^{\perp} = q_{\text{sea}}^{\perp} + \delta_{\text{sea}} \tag{6.18}$$

as the energy jump condition at the ice shelf base. Since the basal temperature is already determined by (6.11), Eq. (6.18) is not required as a boundary condition for the temperature field. However, provided that $q_{\text{sea}}^{\perp}$ and δ_{sea} are known (for instance, from measurements), it determines the basal melting-freezing rate $a_{\text{b}}^{\perp}$. Alternatively, if $a_{\text{b}}^{\perp}$ is known, Eq. (6.18) can be used to compute the total heat input $q_{\text{sea}}^{\perp} + \delta_{\text{sea}}$ from the sea water below the ice shelf.

Concerning the effect of the basal shear stress $\boldsymbol{\tau}_{\text{sea}}$, an estimate with $C_{\text{wi}} = 2.5 \times 10^{-3}$, $\rho_{\text{sw}} = 1028\,\text{kg}\,\text{m}^{-3}$ and $|\mathbf{v}_{\text{sea}}| = 0.1\,\text{m}\,\text{s}^{-1}$ gives the values $|\boldsymbol{\tau}_{\text{sea}}| \approx 0.025\,\text{Pa}$ and $\delta_{\text{sea}} \approx 2.5\,\text{mW}\,\text{m}^{-2}$. Evidently, the shear stress itself is extremely small, and therefore its contribution to the stress condition (6.10) is negligible. By contrast, the frictional dissipation resulting from it is significant and should be taken into account in the energy jump condition (6.18).

6.1.3 Boundary Conditions at the Grounding Line and Calving Front

Boundary conditions are also required at the lateral margins, that is, the grounding line (in contact with the attached ice sheet) and the calving front (in contact with the surrounding ocean). At the grounding line, the inflow $\mathbf{v}_{\text{gl}}$ and temperature T_{gl} from the ice sheet are usually prescribed directly. At the calving front, the temperature can also be prescribed directly as T_{cf}. The hydrostatic pressure distribution p_{sw} of the sea water provides a boundary condition for the stress vector $\mathsf{t}|_{\text{cf}} \cdot \mathbf{n}$,

$$\mathsf{t}|_{\text{cf}} \cdot \mathbf{n} = -p_{\text{sw}} \mathbf{n} \,, \tag{6.19}$$

where $\mathbf{n}$ is the unit normal vector which points in a horizontal direction away from the calving front (assumed to be vertical). The sea water pressure is given by

$$p_{\text{sw}} = \begin{cases} 0\,, & \text{for } z \geq z_{\text{sl}}\,, \\ \rho_{\text{sw}} g (z_{\text{sl}} - z)\,, & \text{for } z \leq z_{\text{sl}} \end{cases} \tag{6.20}$$

[compare Eq. (6.10) for the ice shelf base]. This relation is illustrated in Fig. 6.1.

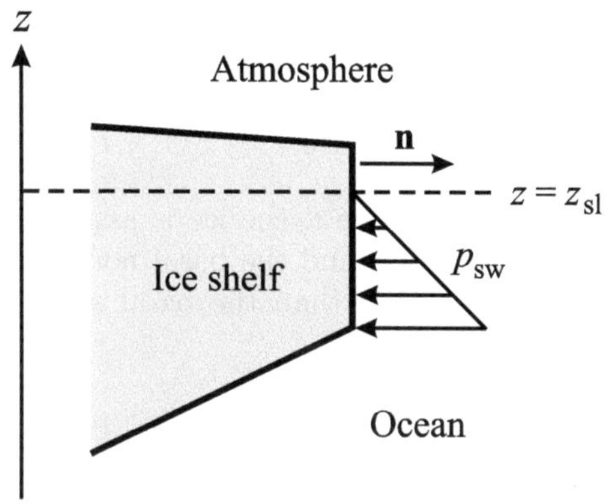

Fig. 6.1. Stress condition at the calving front of an ice shelf.

For the positions of the grounding line and calving front, let us assume that they are described by implicit functions

$$F_{\rm gl}(x, y, t) = 0 \tag{6.21}$$

and

$$F_{\rm cf}(x, y, t) = 0\,, \tag{6.22}$$

respectively. Since both boundaries are vertical faces, these functions do not depend on the vertical coordinate z. The unit normal vectors are the normalised gradients,

$$\mathbf{n} = \frac{\operatorname{grad} F_{\rm gl/cf}}{|\operatorname{grad} F_{\rm gl/cf}|} = \frac{1}{N_{\rm gl/cf}} \begin{pmatrix} \dfrac{\partial F_{\rm gl/cf}}{\partial x} \\ \dfrac{\partial F_{\rm gl/cf}}{\partial y} \\ 0 \end{pmatrix}, \tag{6.23}$$

with the gradient norm

$$N_{\rm gl/cf} = |\operatorname{grad} F_{\rm gl/cf}| = \left(\left(\frac{\partial F_{\rm gl/cf}}{\partial x} \right)^2 + \left(\frac{\partial F_{\rm gl/cf}}{\partial y} \right)^2 \right)^{1/2}. \tag{6.24}$$

We assume that the functions $F_{\rm gl/cf}$ are chosen such that $\mathbf{n}$ points away from the ice shelf.

Formulating a suitable condition which describes the position of the grounding line (evolution of the function $F_{\rm gl}$) constitutes a major problem in recent glaciological research. The easiest possibility is to simply prescribe its position. If this is not possible due to a lack of observational data, the position of the grounding line can be found by solving a contact problem for the coupled ice-sheet/ice-shelf problem, according to a recent suggestion by

Durand et al. (2009). This is based on the topological condition that the ice base cannot penetrate the lithosphere surface, that is,

$$\forall x, y, t: \quad b(x, y, t) \geq z_1(x, y, t)\,. \tag{6.25}$$

For any given point (x, y) and time t, the ice is assumed to be grounded if it touches the lithosphere surface, and the basal normal stress N_b (which is positive for compression) is larger than the basal sea water pressure $p_\mathrm{sw,b}$ $[= \rho_\mathrm{sw} g(z_\mathrm{sl} - b)$, cf. Eq. (6.20)]:

$$b(x, y, t) = z_1(x, y, t) \;\text{ and }\; N_\mathrm{b}(x, y, t) > p_\mathrm{sw,b}(x, y, t)\,. \tag{6.26}$$

By contrast, the ice is assumed to be floating if the ice base is above the lithosphere surface, or if it touches the lithosphere surface, but the basal normal stress is smaller than or equal to the basal sea water pressure:

$$\begin{aligned} & b(x, y, t) > z_1(x, y, t)\,, \\ \text{or} \quad & b(x, y, t) = z_1(x, y, t) \;\text{ and }\; N_\mathrm{b}(x, y, t) \leq p_\mathrm{sw,b}(x, y, t)\,. \end{aligned} \tag{6.27}$$

The position of the grounding line is the boundary of the two domains (grounded vs. floating ice) which result from the conditions (6.26) and (6.27). A further possibility will be discussed below (Sect. 6.3).

For determining the position of the calving front, a kinematic condition is well suited. As a direct consequence of Eq. (6.22), the time derivative of F_cf following the motion of the calving front with velocity $\mathbf{w} = w_x \mathbf{e}_x + w_y \mathbf{e}_y$ must vanish,

$$\frac{\mathrm{d}_w F_\mathrm{cf}}{\mathrm{d}t} = \frac{\partial F_\mathrm{cf}}{\partial t} + (\operatorname{grad} F_\mathrm{cf}) \cdot \mathbf{w} = 0 \tag{6.28}$$

[compare Eq. (5.18)]. Let $\mathbf{v}_\mathrm{h}$ be the horizontal component of the ice velocity at the calving front. We can then introduce the *calving rate* $c^\perp$,

$$c^\perp = (\mathbf{v}_\mathrm{h} - \mathbf{w}) \cdot \mathbf{n}\,, \tag{6.29}$$

and rewrite Eq. (6.28) as

$$\frac{\partial F_\mathrm{cf}}{\partial t} + (\operatorname{grad} F_\mathrm{cf}) \cdot \mathbf{v}_\mathrm{h} = N_\mathrm{cf} c^\perp\,. \tag{6.30}$$

This equation represents an evolution equation for the calving front, which can be solved provided that the calving rate $c^\perp$ is prescribed.

6.2 Hydrostatic Approximation

The hydrostatic approximation works the same way as for ice sheets. By neglecting the shear stresses t_{xz} and t_{yz} in the vertical momentum balance, one finds the hydrostatic distribution (5.59) for the vertical normal stress

t_{zz}, and the relation (5.60) for the pressure p. This leads to the expressions (5.61) for the horizontal normal stresses t_{xx} and t_{yy}. By inserting them in the horizontal components of the momentum balance and applying Glen's flow law, one obtains the equations of motion (5.64), which replace the full-Stokes-flow equation (5.11).

6.3 Shallow Shelf Approximation

Similar to ice sheets, the interior regions of an ice shelf (that is, some 10 km away from the grounding line and the calving front) are characterised by a small aspect ratio [see Eq. (6.2)] and small slopes of the free surface and the ice base. The latter can be expressed as

$$\begin{aligned} \frac{\partial h}{\partial x}, \frac{\partial h}{\partial y} &\sim \frac{[H]}{[L]} = \varepsilon\,, \\ \frac{\partial b}{\partial x}, \frac{\partial b}{\partial y} &\sim \frac{[H]}{[L]} = \varepsilon\,. \end{aligned} \tag{6.31}$$

Since at the base of an ice shelf significant shear stresses cannot be maintained [see the stress condition (6.10); as mentioned at the end of Sect. 6.1, the contribution from the water-flow-induced shear stress $\boldsymbol{\tau}_{\mathrm{sea}}$ is negligibly small in that context], the regime of bed-parallel shear flow found in grounded ice sheets cannot exist in floating ice shelves. By contrast, the typical situation is that of *plug flow*, whereby the horizontal velocities are essentially constant over depth. Naturally, in the immediate vicinity of the grounding line, a transitional state between these two limits will develop. The situation is illustrated in Fig. 6.2.

These considerations will now be used for deriving a further simplification of the hydrostatic approximation which is called the *shallow shelf approximation* (SSA) (Morland 1987). Mathematically, the regime of plug flow can be defined by

$$\frac{\partial v_x}{\partial z} \approx 0\,, \quad \frac{\partial v_y}{\partial z} \approx 0\,. \tag{6.32}$$

In other words, the horizontal velocities only depend on the horizontal coordinates and the time,

$$v_x = v_x(x,y,t)\,, \quad v_y = v_y(x,y,t)\,. \tag{6.33}$$

We go back to the horizontal components of the momentum balance of the hydrostatic approximation in the form (5.62). In order to make use of the fact that the shear stresses vanish at the surface and at the base, these equations are now integrated over depth:

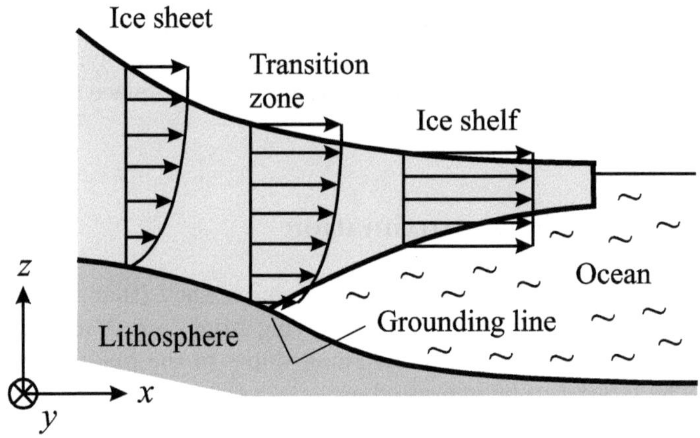

Fig. 6.2. Flow regimes in an ice-sheet/ice-shelf system. In the grounded ice sheet, shear flow prevails, whereas in the floating ice shelf plug flow is present. The vicinity of the grounding line shows a transitional flow pattern.

$$2\int_b^h \frac{\partial t^{\mathrm{D}}_{xx}}{\partial x}\,\mathrm{d}z + \int_b^h \frac{\partial t^{\mathrm{D}}_{yy}}{\partial x}\,\mathrm{d}z + \int_b^h \frac{\partial t_{xy}}{\partial y}\,\mathrm{d}z + \int_b^h \frac{\partial t_{xz}}{\partial z}\,\mathrm{d}z = \rho g H \frac{\partial h}{\partial x}\,, \tag{6.34a}$$

$$2\int_b^h \frac{\partial t^{\mathrm{D}}_{yy}}{\partial y}\,\mathrm{d}z + \int_b^h \frac{\partial t^{\mathrm{D}}_{xx}}{\partial y}\,\mathrm{d}z + \int_b^h \frac{\partial t_{xy}}{\partial x}\,\mathrm{d}z + \int_b^h \frac{\partial t_{yz}}{\partial z}\,\mathrm{d}z = \rho g H \frac{\partial h}{\partial y}\,. \tag{6.34b}$$

The first three integrals of each equation can be modified by using Leibniz's rule (see also Sect. 5.1.3), whereas the last one simply gives the difference of t_{xz} or t_{yz} at the surface and at the base. One obtains

$$\begin{aligned}
&2\frac{\partial}{\partial x}\int_b^h t^{\mathrm{D}}_{xx}\,\mathrm{d}z - 2t^{\mathrm{D}}_{xx}|_{z=h}\frac{\partial h}{\partial x} + 2t^{\mathrm{D}}_{xx}|_{z=b}\frac{\partial b}{\partial x}\\
&+\frac{\partial}{\partial x}\int_b^h t^{\mathrm{D}}_{yy}\,\mathrm{d}z - t^{\mathrm{D}}_{yy}|_{z=h}\frac{\partial h}{\partial x} + t^{\mathrm{D}}_{yy}|_{z=b}\frac{\partial b}{\partial x}\\
&+\frac{\partial}{\partial y}\int_b^h t_{xy}\,\mathrm{d}z - t_{xy}|_{z=h}\frac{\partial h}{\partial y} + t_{xy}|_{z=b}\frac{\partial b}{\partial y}\\
&\qquad + t_{xz}|_{z=h} - t_{xz}|_{z=b} = \rho g H\frac{\partial h}{\partial x}\,,
\end{aligned} \tag{6.35a}$$

$$
\begin{aligned}
&2\frac{\partial}{\partial y}\int_b^h t_{yy}^{\mathrm{D}}\,\mathrm{d}z - 2t_{yy}^{\mathrm{D}}|_{z=h}\frac{\partial h}{\partial y} + 2t_{yy}^{\mathrm{D}}|_{z=b}\frac{\partial b}{\partial y}\\
&+\frac{\partial}{\partial y}\int_b^h t_{xx}^{\mathrm{D}}\,\mathrm{d}z - t_{xx}^{\mathrm{D}}|_{z=h}\frac{\partial h}{\partial y} + t_{xx}^{\mathrm{D}}|_{z=b}\frac{\partial b}{\partial y}\\
&+\frac{\partial}{\partial x}\int_b^h t_{xy}\,\mathrm{d}z - t_{xy}|_{z=h}\frac{\partial h}{\partial x} + t_{xy}|_{z=b}\frac{\partial b}{\partial x}\\
&\qquad\qquad\qquad + t_{yz}|_{z=h} - t_{yz}|_{z=b} = \rho g H\frac{\partial h}{\partial y}\,.
\end{aligned}
\tag{6.35b}
$$

By introducing the *membrane stress* N as the vertically integrated deviatoric stress,

$$
\mathsf{N} = \int_b^h \mathrm{t}^{\mathrm{D}}\,\mathrm{d}z\,,\qquad\text{or}\qquad N_{ij} = \int_b^h t_{ij}^{\mathrm{D}}\,\mathrm{d}z\,,
\tag{6.36}
$$

and rearranging terms, this yields

$$
\begin{aligned}
&2\frac{\partial N_{xx}}{\partial x} + \frac{\partial N_{yy}}{\partial x} + \frac{\partial N_{xy}}{\partial y}\\
&- 2t_{xx}^{\mathrm{D}}|_{z=h}\frac{\partial h}{\partial x} - t_{yy}^{\mathrm{D}}|_{z=h}\frac{\partial h}{\partial x} - t_{xy}|_{z=h}\frac{\partial h}{\partial y} + t_{xz}|_{z=h}\\
&+ 2t_{xx}^{\mathrm{D}}|_{z=b}\frac{\partial b}{\partial x} + t_{yy}^{\mathrm{D}}|_{z=b}\frac{\partial b}{\partial x} + t_{xy}|_{z=b}\frac{\partial b}{\partial y} - t_{xz}|_{z=b} = \rho g H\frac{\partial h}{\partial x}\,,
\end{aligned}
\tag{6.37a}
$$

$$
\begin{aligned}
&2\frac{\partial N_{yy}}{\partial y} + \frac{\partial N_{xx}}{\partial y} + \frac{\partial N_{xy}}{\partial x}\\
&- 2t_{yy}^{\mathrm{D}}|_{z=h}\frac{\partial h}{\partial y} - t_{xx}^{\mathrm{D}}|_{z=h}\frac{\partial h}{\partial y} - t_{xy}|_{z=h}\frac{\partial h}{\partial x} + t_{yz}|_{z=h}\\
&+ 2t_{yy}^{\mathrm{D}}|_{z=b}\frac{\partial b}{\partial y} + t_{xx}^{\mathrm{D}}|_{z=b}\frac{\partial b}{\partial y} + t_{xy}|_{z=b}\frac{\partial b}{\partial x} - t_{yz}|_{z=b} = \rho g H\frac{\partial h}{\partial y}\,.
\end{aligned}
\tag{6.37b}
$$

In order to further simplify this system, the stress conditions at the surface (5.23) and at the base (6.10) must be considered. With the unit normal vector (5.16), condition (5.23) reads, in component form,

$$
\begin{aligned}
&\frac{1}{N_{\mathrm{s}}}\left(-t_{xx}|_{z=h}\frac{\partial h}{\partial x} - t_{xy}|_{z=h}\frac{\partial h}{\partial y} + t_{xz}|_{z=h}\right) = 0\,,\\
&\frac{1}{N_{\mathrm{s}}}\left(-t_{yy}|_{z=h}\frac{\partial h}{\partial y} - t_{xy}|_{z=h}\frac{\partial h}{\partial x} + t_{yz}|_{z=h}\right) = 0\,,\\
&\frac{1}{N_{\mathrm{s}}}\left(-t_{xz}|_{z=h}\frac{\partial h}{\partial x} - t_{yz}|_{z=h}\frac{\partial h}{\partial y} + t_{zz}|_{z=h}\right) = 0\,.
\end{aligned}
\tag{6.38}
$$

In the last component, the shear stresses are negligible compared to the vertical normal stress, so that it simplifies to

$$t_{zz}|_{z=h} = 0\,, \tag{6.39}$$

in agreement with the hydrostatic distribution (5.59). In the first two components, we replace the normal stresses t_{xx} and t_{yy} by using the expressions (5.61) evaluated for $z = h$,

$$\begin{aligned}
-2t^{\mathrm{D}}_{xx}|_{z=h}\frac{\partial h}{\partial x} - t^{\mathrm{D}}_{yy}|_{z=h}\frac{\partial h}{\partial x} - t_{xy}|_{z=h}\frac{\partial h}{\partial y} + t_{xz}|_{z=h} = 0\,,\\
-2t^{\mathrm{D}}_{yy}|_{z=h}\frac{\partial h}{\partial y} - t^{\mathrm{D}}_{xx}|_{z=h}\frac{\partial h}{\partial y} - t_{xy}|_{z=h}\frac{\partial h}{\partial x} + t_{yz}|_{z=h} = 0\,.
\end{aligned} \tag{6.40}$$

As a consequence, the second and fifth lines of the vertically integrated momentum balance (6.37) vanish.

At the ice base, the component form of the stress condition (6.10), with the unit normal vector (5.26) and upon neglecting the shear stress $\boldsymbol{\tau}_{\mathrm{sea}}$, becomes

$$\begin{aligned}
\frac{1}{N_{\mathrm{b}}}\left(t_{xx}|_{z=b}\frac{\partial b}{\partial x} + t_{xy}|_{z=b}\frac{\partial b}{\partial y} - t_{xz}|_{z=b}\right) &= -\frac{\rho_{\mathrm{sw}} g}{N_{\mathrm{b}}}(z_{\mathrm{sl}} - b)\,\frac{\partial b}{\partial x}\,,\\
\frac{1}{N_{\mathrm{b}}}\left(t_{yy}|_{z=b}\frac{\partial b}{\partial y} + t_{xy}|_{z=b}\frac{\partial b}{\partial x} - t_{yz}|_{z=b}\right) &= -\frac{\rho_{\mathrm{sw}} g}{N_{\mathrm{b}}}(z_{\mathrm{sl}} - b)\,\frac{\partial b}{\partial y}\,,\\
\frac{1}{N_{\mathrm{b}}}\left(t_{xz}|_{z=b}\frac{\partial b}{\partial x} + t_{yz}|_{z=b}\frac{\partial b}{\partial y} - t_{zz}|_{z=b}\right) &= \frac{\rho_{\mathrm{sw}} g}{N_{\mathrm{b}}}(z_{\mathrm{sl}} - b)\,.
\end{aligned} \tag{6.41}$$

Again, in the last component the shear stresses can be neglected compared to the vertical normal stress, which yields

$$t_{zz}|_{z=b} = -\rho_{\mathrm{sw}} g (z_{\mathrm{sl}} - b)\,. \tag{6.42}$$

However, we can also obtain the value of $t_{zz}|_{z=b}$ by evaluating the hydrostatic distribution (5.59) for $z = b$,

$$t_{zz}|_{z=b} = -\rho g H\,. \tag{6.43}$$

Equating (6.42) and (6.43) relates the ice thickness H to the water column $z_{\mathrm{sl}} - b$,

$$\rho H = \rho_{\mathrm{sw}}(z_{\mathrm{sl}} - b)\,. \tag{6.44}$$

This result is called the *floating condition*. By replacing $b = h - H$, it can be rearranged into a relation for the *freeboard*, $h - z_{\mathrm{sl}}$, (part of the ice shelf above the water level),

$$\rho H = \rho_{\mathrm{sw}}(z_{\mathrm{sl}} - h + H) \;\Rightarrow\; h - z_{\mathrm{sl}} = H - \frac{\rho}{\rho_{\mathrm{sw}}}H = \frac{\rho_{\mathrm{sw}} - \rho}{\rho_{\mathrm{sw}}}H\,. \tag{6.45}$$

With the values $\rho = 910\,\mathrm{kg\,m^{-3}}$ and $\rho_{\mathrm{sw}} = 1028\,\mathrm{kg\,m^{-3}}$, the freeboard is approximately equal to 11.5% of the ice shelf thickness, so that the remaining 88.5% will be below the water level.

In the first two components of (6.41), we replace t_{xx} and t_{yy} again by using (5.61), now evaluated for $z = b$,

$$\begin{aligned}
&2t^{\mathrm{D}}_{xx}|_{z=b}\frac{\partial b}{\partial x} + t^{\mathrm{D}}_{yy}|_{z=b}\frac{\partial b}{\partial x} - \rho g H\frac{\partial b}{\partial x} + t_{xy}|_{z=b}\frac{\partial b}{\partial y} - t_{xz}|_{z=b} \\
&\qquad\qquad\qquad\qquad = -\rho_{\mathrm{sw}} g(z_{\mathrm{sl}} - b)\,\frac{\partial b}{\partial x}\,, \\
&2t_{yy}|_{z=b}\frac{\partial b}{\partial y} + t_{xx}|_{z=b}\frac{\partial b}{\partial y} - \rho g H\frac{\partial b}{\partial y} + t_{xy}|_{z=b}\frac{\partial b}{\partial x} - t_{yz}|_{z=b} \\
&\qquad\qquad\qquad\qquad = -\rho_{\mathrm{sw}} g(z_{\mathrm{sl}} - b)\,\frac{\partial b}{\partial y}\,.
\end{aligned} \tag{6.46}$$

Because of the floating condition (6.44), this simplifies to

$$\begin{aligned}
2t^{\mathrm{D}}_{xx}|_{z=b}\frac{\partial b}{\partial x} + t^{\mathrm{D}}_{yy}|_{z=b}\frac{\partial b}{\partial x} + t_{xy}|_{z=b}\frac{\partial b}{\partial y} - t_{xz}|_{z=b} = 0\,, \\
2t_{yy}|_{z=b}\frac{\partial b}{\partial y} + t_{xx}|_{z=b}\frac{\partial b}{\partial y} + t_{xy}|_{z=b}\frac{\partial b}{\partial x} - t_{yz}|_{z=b} = 0\,.
\end{aligned} \tag{6.47}$$

Thus, the terms on the left hand sides in the third and sixth lines of (6.37) vanish as well.

Application of the conditions (6.40) and (6.47) in the vertically integrated momentum balance (6.37) provides

$$\begin{aligned}
2\frac{\partial N_{xx}}{\partial x} + \frac{\partial N_{yy}}{\partial x} + \frac{\partial N_{xy}}{\partial y} = \rho g H\frac{\partial h}{\partial x}\,, \\
2\frac{\partial N_{yy}}{\partial y} + \frac{\partial N_{xx}}{\partial y} + \frac{\partial N_{xy}}{\partial x} = \rho g H\frac{\partial h}{\partial y}\,,
\end{aligned} \tag{6.48}$$

where h and H are related by the floating condition in the form (6.45). Vertically integrating the flow law (4.21) gives

$$\begin{aligned}
N_{xx} &= 2\bar{\eta}\, D_{xx} = 2\bar{\eta}\,\frac{\partial v_x}{\partial x}\,, \\
N_{yy} &= 2\bar{\eta}\, D_{yy} = 2\bar{\eta}\,\frac{\partial v_y}{\partial y}\,, \\
N_{xy} &= 2\bar{\eta}\, D_{xy} = \bar{\eta}\left(\frac{\partial v_x}{\partial y} + \frac{\partial v_y}{\partial x}\right),
\end{aligned} \tag{6.49}$$

where the independence of v_x and v_y on z, Eq. (6.33), and the definition of the membrane stresses, Eq. (6.36), have been used, and the depth-integrated viscosity,

$$\bar{\eta} = \int_b^h \eta \, \mathrm{d}z = \int_b^h \frac{1}{2} B(T') \, d_\mathrm{e}^{-(1-1/n)} \, \mathrm{d}z \,, \tag{6.50}$$

has been introduced. The effective strain rate d_e, of which the general component form is given in Eq. (5.66), can be simplified for the plug flow regime [Eq. (6.32)] to

$$\begin{aligned} d_\mathrm{e} = \Bigg\{ & \left(\frac{\partial v_x}{\partial x}\right)^2 + \left(\frac{\partial v_y}{\partial y}\right)^2 + \frac{\partial v_x}{\partial x}\frac{\partial v_y}{\partial y} \\ & + \frac{1}{4}\left(\frac{\partial v_x}{\partial y} + \frac{\partial v_y}{\partial x}\right)^2 + \frac{1}{4}\left(\frac{\partial v_z}{\partial x}\right)^2 + \frac{1}{4}\left(\frac{\partial v_z}{\partial y}\right)^2 \Bigg\}^{1/2} . \end{aligned} \tag{6.51}$$

With the scaling (6.1) and the aspect ratio (6.2), we can estimate the orders of magnitude of the terms $(\partial v_x/\partial x)^2$ and $(\partial v_z/\partial x)^2$,

$$\left(\frac{\partial v_x}{\partial x}\right)^2 \sim \frac{[U]^2}{[L]^2} , \qquad \left(\frac{\partial v_z}{\partial x}\right)^2 \sim \frac{[W]^2}{[L]^2} = \varepsilon^2 \frac{[U]^2}{[L]^2} . \tag{6.52}$$

Consequently, the term $(\partial v_z/\partial x)^2$ is negligible compared to $(\partial v_x/\partial x)^2$. The same holds for $(\partial v_z/\partial y)^2$ compared to $(\partial v_y/\partial y)^2$, and so we obtain a further simplified expression for the effective strain rate,

$$d_\mathrm{e} = \left\{ \left(\frac{\partial v_x}{\partial x}\right)^2 + \left(\frac{\partial v_y}{\partial y}\right)^2 + \frac{\partial v_x}{\partial x}\frac{\partial v_y}{\partial y} + \frac{1}{4}\left(\frac{\partial v_x}{\partial y} + \frac{\partial v_y}{\partial x}\right)^2 \right\}^{1/2} . \tag{6.53}$$

This relation is independent of the vertical coordinate z, which reduces the depth-integrated viscosity (6.50) to

$$\bar{\eta} = \frac{1}{2} d_\mathrm{e}^{-(1-1/n)} \int_b^h B(T') \, \mathrm{d}z \,. \tag{6.54}$$

Inserting the vertically integrated flow law, Eq. (6.49), in the vertically integrated momentum balance, Eq. (6.48), finally yields

$$\begin{aligned} 4\frac{\partial}{\partial x}\left(\bar{\eta}\frac{\partial v_x}{\partial x}\right) + 2\frac{\partial}{\partial x}\left(\bar{\eta}\frac{\partial v_y}{\partial y}\right) + \frac{\partial}{\partial y}\left(\bar{\eta}\left(\frac{\partial v_x}{\partial y} + \frac{\partial v_y}{\partial x}\right)\right) &= \rho g H \frac{\partial h}{\partial x} , \\ 4\frac{\partial}{\partial y}\left(\bar{\eta}\frac{\partial v_y}{\partial y}\right) + 2\frac{\partial}{\partial y}\left(\bar{\eta}\frac{\partial v_x}{\partial x}\right) + \frac{\partial}{\partial x}\left(\bar{\eta}\left(\frac{\partial v_x}{\partial y} + \frac{\partial v_y}{\partial x}\right)\right) &= \rho g H \frac{\partial h}{\partial y} , \end{aligned} \tag{6.55}$$

which is a system of non-linear elliptical differential equations for the horizontal velocities v_x and v_y.

This system must be complemented by boundary conditions along the lateral margin of the ice shelf. At the grounding line, the inflow $\mathbf{v}_\mathrm{gl}$ from the nourishing ice sheet is usually prescribed, which serves as a dynamic condition. The position of the grounding line can either be prescribed or determined by solving a contact problem (see Sect. 6.1.3). The shallow shelf approximation allows a simpler formulation of the contact problem by using the floating condition (6.44), which can be rearranged to

$$b = z_{\mathrm{sl}} - \frac{\rho}{\rho_{\mathrm{sw}}} H \,. \tag{6.56}$$

Combining this with the inequality (6.25) yields

$$z_{\mathrm{sl}} - \frac{\rho}{\rho_{\mathrm{sw}}} H \geq z_{\mathrm{l}} \quad \Rightarrow \quad H \leq \frac{\rho_{\mathrm{sw}}}{\rho}(z_{\mathrm{sl}} - z_{\mathrm{l}}) \,. \tag{6.57}$$

For any point (x, y) and time t, the ice is assumed to be floating if the inequality (6.57) is fulfilled. Otherwise, for the case

$$H > \frac{\rho_{\mathrm{sw}}}{\rho}(z_{\mathrm{sl}} - z_{\mathrm{l}}) \,, \tag{6.58}$$

the ice is assumed to be grounded, and the position of the grounding line is given by the boundary of the two domains.

At the calving front, the stress condition needs to be evaluated. Equation (6.19) reads in component form

$$\begin{aligned} (-p + t^{\mathrm{D}}_{xx})_{\mathrm{cf}}\, n_x + t_{xy}|_{\mathrm{cf}}\, n_y &= -p_{\mathrm{sw}}\, n_x \,, \\ t_{xy}|_{\mathrm{cf}}\, n_x + (-p + t^{\mathrm{D}}_{yy})_{\mathrm{cf}}\, n_y &= -p_{\mathrm{sw}}\, n_y \,, \\ t_{xz}|_{\mathrm{cf}}\, n_x + t_{yz}|_{\mathrm{cf}}\, n_y &= 0 \,. \end{aligned} \tag{6.59}$$

With the relation (5.60) for the pressure p in the hydrostatic approximation, Eqs. $(6.59)_{1,2}$ yield

$$\begin{aligned} (2t^{\mathrm{D}}_{xx} + t^{\mathrm{D}}_{yy})_{\mathrm{cf}}\, n_x + t_{xy}|_{\mathrm{cf}}\, n_y &= -p_{\mathrm{sw}}\, n_x + \rho g(h - z) n_x \,, \\ t_{xy}|_{\mathrm{cf}}\, n_x + (2t^{\mathrm{D}}_{yy} + t^{\mathrm{D}}_{xx})_{\mathrm{cf}}\, n_y &= -p_{\mathrm{sw}}\, n_y + \rho g(h - z) n_y \,. \end{aligned} \tag{6.60}$$

Using Eq. (6.20) for the sea-water pressure and definition (6.36) for the membrane stresses, this is readily integrated,

$$\begin{aligned} &(2N_{xx} + N_{yy})_{\mathrm{cf}}\, n_x + N_{xy}|_{\mathrm{cf}}\, n_y \\ &\quad = \left(\int_b^h \rho g (h - z)\, \mathrm{d}z - \int_b^{z_{\mathrm{sl}}} \rho_{\mathrm{sw}} g (z_{\mathrm{sl}} - z)\, \mathrm{d}z \right) n_x \,, \\ &N_{xy}|_{\mathrm{cf}}\, n_x + (2N_{yy} + N_{xx})_{\mathrm{cf}}\, n_y \\ &\quad = \left(\int_b^h \rho g (h - z)\, \mathrm{d}z - \int_b^{z_{\mathrm{sl}}} \rho_{\mathrm{sw}} g (z_{\mathrm{sl}} - z)\, \mathrm{d}z \right) n_y \,. \end{aligned} \tag{6.61}$$

The integral terms on the right hand side yield

$$
\begin{aligned}
\Big(\cdots\Big) &= \frac{\rho g H^2}{2} - \frac{\rho_{\rm sw} g (z_{\rm sl}-b)^2}{2} \\
&\overset{(6.44)}{=} \frac{\rho g H^2}{2} - \left(\frac{\rho}{\rho_{\rm sw}}\right)^2 \frac{\rho_{\rm sw} g H^2}{2} \\
&= \frac{\rho}{\rho_{\rm sw}}(\rho_{\rm sw}-\rho)\frac{gH^2}{2}\,, \qquad (6.62)
\end{aligned}
$$

so that for the stress condition at the calving front we obtain

$$
\begin{aligned}
(2N_{xx}+N_{yy})_{\rm cf}\, n_x + N_{xy}|_{\rm cf}\, n_y &= \frac{\rho}{\rho_{\rm sw}}(\rho_{\rm sw}-\rho)\frac{gH^2}{2}\, n_x\,, \\
N_{xy}|_{\rm cf}\, n_x + (2N_{yy}+N_{xx})_{\rm cf}\, n_y &= \frac{\rho}{\rho_{\rm sw}}(\rho_{\rm sw}-\rho)\frac{gH^2}{2}\, n_y\,.
\end{aligned}
\qquad (6.63)
$$

Inserting the vertically integrated flow law (6.49) yields the corresponding boundary condition for the velocities,

$$
\begin{aligned}
&4\bar\eta\left.\frac{\partial v_x}{\partial x}\right|_{\rm cf} n_x + 2\bar\eta\left.\frac{\partial v_y}{\partial y}\right|_{\rm cf} n_x + \bar\eta\left(\frac{\partial v_x}{\partial y}+\frac{\partial v_y}{\partial x}\right)_{\rm cf} n_y \\
&\qquad = \frac{\rho}{\rho_{\rm sw}}(\rho_{\rm sw}-\rho)\frac{gH^2}{2}\, n_x\,, \\
&\bar\eta\left(\frac{\partial v_x}{\partial y}+\frac{\partial v_y}{\partial x}\right)_{\rm cf} n_x + 4\bar\eta\left.\frac{\partial v_y}{\partial y}\right|_{\rm cf} n_y + 2\bar\eta\left.\frac{\partial v_x}{\partial x}\right|_{\rm cf} n_y \\
&\qquad = \frac{\rho}{\rho_{\rm sw}}(\rho_{\rm sw}-\rho)\frac{gH^2}{2}\, n_y\,.
\end{aligned}
\qquad (6.64)
$$

The position of the calving front is governed by the kinematic condition (6.30), which does not simplify in the shallow shelf approximation. Note that it requires the calving rate $c^{\perp}$ as an input quantity.

The geometry and structure of the elliptical boundary-value problem for the horizontal velocity field, which consists of the system of differential equations (6.55) and its associated boundary conditions at the grounding line and at the calving front, are illustrated in Fig. 6.3.

The system of model equations in the shallow shelf approximation is completed by the vertically integrated mass balance, the ice thickness equation and the temperature evolution equation. The vertically integrated mass balance is essentially the same as Eq. (5.95); however, we formulate it with the sea level instead of the ice base as the reference horizon,

$$
\begin{aligned}
v_z &= v_z|_{z=z_{\rm sl}} - \int_{z_{\rm sl}}^{z}\left(\frac{\partial v_x}{\partial x}+\frac{\partial v_y}{\partial y}\right)\mathrm{d}\bar z \\
&= v_z|_{z=z_{\rm sl}} - (z-z_{\rm sl})\left(\frac{\partial v_x}{\partial x}+\frac{\partial v_y}{\partial y}\right). \qquad (6.65)
\end{aligned}
$$

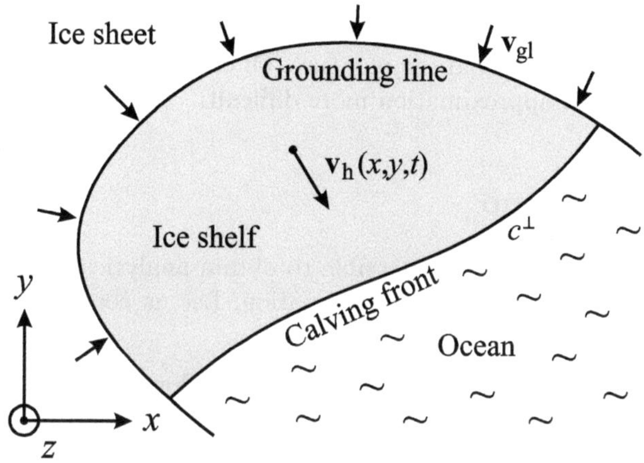

Fig. 6.3. Ice shelf geometry in the horizontal map plane. The horizontal velocity field $\mathbf{v}_\mathrm{h}$ is a solution of the elliptical differential equations (6.55) and the boundary conditions at the grounding line and at the calving front. Input quantities are the inflow velocity at the grounding line, $\mathbf{v}_\mathrm{gl}$, and the calving rate at the calving front, $c^{\perp}$.

Hence the vertical velocity is linear over depth. The ice thickness equation results from Eq. (5.55), with $N_\mathrm{s} \approx 1$ and $N_\mathrm{b} \approx 1$, so that due to (6.31) $a_\mathrm{s} \approx a_\mathrm{s}^{\perp}$ and $a_\mathrm{b} \approx a_\mathrm{b}^{\perp}$,

$$\begin{aligned}\frac{\partial H}{\partial t} &= -\mathrm{div}\,\mathbf{Q} + a_\mathrm{s} - a_\mathrm{b} \\ &= -\mathrm{div}\,(H\mathbf{v}_\mathrm{h}) + a_\mathrm{s} - a_\mathrm{b}\,. \end{aligned} \tag{6.66}$$

The temperature evolution equation [see Eq. (5.105)] is

$$\rho c\Big(\frac{\partial T}{\partial t} + v_x\frac{\partial T}{\partial x} + v_y\frac{\partial T}{\partial y} + v_z\frac{\partial T}{\partial z}\Big) = \frac{\partial}{\partial z}\Big(\kappa\frac{\partial T}{\partial z}\Big) + 2A(T')\,\sigma_\mathrm{e}^{n+1}\,, \tag{6.67}$$

and is given simple Dirichlet boundary conditions (prescribed temperatures) at the surface, the ice base, the grounding line and the calving front as described above (Sect. 6.1).

It is instructive to compare the structure of the model equations of the shallow shelf approximation with that of the shallow ice approximation (Sect. 5.4). In the shallow ice approximation for ice sheets, the stresses, the horizontal velocity and the volume flux are *local* functions of the ice geometry and temperature. This means that information about the state of the ice sheet at other positions is not required in order to compute them. This is in strong contrast to the shallow shelf approximation, where the stresses, the horizontal velocity and the volume flux are determined by differential equations for the entire area of the ice shelf. Therefore, at any position they are influenced by the state of the whole ice shelf, or, in other words, they depend in a *non-local* way on the ice geometry and temperature. Of course, this makes the solution of the shallow shelf approximation more difficult.

6.4 Ice Shelf Ramp

Under certain conditions, it is possible to obtain analytical solutions for the equations of the shallow shelf approximation. Let us consider the following problem:

- Plane strain approximation (two-dimensional flow in the x-z plane).
- Ice shelf extent between $x = 0$ and $x = L$, grounding line at $x = 0$, calving front at $x = L$.
- Thickness of the ice shelf linearly decreasing from $H = H_{\mathrm{gl}}$ at the grounding line to $H = H_{\mathrm{cf}}$ at the calving front.
- Mean sea level at $z = z_{\mathrm{sl}} = 0$.
- Inflow from attached ice sheet with velocity $\mathbf{v}_{\mathrm{gl}} = v_{\mathrm{gl}}\,\mathbf{e}_x$.
- Steady-state conditions: $\partial(\cdot)/\partial t = 0$ for all field quantities.
- Constant rate factor: $A(T') = A = \mathrm{const.}$

The geometry of this ice shelf ramp is illustrated in Fig. 6.4. The thickness $H(x)$ can be expressed as

$$H = H_{\mathrm{gl}} - \frac{H_{\mathrm{gl}} - H_{\mathrm{cf}}}{L}\,x\,. \tag{6.68}$$

With the floating condition in the alternative forms (6.44) and (6.45), the free surface and the ice base are given by

$$h = \frac{\rho_{\mathrm{sw}} - \rho}{\rho_{\mathrm{sw}}}\,H\,, \qquad b = -\frac{\rho}{\rho_{\mathrm{sw}}}\,H\,. \tag{6.69}$$

The velocity field consists of the two functions $v_x(x)$ and $v_z(x,z)$, so that the effective strain rate (6.53) is simply

$$d_{\mathrm{e}} = \left|\frac{\mathrm{d}v_x}{\mathrm{d}x}\right| = \frac{\mathrm{d}v_x}{\mathrm{d}x}\,. \tag{6.70}$$

The absolute value bars can be dropped because $\mathrm{d}v_x/\mathrm{d}x > 0$ for the ice shelf ramp. With this result, the depth-integrated viscosity (6.54) reads

$$\bar{\eta} = \frac{BH}{2}\left(\frac{\mathrm{d}v_x}{\mathrm{d}x}\right)^{-(1-1/n)}\,, \tag{6.71}$$

where the constant $B = A^{-1/n}$ (associated rate factor) has been introduced.

The horizontal velocity v_x is governed by Eq. $(6.55)_1$, whereas $(6.55)_2$ vanishes identically due to the plane strain approximation. Equation $(6.55)_1$ simplifies to

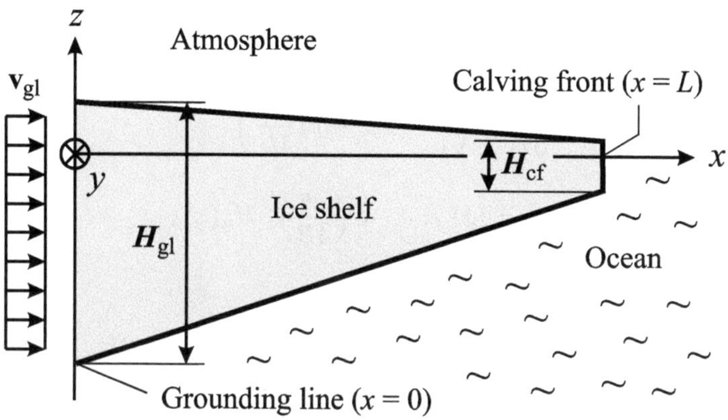

Fig. 6.4. Ice shelf ramp of length L. The thickness is linearly decreasing from $H = H_{gl}$ at the grounding line ($x = 0$) to $H = H_{cf}$ at the calving front ($x = L$).

$$4\frac{\mathrm{d}}{\mathrm{d}x}\left(\bar{\eta}\frac{\mathrm{d}v_x}{\mathrm{d}x}\right) = \rho g H \frac{\mathrm{d}h}{\mathrm{d}x}, \tag{6.72}$$

and by inserting Eqs. $(6.69)_1$ and (6.71), we obtain

$$\begin{aligned} 2B\frac{\mathrm{d}}{\mathrm{d}x}\left[H\left(\frac{\mathrm{d}v_x}{\mathrm{d}x}\right)^{1/n}\right] &= \frac{\rho}{\rho_{sw}}(\rho_{sw} - \rho) g H \frac{\mathrm{d}H}{\mathrm{d}x} \\ &= \frac{\rho}{\rho_{sw}}(\rho_{sw} - \rho) g \frac{\mathrm{d}}{\mathrm{d}x}\left(\frac{H^2}{2}\right). \end{aligned} \tag{6.73}$$

The first integral of this equation is

$$2BH\left(\frac{\mathrm{d}v_x}{\mathrm{d}x}\right)^{1/n} = \frac{\rho}{\rho_{sw}}(\rho_{sw} - \rho)\frac{gH^2}{2} + C_1. \tag{6.74}$$

The integration constant C_1 can be determined by using the boundary condition (6.64) at the calving front. With $n_x = 1$ and $n_y = 0$, it yields

$$4\bar{\eta}\left.\frac{\mathrm{d}v_x}{\mathrm{d}x}\right|_{cf} = \frac{\rho}{\rho_{sw}}(\rho_{sw} - \rho)\frac{gH^2}{2}, \tag{6.75}$$

or with the depth-integrated viscosity (6.71),

$$2BH\left.\left(\frac{\mathrm{d}v_x}{\mathrm{d}x}\right)^{1/n}\right|_{cf} = \frac{\rho}{\rho_{sw}}(\rho_{sw} - \rho)\frac{gH^2}{2}. \tag{6.76}$$

Comparison of Eq. (6.74), taken at the calving front, and the boundary condition (6.76) yields $C_1 = 0$.

We proceed by solving Eq. (6.74) for the velocity derivative and integrating the result from the grounding line $x = 0$ to an arbitrary position x,

$$\frac{\mathrm{d}v_x}{\mathrm{d}x} = \left(\frac{\rho}{\rho_{\mathrm{sw}}}(\rho_{\mathrm{sw}} - \rho)\frac{gH}{4B}\right)^n$$

$$\Rightarrow\ v_x(x) = v_{\mathrm{gl}} + \left(\frac{\varrho g}{4B}\right)^n \mathcal{H}_n(x)\,, \tag{6.77}$$

with the abbreviations

$$\varrho = \frac{\rho}{\rho_{\mathrm{sw}}}(\rho_{\mathrm{sw}} - \rho)\,, \qquad \mathcal{H}_n(x) = \int_0^x H^n(\tilde{x})\,\mathrm{d}\tilde{x}\,. \tag{6.78}$$

For the ice ramp defined by Eq. (6.68), the thickness integral reads

$$\begin{aligned}\mathcal{H}_n(x) &= \left[-\frac{L}{(n+1)(H_{\mathrm{gl}} - H_{\mathrm{cf}})}\left(H_{\mathrm{gl}} - \frac{H_{\mathrm{gl}} - H_{\mathrm{cf}}}{L}\,\tilde{x}\right)^{n+1}\right]_0^x \\ &= \frac{L}{(n+1)(H_{\mathrm{gl}} - H_{\mathrm{cf}})} \\ &\quad \times H_{\mathrm{gl}}^{n+1}\left[1 - \left(1 - \frac{H_{\mathrm{gl}} - H_{\mathrm{cf}}}{LH_{\mathrm{gl}}}\,x\right)^{n+1}\right].\end{aligned} \tag{6.79}$$

Equations (6.77) and (6.79) describe the horizontal velocity for the ice shelf ramp. For the stress exponent $n = 3$, it is a quartic function of the position x.

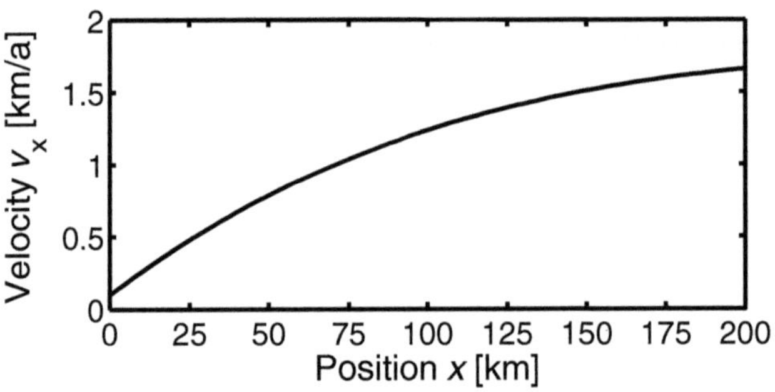

Fig. 6.5. Ice shelf ramp: Horizontal velocity v_x according to Eqs. (6.77) and (6.79), for $H_{\mathrm{gl}} = 400\,\mathrm{m}$, $H_{\mathrm{cf}} = 200\,\mathrm{m}$, $L = 200\,\mathrm{km}$, $v_{\mathrm{gl}} = 100\,\mathrm{m\,a^{-1}}$, $n = 3$, $A = 4.9 \times 10^{-25}\,\mathrm{s^{-1}\,Pa^{-3}}$ (value for $T' = -10°\mathrm{C}$), $\rho = 910\,\mathrm{kg\,m^{-3}}$, $\rho_{\mathrm{sw}} = 1028\,\mathrm{kg\,m^{-3}}$ and $g = 9.81\,\mathrm{m\,s^{-2}}$.

An example is shown in Fig. 6.5. Even though the set-up is that of a rather small and thin ice shelf (see figure caption), a maximum velocity as large as

$1.66\,\mathrm{km\,a^{-1}}$ is reached at the calving front. This is due to the fact that our two-dimensional, plane strain ice shelf does not feel any lateral drag, which would reduce the flow speed in a natural ice shelf.

The vertical velocity can be computed by inserting the horizontal velocity (6.77) into the vertically integrated mass balance (6.65),

$$\begin{aligned} v_z &= v_z|_{z=0} - \frac{\mathrm{d}v_x}{\mathrm{d}x}\,z = v_z|_{z=0} - \left(\frac{\varrho g}{4B}\right)^n \frac{\mathrm{d}\mathcal{H}_n}{\mathrm{d}x}\,z \\ &= v_z|_{z=0} - \left(\frac{\varrho g H}{4B}\right)^n z\,. \end{aligned} \tag{6.80}$$

The value of the integration constant $v_z|_{z=0}$ (vertical velocity at sea level) can be determined only if the distributions of the surface mass balance a_s and the basal mass balance a_b are known. For simplicity we assume instead that the ratio of a_s and a_b is such that $v_z|_{z=0} = 0$. The result for the vertical velocity is then

$$v_z(x,z) = -\left(\frac{\varrho g H(x)}{4B}\right)^n z\,. \tag{6.81}$$

It is illustrated in Fig. 6.6 for the center of the ice shelf described in the legend of Fig. 6.5.

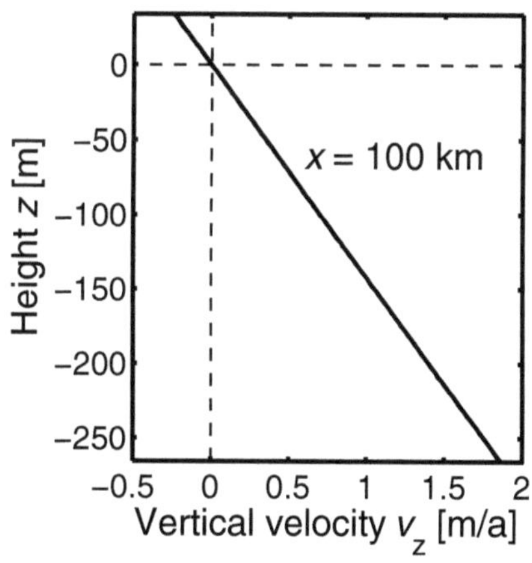

Fig. 6.6. Ice shelf ramp: Profile of the vertical velocity v_z in the center of the ice shelf ($x = L/2 = 100\,\mathrm{km}$) according to Eq. (6.81). Parameters as in Fig. 6.5.

The total mass balance at the upper and lower interface, $a_\mathrm{s} - a_\mathrm{b}$ (where supply is positive and loss negative), can be obtained from the ice thickness equation (6.66) by employing the steady-state assumption,

$$a_\mathrm{s} - a_\mathrm{b} = \frac{\mathrm{d}(Hv_x)}{\mathrm{d}x}\,. \tag{6.82}$$

With the horizontal velocity (6.77), this yields

$$\begin{aligned} a_\mathrm{s} - a_\mathrm{b} &= H\frac{\mathrm{d}v_x}{\mathrm{d}x} + v_x\frac{\mathrm{d}H}{\mathrm{d}x} \\ &= H\left(\frac{\varrho g}{4B}\right)^n \frac{\mathrm{d}\mathcal{H}_n}{\mathrm{d}x} + \frac{\mathrm{d}H}{\mathrm{d}x}\left[v_\mathrm{gl} + \left(\frac{\varrho g}{4B}\right)^n \mathcal{H}_n\right] \\ &= H\left(\frac{\varrho g H}{4B}\right)^n + \frac{\mathrm{d}H}{\mathrm{d}x}\left[v_\mathrm{gl} + \left(\frac{\varrho g}{4B}\right)^n \mathcal{H}_n\right], \end{aligned} \tag{6.83}$$

which is depicted in Fig. 6.7 for the set-up of Fig. 6.5. In this example, the total mass balance is positive in slightly more than the inner half of the ice shelf (until $x \approx 128\,\mathrm{km}$) and negative in the outer part towards the calving front.

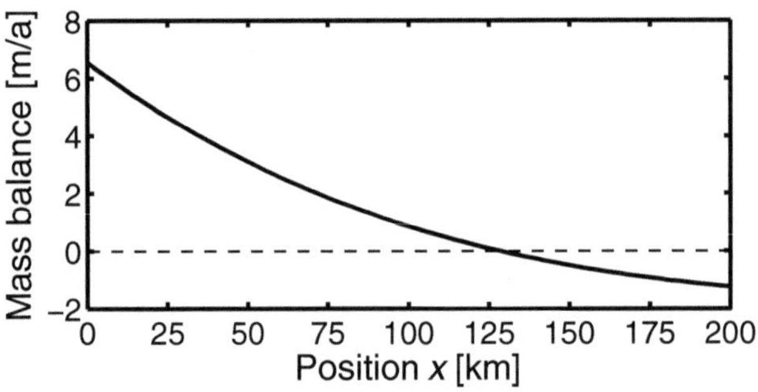

Fig. 6.7. Ice shelf ramp: Total mass balance at the upper and lower interface, $a_\mathrm{s} - a_\mathrm{b}$, according to Eq. (6.83). Parameters as in Fig. 6.5.

It is also instructive to discuss the degenerate case of an *ice shelf with constant thickness*, which can be realised by setting $H_\mathrm{gl} = H_\mathrm{cf} = H = \mathrm{const}$. One may intuitively conclude that for this situation there is no ice flow, because the forcing term $\rho g H(\mathrm{d}h/\mathrm{d}x)$ on the right-hand side of the momentum equation (6.72) vanishes. However, this is not the case. For a constant thickness H, the thickness integral $\mathcal{H}_n(x)$ defined in Eq. $(6.78)_2$ is given by $\mathcal{H}_n(x) = H^n x$. Inserting this into the velocity equation (6.77) yields the linearly increasing velocity distribution

$$v_x(x) = v_\mathrm{gl} + \left(\frac{\varrho g H}{4B}\right)^n x\,, \tag{6.84}$$

which corresponds to the constant horizontal dilatation rate

$$D_{xx} = \frac{\mathrm{d}v_x}{\mathrm{d}x} = \left(\frac{\varrho g H}{4B}\right)^n. \tag{6.85}$$

The vertical compression rate [due to Eq. (6.81)] is then

$$D_{zz} = \frac{\mathrm{d}v_z}{\mathrm{d}z} = -\left(\frac{\varrho g H}{4B}\right)^n . \tag{6.86}$$

Of course, the two strain rates add up to zero, which is a requirement of incompressibility.

The reason for this counterintuitive behaviour of horizontal stretching and vertical compression of the ice, despite constant ice thickness, is the existence of the calving front as a boundary of the system. Beyond the calving front, there is no more ice which could hinder the extension of the ice shelf, and the sea-water pressure at the calving front is not strong enough to buttress the ice shelf completely.

6.5 Numerical Methods

Of course, the equations of the shallow shelf approximation derived in Sect. 6.3 can be solved numerically by finite difference methods, which have been discussed for the shallow ice approximation of ice sheets in Sect. 5.7. However, the two-dimensional, elliptical boundary-value problem (6.55) for the horizontal velocities v_x and v_y, along with its boundary conditions at the grounding line and the calving front, is equally well suited for a different type of numerical solution, namely the *finite element method* [see, e.g., Reddy (2006) for a general introduction]. The method now described follows Weis (2001) closely.

6.5.1 Mechanical Ice Shelf Problem

Let us assume that, for some given time t, the geometry and temperature field of an ice shelf are known, and the horizontal velocity field $\mathbf{v}_\mathrm{h} = (v_x, v_y)$ is computed by solving Eqs. (6.55). By introducing the symmetric, two-dimensional tensor

$$\mathsf{M} = \begin{pmatrix} 2\dfrac{\partial v_x}{\partial x} + \dfrac{\partial v_y}{\partial y} & \dfrac{1}{2}\left(\dfrac{\partial v_x}{\partial y} + \dfrac{\partial v_y}{\partial x}\right) \\[2ex] \dfrac{1}{2}\left(\dfrac{\partial v_x}{\partial y} + \dfrac{\partial v_y}{\partial x}\right) & \dfrac{\partial v_x}{\partial x} + 2\dfrac{\partial v_y}{\partial y} \end{pmatrix} \tag{6.87}$$

and inserting the floating condition (6.45), they can be written in abbreviated form as

$$\begin{aligned} 2\frac{\partial(\bar{\eta} M_{xx})}{\partial x} + 2\frac{\partial(\bar{\eta} M_{xy})}{\partial y} &= \frac{\rho}{\rho_\mathrm{sw}}(\rho_\mathrm{sw} - \rho) g H \frac{\partial H}{\partial x} , \\ 2\frac{\partial(\bar{\eta} M_{xy})}{\partial x} + 2\frac{\partial(\bar{\eta} M_{yy})}{\partial y} &= \frac{\rho}{\rho_\mathrm{sw}}(\rho_\mathrm{sw} - \rho) g H \frac{\partial H}{\partial y} . \end{aligned} \tag{6.88}$$

The combination of density terms on the right-hand sides is equal to the quantity ϱ introduced in Eq. $(6.78)_1$, and so we can assemble the compact form

$$2\frac{\partial(\bar{\eta}M_{ij})}{\partial x_j} = \varrho g H \frac{\partial H}{\partial x_i}, \tag{6.89}$$

where the indices i and j can take the values 1 and 2 (note the summation over j by Einstein's summation convention).

As for the boundary conditions at the grounding line, the inflow velocity is prescribed,

$$\mathbf{v}_\mathrm{h}\big|_\mathrm{gl} = \mathbf{v}_\mathrm{gl} . \tag{6.90}$$

At the calving front, Eq. (6.64) holds, which can be written as

$$\begin{aligned} 2\bar{\eta}M_{xx}\big|_\mathrm{cf}\, n_x + 2\bar{\eta}M_{xy}\big|_\mathrm{cf}\, n_y &= \frac{\varrho g H^2}{2}\, n_x , \\ 2\bar{\eta}M_{xy}\big|_\mathrm{cf}\, n_x + 2\bar{\eta}M_{yy}\big|_\mathrm{cf}\, n_y &= \frac{\varrho g H^2}{2}\, n_y , \end{aligned} \tag{6.91}$$

or, in compact index notation,

$$2\bar{\eta}M_{ij}\big|_\mathrm{cf}\, n_j = \frac{\varrho g H^2}{2}\, n_i . \tag{6.92}$$

In the following, we will refer to the boundary-value problem consisting of Eqs. (6.89), (6.90) and (6.92) as the *mechanical ice shelf problem.*

6.5.2 Weak Formulation

Let ω be the domain covered by the ice shelf in the horizontal plane, and $\partial\omega$ its margin, which consists of the grounding line, $\partial\omega_\mathrm{gl}$, and the calving front, $\partial\omega_\mathrm{cf}$ (that is, $\partial\omega = \partial\omega_\mathrm{gl} \cup \partial\omega_\mathrm{cf}$). Since the second order system of differential equations (6.89) is valid at any point within the domain ω, we can also multiply it by an arbitrary smooth function $w(x,y)$ and integrate it over the entire domain,

$$\int_\omega w \left[2\frac{\partial(\bar{\eta}M_{ij})}{\partial x_j} - \varrho g H \frac{\partial H}{\partial x_i} \right] \mathrm{d}x\,\mathrm{d}y = 0 . \tag{6.93}$$

The function $w(x,y)$ is called the *weight function,* and Eq. (6.93) is the *weighted-integral form* of Eq. (6.89).

We proceed by employing the two-dimensional version of the divergence theorem (2.63) as follows,

$$\begin{aligned} &\int_\omega \frac{\partial(w\bar{\eta}M_{ij})}{\partial x_j}\,\mathrm{d}x\,\mathrm{d}y = \int_{\partial\omega} w\bar{\eta}M_{ij}n_j\,\mathrm{d}s \\ \Rightarrow\; &\int_\omega w\frac{\partial(\bar{\eta}M_{ij})}{\partial x_j}\,\mathrm{d}x\,\mathrm{d}y = \int_{\partial\omega} w\bar{\eta}M_{ij}n_j\,\mathrm{d}s - \int_\omega \frac{\partial w}{\partial x_j}\bar{\eta}M_{ij}\,\mathrm{d}x\,\mathrm{d}y , \end{aligned} \tag{6.94}$$

where ds is a line element along the boundary $\partial\omega$. Then the weighted-integral statement (6.93) becomes

$$\int_{\omega} \left[2\bar{\eta} M_{ij} \frac{\partial w}{\partial x_j} + w \varrho g H \frac{\partial H}{\partial x_i} \right] \mathrm{d}x\,\mathrm{d}y - \int_{\partial\omega} 2w\bar{\eta} M_{ij} n_j \,\mathrm{d}s = 0\,. \tag{6.95}$$

By applying the decomposition $\partial\omega = \partial\omega_{\mathrm{gl}} \cup \partial\omega_{\mathrm{cf}}$ and inserting the boundary condition (6.92) for the calving front, the boundary integral yields

$$\begin{aligned} \int_{\partial\omega} 2w\bar{\eta} M_{ij} n_j \,\mathrm{d}s &= \int_{\partial\omega_{\mathrm{gl}}} 2w\bar{\eta} M_{ij} n_j \,\mathrm{d}s + \int_{\partial\omega_{\mathrm{cf}}} 2w\bar{\eta} M_{ij} n_j \,\mathrm{d}s \\ &= \int_{\partial\omega_{\mathrm{gl}}} 2w\bar{\eta} M_{ij} n_j \,\mathrm{d}s + \int_{\partial\omega_{\mathrm{cf}}} w \frac{\varrho g H^2}{2} n_i \,\mathrm{d}s\,, \end{aligned} \tag{6.96}$$

and so from (6.95) we obtain

$$\begin{aligned} &\int_{\omega} \left[2\bar{\eta} M_{ij} \frac{\partial w}{\partial x_j} + w \varrho g H \frac{\partial H}{\partial x_i} \right] \mathrm{d}x\,\mathrm{d}y \\ &\qquad - \int_{\partial\omega_{\mathrm{gl}}} 2w\bar{\eta} M_{ij} n_j \,\mathrm{d}s - \int_{\partial\omega_{\mathrm{cf}}} w \frac{\varrho g H^2}{2} n_i \,\mathrm{d}s = 0\,. \end{aligned} \tag{6.97}$$

The remaining boundary condition (6.90) for the grounding line reads, in component form,

$$v_i|_{\mathrm{gl}} = (v_i)_{\mathrm{gl}}\,, \tag{6.98}$$

where $(v_i)_{\mathrm{gl}}$ is a prescribed function along the grounding line.

Equations (6.97) and (6.98) constitute the *weak formulation* of the mechanical ice shelf problem. Note that the second derivatives of the velocity components in the original differential equation (6.89) and its weighted-integral form (6.93) have been replaced by first derivatives of the velocity components (hidden in M) and the weight function. The boundary condition at the calving front is automatically fulfilled by the integral statement (6.97) ("natural boundary condition"), whereas the boundary condition (6.98) at the grounding line needs to be accounted for separately ("essential boundary condition").

6.5.3 Discretisation of the Ice Shelf Domain

In order to convert the weak formulation of the mechanical ice shelf problem into a numerical scheme, the ice shelf domain ω is now divided into non-overlapping elements ω_e ("finite elements"), such that adjacent elements share common edges and vertices ("nodes"). The entire set of finite elements is referred to as the *finite element mesh*.

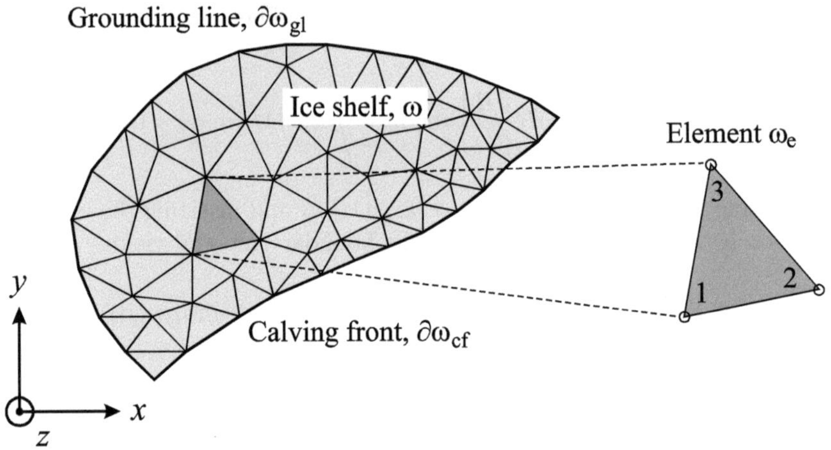

Fig. 6.8. Triangulation of an ice shelf domain ω. The boundary $\partial\omega$ consists of the grounding line $\partial\omega_{\mathrm{gl}}$ and the calving front $\partial\omega_{\mathrm{cf}}$. A typical triangular element ω_{e} with its three element nodes, numbered counterclockwise from 1 to 3, is highlighted.

The decomposition of the domain into finite elements can be achieved in many different ways, of which we only consider the simplest one, namely a *triangulation* (Fig. 6.8). This is to say, the elements are triangles which cover the domain, and consequently the margin of the domain is approximated by a closed polygon. The elements are assigned a consecutive numbering $e = 1 \ldots E$, where E is the number of elements, and each element has three nodes, numbered by $l = 1, 2, 3$. We also introduce a global (element-independent) numbering of the nodes, $n = 1 \ldots N$, where N is the number of nodes of the entire domain (including the boundary). It is of course important to do careful bookkeeping for the global and element-wise numbering of the nodes by establishing the relation $n(e, l)$.

Carrying out a triangulation for a given ice shelf domain is a non-trivial problem. A commonly employed method that has favourable properties for the finite element method is the *Delaunay triangulation*. However, this is not discussed here, and we refer the reader to the numerous literature on the subject [e.g., de Berg et al. (2008)].

For the computations required in the following, it is convenient to introduce a *master element* Ω to which the elements ω_{e} are mapped. As shown in Fig. 6.9, the master element is the isosceles unit triangle with area 1/2 in the transformed coordinates ξ and η. The transformation is given by

$$\begin{pmatrix} x \\ y \end{pmatrix} = \begin{pmatrix} x_2 - x_1 & x_3 - x_1 \\ y_2 - y_1 & y_3 - y_1 \end{pmatrix} \cdot \begin{pmatrix} \xi \\ \eta \end{pmatrix} + \begin{pmatrix} x_1 \\ y_1 \end{pmatrix}, \tag{6.99}$$

and its inverse reads

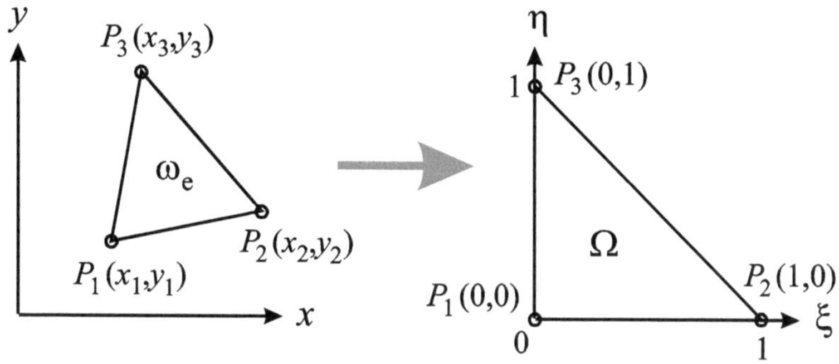

Fig. 6.9. Mapping of an arbitrary triangular element ω_e with nodes P_1, P_2 and P_3 (element numbering) to the master element Ω.

$$\begin{pmatrix} \xi \\ \eta \end{pmatrix} = \frac{1}{2A_e} \begin{pmatrix} y_3 - y_1 & -(x_3 - x_1) \\ -(y_2 - y_1) & x_2 - x_1 \end{pmatrix} \cdot \begin{pmatrix} x - x_1 \\ y - y_1 \end{pmatrix} , \qquad (6.100)$$

where $2A_e$ is the determinant of the transformation matrix,

$$2A_e = (x_2 - x_1)(y_3 - y_1) - (x_3 - x_1)(y_2 - y_1) . \qquad (6.101)$$

Evaluation of the integral

$$\int\limits_{\omega_e} \mathrm{d}x\,\mathrm{d}y = \int\limits_{\omega_e} 2A_e\,\mathrm{d}\xi\,\mathrm{d}\eta = 2A_e \int\limits_{\omega_e} \mathrm{d}\xi\,\mathrm{d}\eta = 2A_e \times \frac{1}{2} = A_e \qquad (6.102)$$

shows that A_e denotes the area of the element ω_e.

6.5.4 Galerkin Finite Element Method

We return to the weak formulation (6.97) and (6.98) of the mechanical ice shelf problem. The horizontal velocity field is now approximated by a set of *basis functions* Φ_n, $n = 1 \ldots N$, as follows,

$$\begin{aligned} v_x(x,y) &\sim \sum_{n=1}^{N} (v_x)_n\, \Phi_n(x,y) , \\ v_y(x,y) &\sim \sum_{n=1}^{N} (v_y)_n\, \Phi_n(x,y) . \end{aligned} \qquad (6.103)$$

For the tensor M, which was defined in Eq. (6.87) and can be expressed in index notation as

$$M_{ij} = \frac{1}{2}\left(\frac{\partial v_i}{\partial x_j} + \frac{\partial v_j}{\partial x_i}\right) + \frac{\partial v_k}{\partial x_k}\delta_{ij} \qquad (i,j,k = 1,2;\ \text{sum over } k) , \qquad (6.104)$$

we obtain

$$M_{ij} \sim \sum_{n=1}^{N} \left[\frac{1}{2}(v_i)_n \frac{\partial \Phi_n}{\partial x_j} + \frac{1}{2}(v_j)_n \frac{\partial \Phi_n}{\partial x_i} + (v_k)_n \frac{\partial \Phi_n}{\partial x_k} \delta_{ij} \right] . \tag{6.105}$$

The basis functions Φ_n must be linearly independent, continuously differentiable within the elements and continuous across element boundaries. A central idea of the finite element method is to employ basis functions which are non-zero only in a small sub-domain of the discretised ice shelf area. Specifically, we choose functions with the following properties:

- The basis function Φ_n is non-zero only in the elements adjacent to the node n (global numbering, that is, $n = 1 \dots N$) .
- $\Phi_n(x_n, y_n) = 1$, $\Phi_n(x_m, y_m) = 0$ for nodes $m \neq n$.
- In the elements adjacent to the node n, Φ_n is a linear function of x and y (that is, geometrically a plane).

This yields piecewise linear functions, an example of which is depicted in Fig. 6.10. Note that the coefficients $(v_x)_n$ and $(v_y)_n$ in Eq. (6.103) denote the values of v_x and v_y at the element nodes.

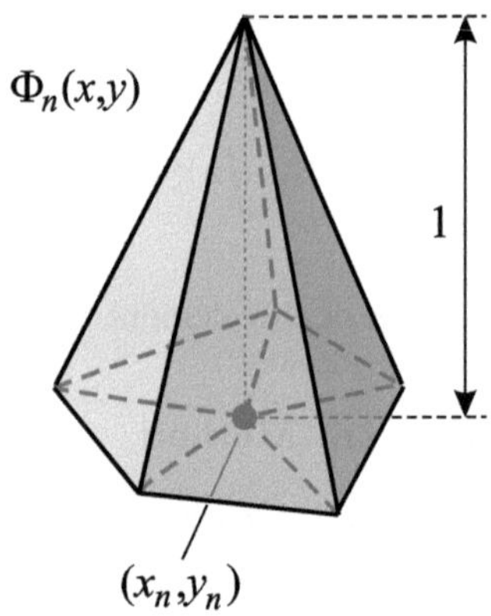

Fig. 6.10. Piecewise linear basis function $\Phi_n(x, y)$ associated with the node with the global number n and the position (x_n, y_n).

Since in real-world problems the ice thickness H is never known exactly at every point (x, y) in the domain continuum, but only for a finite number of points (either from observational data or from simulation results), we express it also in terms of the basis functions,

$$H(x, y) \sim \sum_{n=1}^{N} H_n \, \Phi_n(x, y) , \tag{6.106}$$

so that only the nodal values H_n, $n = 1 \dots N$, need to be known. The same procedure is applied for the depth-integrated viscosity $\bar{\eta}$,

$$\bar{\eta}(x, y) \sim \sum_{n=1}^{N} \bar{\eta}_n \, \Phi_n(x, y) \,. \tag{6.107}$$

Even though the depth-integrated viscosity depends on the velocity components [see Eqs. (6.53), (6.54)] and is therefore part of the solution, let us assume for the moment that its nodal values $\bar{\eta}_n$ are also known. We will come back to this point in Sect. 6.5.5.

The approximations (6.105), (6.106) and (6.107) are inserted into the integral statement (6.97) of the weak formulation. In order to turn it into a set of equations for the unknown nodal values $(v_x)_n$ and $(v_y)_n$, the weight functions w must be prescribed. It is convenient (but not the only possibility) to choose the weight functions as the set of basis functions Φ_n. This is called the *Galerkin finite element method.* Specifically, we choose

$$w = \Phi_m \,, \quad m = 1 \dots N - N_{\mathrm{gl}} \,, \tag{6.108}$$

where N_{gl} is the number of nodes on the grounding line, which are excluded because the velocity on the grounding line is already prescribed by the boundary condition (6.98).

For any node m, we define ω_m as the area of all adjacent elements, in which the basis function Φ_m does not vanish. The intersections of ω_m with the calving front $\partial\omega_{\mathrm{cf}}$ and the grounding line $\partial\omega_{\mathrm{gl}}$ are $\partial\omega_{\mathrm{cf}} \cap \omega_m$ and $\partial\omega_{\mathrm{gl}} \cap \omega_m$, respectively (Fig. 6.11). For all inner nodes, these are both empty. With these definitions, inserting Eqs. (6.105), (6.106) and (6.107) in the integral statement (6.97), and choosing the weight functions according to Eq. (6.108), yields

$$\begin{aligned}
&\int_{\omega_m} \Bigg\{ 2 \sum_{p=1}^{N} (\bar{\eta}_p \, \Phi_p) \\
&\qquad \times \sum_{n=1}^{N} \left[\frac{1}{2} (v_i)_n \frac{\partial \Phi_n}{\partial x_j} + \frac{1}{2} (v_j)_n \frac{\partial \Phi_n}{\partial x_i} + (v_k)_n \frac{\partial \Phi_n}{\partial x_k} \delta_{ij} \right] \frac{\partial \Phi_m}{\partial x_j} \\
&\qquad + \Phi_m \varrho g \sum_{p=1}^{N} (H_p \, \Phi_p) \sum_{n=1}^{N} \left(H_n \frac{\partial \Phi_n}{\partial x_i} \right) \Bigg\} \, \mathrm{d}x \, \mathrm{d}y \\
&- \int_{\partial\omega_{\mathrm{gl}} \cap \omega_m} 2 \Phi_m \sum_{p=1}^{N} (\bar{\eta}_p \, \Phi_p) \\
&\qquad \times \sum_{n=1}^{N} \left[\frac{1}{2} (v_i)_n \frac{\partial \Phi_n}{\partial x_j} + \frac{1}{2} (v_j)_n \frac{\partial \Phi_n}{\partial x_i} + (v_k)_n \frac{\partial \Phi_n}{\partial x_k} \delta_{ij} \right] n_j \, \mathrm{d}s
\end{aligned}$$

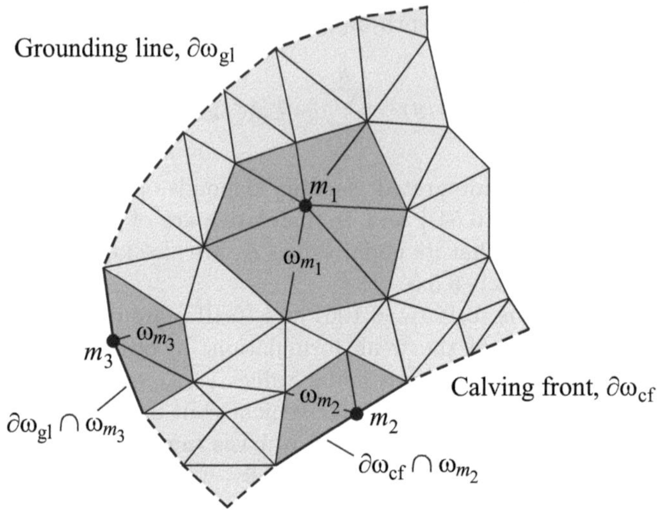

Fig. 6.11. Definition of the sub-domains ω_m adjacent to the nodes m and their intersections with the calving front and the grounding line, $\partial\omega_{cf} \cap \omega_m$ and $\partial\omega_{gl} \cap \omega_m$, for an inner node ($m = m_1$), a node on the calving front ($m = m_2$) and a node on the grounding line ($m = m_3$).

$$- \int\limits_{\partial\omega_{cf} \cap \omega_m} \Phi_m \frac{\varrho g}{2} \sum_{p=1}^{N} (H_p \Phi_p) \sum_{n=1}^{N} (H_n \Phi_n)\, n_i \, ds = 0 \,. \tag{6.109}$$

Since $i = 1, 2$ and $m = 1 \ldots N - N_{gl}$ (all other indices are summation indices), these are $2(N - N_{gl})$ equations. Let us assemble the $2N$ nodal values of the velocities into the column

$$\hat{\mathbf{v}} = \Big((v_x)_1, \, (v_y)_1, \, (v_x)_2, \, (v_y)_2, \, \ldots \, (v_x)_N, \, (v_y)_N \Big)^{\mathrm{T}} , \tag{6.110}$$

or

$$\hat{v}_{2(n-1)+i} = (v_i)_n \quad (i = 1, 2; \; n = 1 \ldots N) \,. \tag{6.111}$$

If the index m in Eq. (6.109) corresponds to an inner node of the ice shelf domain, the intersections $\partial\omega_{gl} \cap \omega_m$ and $\partial\omega_{cf} \cap \omega_m$ are empty. Therefore, the two boundary integrals are automatically equal to zero, and we only have to consider the area integral over ω_m. By applying Einstein's summation convention for the node indices n and p, using (6.111) and re-arranging terms, we obtain, from Eq. (6.109),

$$
\begin{aligned}
&\left(\bar{\eta}_p \int\limits_{\omega_m} \frac{\partial \Phi_m}{\partial x_j}\frac{\partial \Phi_n}{\partial x_j} \Phi_p \,\mathrm{d}x\,\mathrm{d}y \right) \hat{v}_{2(n-1)+i} \\
&+ \left(\bar{\eta}_p \int\limits_{\omega_m} \left(\frac{\partial \Phi_m}{\partial x_j}\frac{\partial \Phi_n}{\partial x_i} + 2\frac{\partial \Phi_m}{\partial x_i}\frac{\partial \Phi_n}{\partial x_j} \right) \Phi_p \,\mathrm{d}x\,\mathrm{d}y \right) \hat{v}_{2(n-1)+j} \\
&= -\varrho g H_n H_p \int\limits_{\omega_m} \Phi_m \frac{\partial \Phi_n}{\partial x_i} \Phi_p \,\mathrm{d}x\,\mathrm{d}y \\
&\qquad\qquad (i = 1, 2;\;\; m = 1 \ldots N - N_{\mathrm{cf}} - N_{\mathrm{gl}}) \,, \qquad (6.112)
\end{aligned}
$$

where N_{cf} and N_{gl} are the numbers of nodes on the calving front and on the grounding line, respectively.

If the index m in Eq. (6.109) corresponds to a node on the calving front, then the intersection $\partial\omega_{\mathrm{cf}} \cap \omega_m$ will not be empty, and the last line of (6.109) provides an additional inhomogeneity,

$$
\begin{aligned}
&\left(\bar{\eta}_p \int\limits_{\omega_m} \frac{\partial \Phi_m}{\partial x_j}\frac{\partial \Phi_n}{\partial x_j} \Phi_p \,\mathrm{d}x\,\mathrm{d}y \right) \hat{v}_{2(n-1)+i} \\
&+ \left(\bar{\eta}_p \int\limits_{\omega_m} \left(\frac{\partial \Phi_m}{\partial x_j}\frac{\partial \Phi_n}{\partial x_i} + 2\frac{\partial \Phi_m}{\partial x_i}\frac{\partial \Phi_n}{\partial x_j} \right) \Phi_p \,\mathrm{d}x\,\mathrm{d}y \right) \hat{v}_{2(n-1)+j} \\
&= -\varrho g H_n H_p \int\limits_{\omega_m} \Phi_m \frac{\partial \Phi_n}{\partial x_i} \Phi_p \,\mathrm{d}x\,\mathrm{d}y \\
&\quad + \frac{\varrho g}{2} H_n H_p \int\limits_{\partial\omega_{\mathrm{cf}} \cap \omega_m} \Phi_m \Phi_n \Phi_p \, n_i \,\mathrm{d}s \\
&\qquad (i = 1, 2;\;\; m = N - N_{\mathrm{cf}} - N_{\mathrm{gl}} + 1 \ldots N - N_{\mathrm{gl}}) \,. \qquad (6.113)
\end{aligned}
$$

The remaining $2N_{\mathrm{gl}}$ equations are provided by evaluating the boundary condition (6.98) at the nodes of the grounding line,

$$
\hat{v}_{2(m-1)+i} = \left[\hat{v}_{2(m-1)+i}\right]_{\mathrm{gl}} \quad (i = 1, 2;\;\; m = N - N_{\mathrm{gl}} + 1 \ldots N) \,. \qquad (6.114)
$$

Equations (6.112), (6.113) and (6.114) constitute a system of $2N$ linear equations for the $2N$ unknown velocity components, which can be written in compact form as

$$
\hat{\mathsf{K}} \cdot \hat{\mathbf{v}} = \hat{\mathbf{r}} \,, \qquad (6.115)
$$

where $\hat{\mathsf{K}}$ is the $2N \times 2N$ coefficient matrix, and $\hat{\mathbf{r}}$ is the column of size $2N$ which contains the inhomogeneities. The integrals in (6.112) and (6.113), which are required to assemble the coefficient matrix and the inhomogeneity column, can be relatively easily computed analytically, because the basis functions are piecewise linear over the elements ω_e, and thus their spatial derivatives are piecewise constant. In order to do so, it is convenient to transform the integrals element-wise to the master element Ω discussed above (Sect. 6.5.3); this is not shown here.

The coefficient matrix $\hat{\mathsf{K}}$ is *sparse*, that is, most of its entries are equal to zero, because the basis functions Φ_m and Φ_n overlap only if m and n are identical or adjacent nodes. However, it lacks the property of symmetry. Since the system (6.115) is usually very large, it is important to choose an efficient numerical solver, for instance, from the set of generalised conjugate gradient methods (Press et al. 1996).

6.5.5 Iteration

The nodal values of the depth-integrated viscosity, $\bar{\eta}_n$ $(n = 1 \ldots N)$ remain to be calculated [see Eq. (6.107)]. This can be done by employing an *iteration method.*

We start with an initial guess of the depth-integrated viscosity, $\bar{\eta}_n^{(1)}$, for instance, a constant value everywhere. Based on this, the coefficient matrix $\hat{\mathsf{K}}^{(1)}$ and the inhomogeneity column $\hat{\mathbf{r}}^{(1)}$ are assembled, and the system

$$\hat{\mathsf{K}}^{(1)} \cdot \hat{\mathbf{v}}^{(1)} = \hat{\mathbf{r}}^{(1)} \tag{6.116}$$

is solved for $\hat{\mathbf{v}}^{(1)}$, the first estimate for the velocity field.

Let $\hat{\mathbf{v}}^{(l)}$ be the l-th iteration of the velocity field. By inserting $\hat{\mathbf{v}}^{(l)}$ into Eqs. (6.53) and (6.54), improved viscosities $\bar{\eta}_n^{(l+1)}$ can be computed. Since we have assumed that the geometry and temperature of the entire ice shelf are known, the integral over the associated rate factor in Eq. (6.54) is also known everywhere. However, due to the choice of piecewise linear basis functions Φ_n, the approximated velocity field (6.103) is not differentiable at the nodes. Therefore, the velocity derivatives in Eq. (6.53), evaluated at the nodes, must be computed by averaging over the respective adjacent elements, where they exist. With the viscosities $\bar{\eta}_n^{(l+1)}$, the updated coefficient matrix $\hat{\mathsf{K}}^{(l+1)}$ and the updated inhomogeneity column $\hat{\mathbf{r}}^{(l+1)}$ can be constructed, and the system

$$\hat{\mathsf{K}}^{(l+1)} \cdot \tilde{\hat{\mathbf{v}}}^{(l+1)} = \hat{\mathbf{r}}^{(l+1)} \tag{6.117}$$

solved for $\tilde{\hat{\mathbf{v}}}^{(l+1)}$. In order to dampen possible numerical oscillations, only a certain part $w \in (0, 1]$ of $\tilde{\hat{\mathbf{v}}}^{(l+1)}$ is used for the updated velocity field $\hat{\mathbf{v}}^{(l+1)}$,

$$\hat{\mathbf{v}}^{(l+1)} = w\,\tilde{\hat{\mathbf{v}}}^{(l+1)} + (1 - w)\,\hat{\mathbf{v}}^{(l)}\,. \tag{6.118}$$

This *relaxation scheme* averages between the new and the old solution, and the relaxation factor w should be chosen such that a good compromise is achieved between speed of convergence and limitation of numerical oscillations.

As the above discussions show, the great advantage of the finite element method is the flexibility in the decomposition of the numerical domain.

Whereas the finite difference method is essentially restricted to a regular grid, a finite element mesh can easily handle irregularly shaped boundaries and local refinements of the resolution. The price to pay is that the finite element method is conceptually more difficult, and, owing to the need to solve large systems of linear equations, computationally very demanding.

6.5.6 Example: The Ross Ice Shelf

An implementation of the numerical solution of the mechanical ice shelf problem as laid down above (Sects. 6.5.1–6.5.5) was realised by Weis (2001) with the model FESSACODE (Finite Element Shallow Shelf Approximation Code). As an example of a numerical ice-shelf-modelling study, we describe briefly its application to the Ross Ice Shelf in Antarctica (Humbert et al. 2005, and references therein).

The Ross Ice Shelf is the largest of the Antarctic ice shelves with an area of approx. 490,000 km^2, situated in a huge embayment of the Antarctic continent (Figs. 1.1, 6.12). Inflow from the ice sheet occurs mainly by five major ice streams from the West-Antarctic Shirase, Siple and Gould Coast (for simplicity, we will refer to it as "Siple Coast" in the following) and a number of glaciers from East Antarctica which enter through the steep coast of the Transantarctic Mountains. The Ross Ice Shelf is well suited for a dynamic/thermodynamic modelling study because of the good data coverage: The geometry, ice thickness distribution, surface temperature and inflow velocities at the grounding line are reasonably well known. The three-dimensional temperature field can be interpolated (with some uncertainty) by assuming that the basal temperature is equal to the freezing temperature of the sea water below the ice shelf, and that the shape of the vertical profile is everywhere the same as that of an existing ice core measurement (for details see Humbert et al. 2005). Therefore, all the required input data for the model FESSACODE are available with acceptable accuracy.

Figure 6.12 shows a triangulation of the Ross Ice Shelf. It consists of 2290 elements with 1398 nodes, and the size of the elements varies from 0.71 km to 49.8 km (minimum and maximum side length, respectively) depending on the position in the ice shelf. In order to facilitate comparisons between simulated velocities and field data, the mesh has been created such that the locations of stations of the Ross Ice Shelf Geophysical and Glaciological Survey ("RIGGS") campaign (Thomas et al. 1984), for which velocities have been measured, fall on the nodes of the mesh.

In order to account for the fact that the viscosity of the in-situ ice of the Ross Ice Shelf may deviate from that of pure, isotropic ice, a stress enhancement factor E_s has been introduced by replacing the associated rate factor in Eq. (6.54) according to

$$B(T') \rightarrow E_s B(T') \tag{6.119}$$

[see also Eq. (4.35)]. A value of E_s greater than unity makes the ice stiffer, whereas a value less than unity makes it softer.

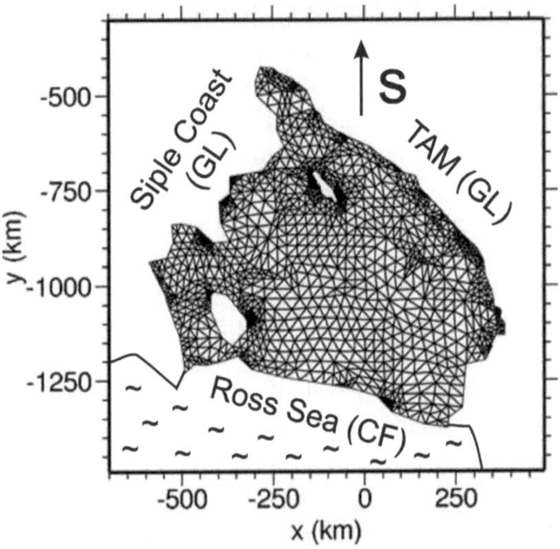

Fig. 6.12. Geometry of the Ross Ice Shelf. TAM: Transantarctic Mountains, GL: grounding line, CF: calving front. The triangular mesh was created by using the ice velocity data of the RIGGS campaign (Thomas et al. 1984) as nodes. Adapted from Humbert et al. (2005), © American Geophysical Union.

For the simulation shown in Fig. 6.13, the stress enhancement factor E_s has been chosen such that the mean difference between measured RIGGS (v_RIGGS) and simulated (v_SIM) velocities is minimised,

$$\left| \frac{1}{N} \sum_{n=1}^{N} (v_{\mathrm{RIGGS},n} - v_{\mathrm{SIM},n}) \right| \stackrel{!}{=} \min\,, \tag{6.120}$$

where $v = (v_x^2 + v_y^2)^{1/2}$ and the index n numbers the N RIGGS-velocity data points (which fall together with the nodes of the numerical grid). This procedure yields the value $E_\mathrm{s} = 0.86$, for which the remaining mean difference is $+8.4\,\mathrm{m\,a^{-1}}$.

The simulation reproduces the general flow pattern and the magnitudes of the flow velocities of the Ross Ice Shelf quite well. The latter generally increase towards the calving front, where values of up to $\sim 1200\,\mathrm{m\,a^{-1}}$ are reached. It is also interesting to see how the two islands (Roosevelt Island close to the calving front, Crary Ice Rise further south) efficiently slow down the coastward ice flow, which may be important for stabilizing the entire ice shelf.

The scatter plot (right panel of Fig. 6.13) reveals a systematic deviation, in that the simulation tends to underestimate small and to overestimate large flow velocities. A number of reasons for this discrepancy are conceivable. The real temperature field likely shows more spatial variability than the constructed one, which affects the distribution of the depth-integrated viscosity.

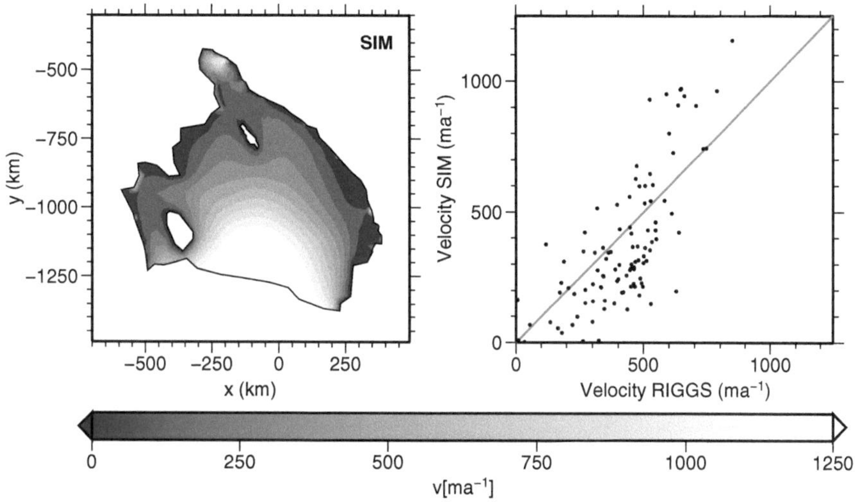

Fig. 6.13. Left: Simulated velocity distribution (absolute values only) of the Ross Ice Shelf. Right: Scatter plot of simulated vs. RIGGS velocities. Reproduced from Humbert et al. (2005), © American Geophysical Union.

Also, the local viscosity can be influenced by several types of impurities as well as anisotropic fabrics of the polycrystalline ice, which are only globally accounted for by the choice of the stress enhancement factor. Another factor is that ice originating from fast-flowing ice streams and glaciers of the adjacent ice sheet may be heavily damaged, thus softening the ice downstream of the main inflow systems. This effect may be further complicated by changing ice stream activities in the past. So it is not surprising that the agreement between measured and simulated velocities is not perfect, and there is still room for improvement.

7

Dynamics of Glacier Flow

7.1 Glaciers Versus Ice Sheets

As mentioned in the introduction (Chapter 1), the size of land ice masses spans several orders of magnitude, from large ice sheets of a few thousand kilometres in diameter down to small glaciers of a few hundreds of metres in length. Consequently, the scaling given for ice sheets in Chapter 5 [Eqs. (5.5) and (5.102)] is not valid for smaller ice caps and glaciers, and needs to be modified. However, the Froude number (5.7) and Coriolis-force-to-pressure-gradient ratio (5.10) are always extremely small compared to unity, and therefore the Stokes flow problem formulated in Sect. 5.1 is applicable to land ice masses of all shapes and sizes. On the other hand, the applicability of the approximations defined in Sects. 5.2 to 5.4 is limited by the size of the ice masses. While the hydrostatic approximation and the first order approximation still provide reasonable accuracy for most problems of glacier flow, the shallow ice approximation is valid for the large-scale description of ice sheets and large ice caps only.

In addition to these dynamical issues, the thermal regime of ice sheets and glaciers shows important differences. Most ice in ice sheets is *cold ice*, i.e., its temperature is below the local pressure melting point. *Temperate ice*, with a temperature at the pressure melting point, exists only at the base or in thin, near-basal layers. While this behaviour is shared by glaciers at high latitudes, in glaciers at lower latitudes temperate ice occurs alongside cold ice, and many of them consist entirely of temperate ice except for a thin cold surface layer in winter. Glaciers that consist entirely of cold ice are called *cold glaciers*, glaciers that consist entirely of temperate ice are called *temperate glaciers*, and glaciers in which significant amounts of both temperate and cold ice occur are called *polythermal glaciers*. We will not expand on glacier thermodynamics in this chapter, but come back to it in Sect. 9.3.

R. Greve, H. Blatter, *Dynamics of Ice Sheets and Glaciers*, Advances in Geophysical and Environmental Mechanics and Mathematics,
DOI 10.1007/978-3-642-03415-2_7, © Springer-Verlag Berlin Heidelberg 2009

7.2 Parallel Sided Slab

Depending on the exact geometry, the central part of a small valley glacier can sometimes be roughly approximated as a parallel sided slab on an inclined bed, similar to the gravity-driven thin-film flow discussed in Sect. 3.4.3. This is sketched in Fig. 7.1.

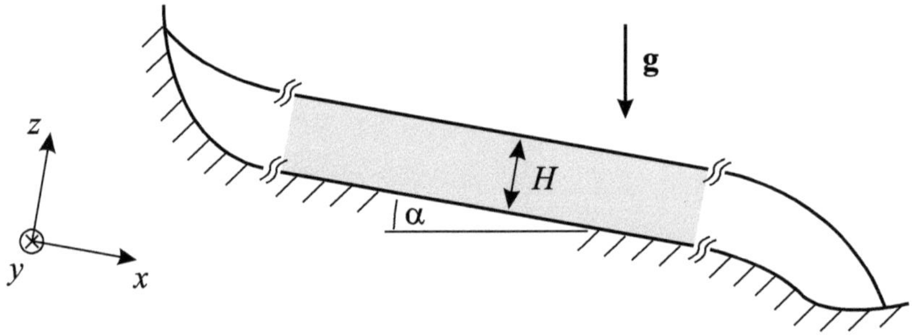

Fig. 7.1. Approximation of the central part of a valley glacier (grey-shaded) as a parallel sided slab.

In order to formulate the ice-flow problem for the parallel sided slab as full Stokes flow, let us make the following assumptions:

- Plane strain approximation: purely two-dimensional flow in the vertical x-z plane, no dependencies on the transverse coordinate y.
- Constant thickness H and inclination angle α.
- Uniformity in the downslope (x) direction: $\partial(\cdot)/\partial x = 0$ for all field quantities.
- Steady-state conditions: $\partial(\cdot)/\partial t = 0$ for all field quantities.
- Flat, rigid bed: $b(x,t) = 0$ (in the inclined coordinate system). The free surface is therefore given by $h(x,t) = H$.
- No surface accumulation ($a_{\mathrm{s}}^{\perp} = 0$).
- No basal melting ($a_{\mathrm{b}}^{\perp} = 0$), no basal sliding ($C_{\mathrm{b}} = 0$).
- Glen's flow law (4.16) with stress exponent $n = 3$.
- Constant rate factor: $A(T') = A = \text{const.}$
- Constant heat conductivity: $\kappa(T) = \kappa = \text{const.}$

Thus, the mass balance (5.1) simplifies to

$$\frac{\partial v_z}{\partial z} = 0\,. \tag{7.1}$$

With the above assumptions, the kinematic conditions (5.21) at the free surface and (5.31) at the glacier bed imply

$$v_z|_{z=H} = 0\,, \quad v_z|_{z=0} = 0\,, \tag{7.2}$$

and so the solution of Eq. (7.1) is simply

$$v_z = 0\,. \tag{7.3}$$

Consequently, the velocity field is entirely described by the profile of the downslope velocity, $v_x(z)$.

With the evident decomposition $\mathbf{g} = g\sin\alpha\,\mathbf{e}_x - g\cos\alpha\,\mathbf{e}_z$, the x-component of the Stokes equation (5.11) reads

$$\begin{aligned}
&-\frac{\partial p}{\partial x} + \eta\left(\frac{\partial^2 v_x}{\partial x^2} + \frac{\partial^2 v_x}{\partial y^2} + \frac{\partial^2 v_x}{\partial z^2}\right) \\
&+ 2\frac{\partial v_x}{\partial x}\frac{\partial \eta}{\partial x} + \left(\frac{\partial v_x}{\partial y} + \frac{\partial v_y}{\partial x}\right)\frac{\partial \eta}{\partial y} + \left(\frac{\partial v_x}{\partial z} + \frac{\partial v_z}{\partial x}\right)\frac{\partial \eta}{\partial z} \\
&+ \rho g \sin\alpha = 0\,,
\end{aligned} \tag{7.4}$$

which simplifies to

$$\begin{aligned}
&\eta\frac{\partial^2 v_x}{\partial z^2} + \frac{\partial v_x}{\partial z}\frac{\partial \eta}{\partial z} + \rho g \sin\alpha = 0 \\
\Rightarrow\quad &\frac{\partial}{\partial z}\left(\eta\frac{\partial v_x}{\partial z}\right) + \rho g \sin\alpha = 0\,.
\end{aligned} \tag{7.5}$$

The first integral of this equation is

$$\eta\frac{\partial v_x}{\partial z} = C_1 - \rho g z \sin\alpha\,, \tag{7.6}$$

where C_1 is an integration constant. Due to Glen's flow law in the form (4.21), the left-hand side is equal to the shear stress t_{xz},

$$t_{xz} = \eta\frac{\partial v_x}{\partial z} = C_1 - \rho g z \sin\alpha\,, \tag{7.7}$$

which vanishes at the free surface ($z = H$) due to the stress-free boundary condition (5.23). Hence, the integration constant can be determined,

$$t_{xz}\big|_{z=H} = C_1 - \rho g H \sin\alpha = 0 \quad\Rightarrow\quad C_1 = \rho g H \sin\alpha\,, \tag{7.8}$$

and we obtain the linear shear stress profile

$$t_{xz} = \eta\frac{\partial v_x}{\partial z} = \rho g (H - z)\sin\alpha\,. \tag{7.9}$$

In order to carry out the second integration, we insert the explicit form of the viscosity which results from Eq. (4.22),

$$\eta = \frac{1}{2}B\,d_\mathrm{e}^{-(1-1/n)}\,, \quad \text{where } B = A^{-1/n} \text{ and } d_\mathrm{e} = \frac{1}{2}\frac{\partial v_x}{\partial z}\,. \tag{7.10}$$

This yields from Eq. (7.9)

$$\begin{aligned}
&\frac{1}{2}A^{-1/n}\left(\frac{1}{2}\frac{\partial v_x}{\partial z}\right)^{-(1-1/n)}\frac{\partial v_x}{\partial z} = \rho g(H-z)\sin\alpha \\
&\Rightarrow\ A^{-1/n}\left(\frac{1}{2}\frac{\partial v_x}{\partial z}\right)^{1/n} = \rho g(H-z)\sin\alpha \\
&\Rightarrow\ \frac{\partial v_x}{\partial z} = 2A(\rho g\sin\alpha)^n(H-z)^n \\
&\Rightarrow\ v_x = -\frac{2A(\rho g\sin\alpha)^n}{n+1}(H-z)^{n+1} + C_2\,. \qquad (7.11)
\end{aligned}$$

The integration constant C_2 results from the no-slip condition $v_x|_{z=0} = 0$ (no basal sliding),

$$\begin{aligned}
v_x|_{z=0} &= -\frac{2A(\rho g\sin\alpha)^n}{n+1}H^{n+1} + C_2 = 0 \\
\Rightarrow\ C_2 &= \frac{2A(\rho g\sin\alpha)^n}{n+1}H^{n+1}\,. \qquad (7.12)
\end{aligned}$$

Therefore, we obtain the velocity profile

$$v_x = \frac{2A(\rho g\sin\alpha)^n}{n+1}\left[H^{n+1} - (H-z)^{n+1}\right], \qquad (7.13)$$

which is a quartic function of z for the stress exponent $n = 3$.

Analogous to Eq. (7.4), the z-component of the Stokes equation (5.11) is

$$\begin{aligned}
&-\frac{\partial p}{\partial z} + \eta\left(\frac{\partial^2 v_z}{\partial x^2} + \frac{\partial^2 v_z}{\partial y^2} + \frac{\partial^2 v_z}{\partial z^2}\right) \\
&+ 2\frac{\partial v_z}{\partial z}\frac{\partial \eta}{\partial z} + \left(\frac{\partial v_x}{\partial z} + \frac{\partial v_z}{\partial x}\right)\frac{\partial \eta}{\partial x} + \left(\frac{\partial v_y}{\partial z} + \frac{\partial v_z}{\partial y}\right)\frac{\partial \eta}{\partial y} \\
&- \rho g\cos\alpha = 0\,, \qquad (7.14)
\end{aligned}$$

and simplifies, for the parallel-sided-slab problem, to

$$\frac{\partial p}{\partial z} = -\rho g\cos\alpha\,. \qquad (7.15)$$

The integral of this equation is

$$p = C_3 - \rho g z\cos\alpha\,. \qquad (7.16)$$

Owing to the stress-free boundary condition at the free surface (5.23), the pressure vanishes there, which yields the integration constant C_3,

$$p|_{z=H} = C_3 - \rho g H \cos\alpha = 0 \quad \Rightarrow \quad C_3 = \rho g H \cos\alpha\,. \tag{7.17}$$

Thus, we obtain the hydrostatic pressure profile

$$p = \rho g (H - z) \cos\alpha\,. \tag{7.18}$$

It is interesting to note that the results for the pressure, Eq. (7.18), and the shear stress, Eq. (7.9), are identical to those for the thin-film flow of an incompressible Newtonian fluid (see Sect. 3.4.3). The nonlinearity of Glen's flow law affects only the velocity profile, which is quartic here [Eq. (7.13)], but parabolic for the Newtonian fluid [Eq. (3.149)].

It is also possible to compute the temperature profile analytically. The temperature equation (5.14) reads

$$\begin{aligned}
&\rho c \left(\frac{\partial T}{\partial t} + v_x \frac{\partial T}{\partial x} + v_y \frac{\partial T}{\partial y} + v_z \frac{\partial T}{\partial z} \right) \\
&= \frac{\partial}{\partial x}\left(\kappa \frac{\partial T}{\partial x}\right) + \frac{\partial}{\partial y}\left(\kappa \frac{\partial T}{\partial y}\right) + \frac{\partial}{\partial z}\left(\kappa \frac{\partial T}{\partial z}\right) + 4\eta d_\mathrm{e}^2\,,
\end{aligned} \tag{7.19}$$

which simplifies to

$$\frac{\partial}{\partial z}\left(\kappa \frac{\partial T}{\partial z}\right) = -4\eta d_\mathrm{e}^2\,. \tag{7.20}$$

By using Eq. (7.10), this can be rewritten as follows,

$$\begin{aligned}
\frac{\partial}{\partial z}\left(\kappa \frac{\partial T}{\partial z}\right) &= -4 \times \frac{1}{2} A^{-1/n} \left(\frac{1}{2}\frac{\partial v_x}{\partial z}\right)^{-(1-1/n)} \times \left(\frac{1}{2}\frac{\partial v_x}{\partial z}\right)^2 \\
&= -(2A)^{-1/n} \left(\frac{\partial v_x}{\partial z}\right)^{1+1/n}\,.
\end{aligned} \tag{7.21}$$

We substitute the term $\partial v_x/\partial z$ from (7.13) and obtain

$$\begin{aligned}
\frac{\partial}{\partial z}\left(\kappa \frac{\partial T}{\partial z}\right) &= -(2A)^{-1/n} \left[2A(\rho g \sin\alpha)^n (H-z)^n\right]^{1+1/n} \\
&= -2A(\rho g \sin\alpha)^{n+1} (H-z)^{n+1}\,.
\end{aligned} \tag{7.22}$$

The first integral of this equation is

$$\kappa \frac{\partial T}{\partial z} = \frac{2A(\rho g \sin\alpha)^{n+1}}{n+2} (H-z)^{n+2} + C_4\,. \tag{7.23}$$

By assuming that the basal temperature does not reach the pressure melting point, the integration constant can be obtained from the basal boundary condition (5.38),

$$
\begin{aligned}
& -\kappa \left.\frac{\partial T}{\partial z}\right|_{z=0} = q_{\text{geo}}^{\perp} \\
\Rightarrow \quad & -\frac{2A(\rho g \sin\alpha)^{n+1}}{n+2} H^{n+2} - C_4 = q_{\text{geo}}^{\perp} \\
\Rightarrow \quad & C_4 = -q_{\text{geo}}^{\perp} - \frac{2A(\rho g \sin\alpha)^{n+1}}{n+2} H^{n+2} ,
\end{aligned}
\tag{7.24}
$$

so that

$$
\kappa \frac{\partial T}{\partial z} = -q_{\text{geo}}^{\perp} - \frac{2A(\rho g \sin\alpha)^{n+1}}{n+2} \left[H^{n+2} - (H-z)^{n+2} \right] . \tag{7.25}
$$

Further integration yields

$$
T = -\frac{q_{\text{geo}}^{\perp}}{\kappa} z - \frac{2A(\rho g \sin\alpha)^{n+1}}{\kappa(n+2)} \left[H^{n+2} z + \frac{(H-z)^{n+3}}{n+3} \right] + C_5 . \tag{7.26}
$$

With the surface boundary condition (5.24), the integration constant can be computed,

$$
\begin{aligned}
& T|_{z=H} = T_{\text{s}} \\
\Rightarrow \quad & -\frac{q_{\text{geo}}^{\perp}}{\kappa} H - \frac{2A(\rho g \sin\alpha)^{n+1}}{\kappa(n+2)} H^{n+3} + C_5 = T_{\text{s}} \\
\Rightarrow \quad & C_5 = T_{\text{s}} + \frac{q_{\text{geo}}^{\perp}}{\kappa} H + \frac{2A(\rho g \sin\alpha)^{n+1}}{\kappa(n+2)} H^{n+3} .
\end{aligned}
\tag{7.27}
$$

Thus, we obtain for the temperature profile the relation

$$
\begin{aligned}
T = T_{\text{s}} &+ \frac{q_{\text{geo}}^{\perp}}{\kappa}(H-z) \\
&+ \frac{2AH^{n+3}(\rho g \sin\alpha)^{n+1}}{\kappa(n+2)} \left[1 - \frac{z}{H} - \frac{1}{n+3} \left(\frac{H-z}{H} \right)^{n+3} \right] .
\end{aligned}
\tag{7.28}
$$

The first line describes a linear profile due to heat conduction only, and the nonlinear modification in the second line is due to the dissipation term $4\eta d_{\text{e}}^2$ in the temperature equation (7.20).

For a glacier of 100 m thickness and 10° inclination, the results for the velocity profile, Eq. (7.13), and the temperature profile, Eq. (7.28), are depicted in Fig. 7.2. The simple-shear profile of the horizontal velocity with the largest shear rates $(\partial v_x/\partial z)$ close to the base and smallest shear rates close to the surface is evident, and the surface velocity reaches a value of $18.6\,\text{m}\,\text{a}^{-1}$. The temperature increases from −10°C at the surface to −4.71°C at the base, and the profile is slightly curved due to viscous dissipation, which is also largest near the base. The importance of dissipation is further highlighted by comparison with the linear profile resulting from heat conduction only, for which the basal temperature reaches a mere −7.62°C.

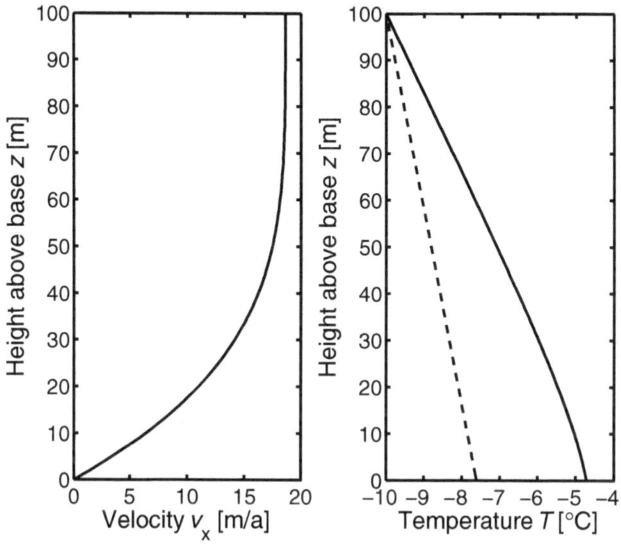

Fig. 7.2. Parallel sided slab: Velocity and temperature profiles according to Eqs. (7.13) and (7.28), for $H = 100\,\mathrm{m}$, $\alpha = 10°$, $T_\mathrm{s} = -10°\mathrm{C}$, $q^{\perp}_{\mathrm{geo}} = 50\,\mathrm{mW\,m^{-2}}$, $n = 3$, $A = 10^{-16}\,\mathrm{a^{-1}\,Pa^{-3}}$, $\rho = 910\,\mathrm{kg\,m^{-3}}$, $\kappa = 2.1\,\mathrm{W\,m^{-1}\,K^{-1}}$ and $g = 9.81\,\mathrm{m\,s^{-2}}$ (solid lines). The dashed line in the temperature panel shows the linear profile that results from heat conduction only (dissipation neglected).

7.3 Scaling Arguments and Hierarchy of Approximations

Basal shear stress is largely independent of the size of an ice mass and hardly exceeds 1 bar in glaciers, ice caps and ice sheets. This is a result of the power-law stress-strain-rate relation whereby the ice softens rapidly as the shear stress exceeds 1 bar. Ice flow velocity also increases rapidly as shear stresses exceed 1 bar and therefore a glacier would expand and thin correspondingly, restricting the shear stress to smaller values. The stress threshold may depend on the ice temperature, however, most ice masses are warmer near the base where most of the shearing takes place.

The existence of such a threshold for basal shear stress has interesting consequences for the aspect ratio of glaciers and ice sheets. For the parallel sided slab, the basal shear stress results from Eq. (7.9) as

$$\tau_\mathrm{b} = \rho g H \sin\alpha\,. \tag{7.29}$$

For a glacier with a geometry as sketched in Fig. 7.3, we infer that the *mean* basal shear stress $\bar{\tau}_\mathrm{b}$ is approximately given by

$$\bar{\tau}_\mathrm{b} \approx \rho g \bar{H} \sin\bar{\alpha}\,, \tag{7.30}$$

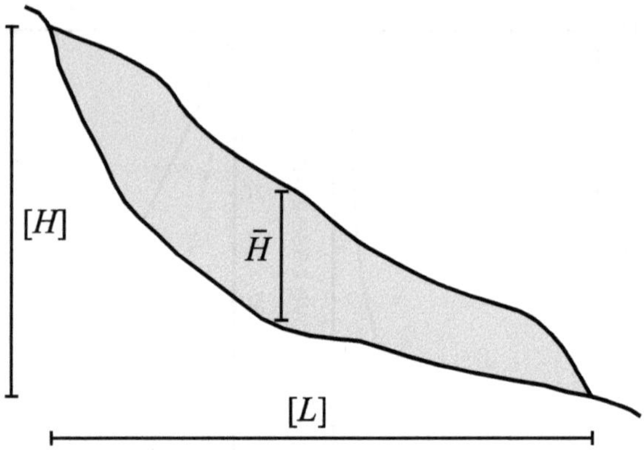

Fig. 7.3. Illustration of the scales: typical vertical extent $[H]$, typical horizontal extent $[L]$ and typical ice thickness $\bar{H}$ of a glacier.

where $\bar{H}$ and $\bar{\alpha}$ are the mean ice thickness and the mean inclination of the ice surface, respectively. The aspect ratio is defined by $\varepsilon = [H]/[L]$, see Eq. (5.6), and thus

$$\varepsilon = \frac{[H]}{[L]} = \tan\bar{\alpha}\,. \tag{7.31}$$

For $\bar{\alpha} \leq 30°$, which is fulfilled by virtually all glaciers on Earth,

$$\sin\bar{\alpha} \approx \tan\bar{\alpha} \tag{7.32}$$

holds (with an error of $\leq 15\%$). From Eqs. (7.30)–(7.32) we find for the aspect ratio

$$\varepsilon = \frac{[H]}{[L]} \approx \frac{\bar{\tau}_\mathrm{b}}{\rho g \bar{H}}\,. \tag{7.33}$$

In the following consideration, we assume $[H] \approx \bar{H}$ (which may be violated for very long and/or very steep glaciers; see Fig. 7.3) and $\bar{\tau}_\mathrm{b} \approx$ const. Equation (7.33) then implies

$$[L] \sim [H]^2\,, \tag{7.34}$$

that is, the horizontal extent of a glacier increases with the square of its vertical extent. Equivalent statements are

$$\varepsilon \sim \frac{1}{[H]} \sim \frac{1}{\sqrt{[L]}}\,. \tag{7.35}$$

Thus, glaciers tend to be the shallower the larger they are. Interestingly, the Vialov profile for ice sheets has the same scaling property [see Eq. (5.117)].

In the shallow ice approximation (Sect. 5.4) ice flow is determined by the local conditions alone, such as ice thickness, surface inclination and temperature. This is justified for large ice sheets where conditions generally vary little

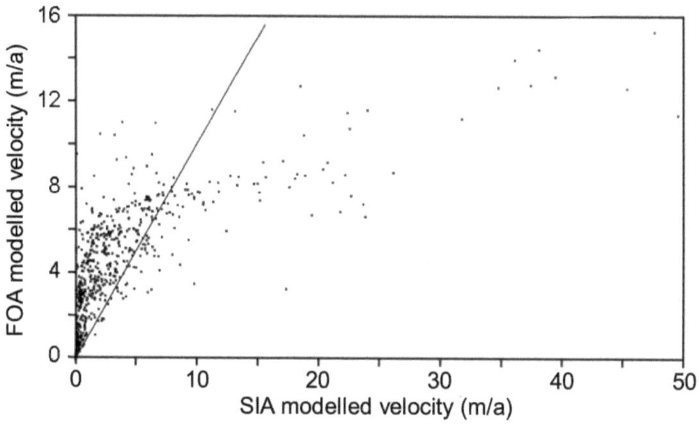

Fig. 7.4. Scatter plot of the horizontal velocity calculated for Haut Glacier d'Arolla in the first order approximation (FOA) versus the horizontal velocity calculated in the shallow ice approximation (SIA) for each grid point of the horizontal grid. The straight line marks equality of the FOA and SIA velocities. (Credit: A. Hubbard, Aberystwyth University, Wales, UK.)

over horizontal distances of 5-10 times the local ice thickness. This is considered to be the distance over which significant horizontal coupling of stresses occurs. Small glaciers may not even be as long as ten times their mean thickness, and their width may be comparable to their maximum ice thickness. Longitudinal and transverse couplings of stresses may thus become important. Such couplings may be enhanced even further if sliding varies over distances comparable to or less than the ice thickness, as occurs frequently in glaciers. For these reasons, the shallow ice approximation is no longer applicable for small glaciers. Depending on their geometry, the first order approximation (Sect. 5.3), the hydrostatic approximation (Sect. 5.2) or the full Stokes flow problem (Sect. 5.1) must be solved to obtain accurate solutions for the flow field in glaciers.

In order to illustrate the inadequacy of the shallow ice approximation for glaciers, Fig. 7.4 shows a scatter plot of horizontal velocity components computed with the shallow ice approximation (SIA) versus horizontal velocity components computed with the first order approximation (FOA) for Haut Glacier d'Arolla in the Swiss Alps. For small velocities, the SIA generally underestimates the velocities compared to the FOA, and vice versa for large velocities.

The relation between $[H]$ and $[L]$ in Eq. (7.34) provides the basis for comparing flow fields of different approximations in glaciers of different size but comparable shape. Starting with the size of the longitudinal section of Storglaciären, Sweden, different glacier sizes are synthesised by multiplying the horizontal scale with a factor Λ and the vertical scale with $\sqrt{\Lambda}$. The hori-

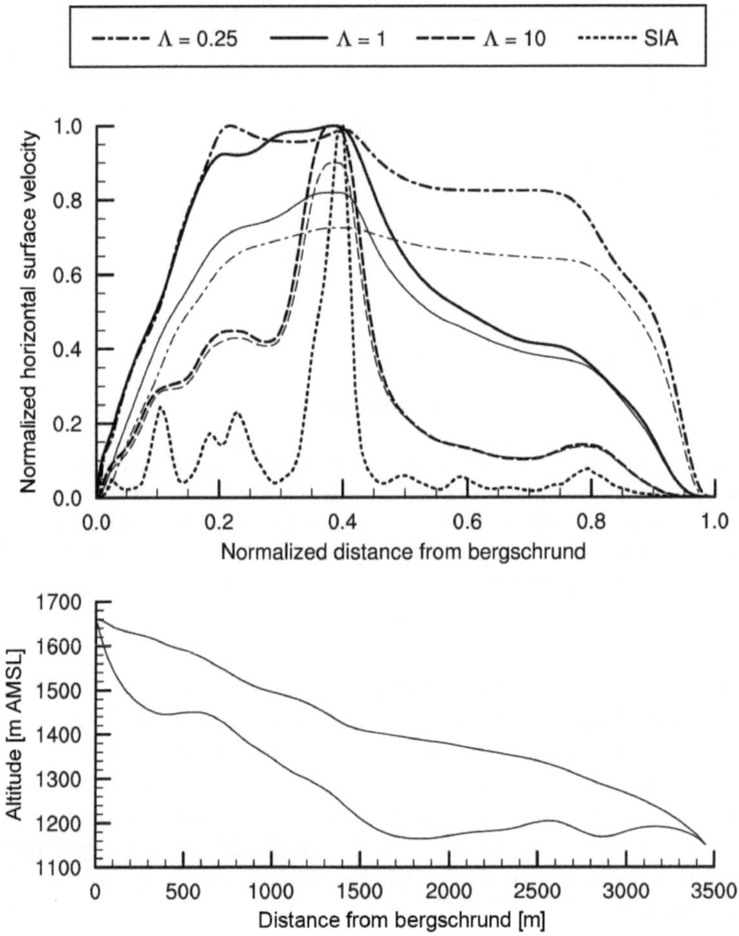

Fig. 7.5. Top panel: Surface velocities for Storglaciären, Sweden, for different scaling factors Λ: dashed lines for $\Lambda = 0.25$, solid lines for $\Lambda = 1$, dash-dotted lines for $\Lambda = 100$. The corresponding upper curves are the solutions of the full Stokes (FS) equations, the lower curves of the first order approximation (FOA). The dotted line shows the shallow ice approximation (SIA), which correspond to $\Lambda = \infty$. The velocity scales of the FS and FOA solutions are normalised to the corresponding maximum velocities of the FS solutions. Bottom panel: Longitudinal section of Storglaciären. (Credit: A. Aschwanden, Swiss Federal Institute of Technology Zurich, Switzerland.)

zontal component of the velocity at the ice surface is computed with the full Stokes equations, the first order approximation and the shallow ice approximation (Fig. 7.5). The results show that a limited area with a steep surface produces locally high velocities compared with shallower regions in large ice masses, whereas the velocity field is strongly smoothed by longitudinal coupling in small ice masses. The differences between the different approximations also become smaller and vanish for very large ice masses.

7.4 First Order Plane Strain Approximation

Small grounded ice masses, such as ice caps, valley glaciers and cirque glaciers may have complex basal and surface topographies. Substantial variations in the ice thickness may occur over horizontal distances comparable to the mean ice thickness, and thus, horizontal coupling of ice flow may be significant. In such cases, the ice flow field is essentially three-dimensional, and a numerical system model to simulate such a glacier must be three-dimensional.

Time scales of climatic variations are of comparable magnitudes as the corresponding time scales of changes in the glacier geometry. Despite this, the flow field of glaciers can be computed as a quasi-stationary field in all situations (see Sect. 5.1.1). However, for cold glaciers, the temperature field is essentially transient. Therefore, initial conditions and time dependent basal and surface boundary conditions for the energy flux or the temperature are required for the time period in consideration. Furthermore, an initial ice surface topography and the variable climatic conditions for the surface mass balance must be prescribed.

There are very few glaciers for which such data is available. Even for well documented glaciers, good coverage of all required information is only available for the past few decades, for which direct monitoring of the conditions on the glacier were carried out. Further back in time, data becomes sparse and indirect reconstruction has to be relied on.

A variety of simplified glacier models have been developed according to the availability of data or the processes to be investigated. The simplifications may concern the physical processes taken into account such as surface evolution, basal sliding and thermomechanical coupling. Other simplifications concern the degree of approximations for the flow and temperature fields from the full Stokes problem to the shallow ice approximation (see Sect. 5). Further simplifications can be introduced by reducing the number of dimensions from a full three-dimensional time dependent model to zero-dimensional steady state considerations.

Many valley glaciers are much longer than their average width, and often they display some degree of symmetry with respect to transverse sections through their ice body. This motivates the application of glacier models in which ice flow, temperature and surface changes are computed along a longitudinal section following a central flow line of the glacier tongue. The basic assumption to this two-dimensional plane strain approximation is that transverse gradients of flow and energy are small or may be considered in a simplified and generalised way (see also Sects. 3.4.3, 5.6.2 and 5.7.2).

For now, let us assume that the temperature field and the geometry (surface and bedrock topography) of the glacier are known for a given time t, and restrict our considerations to the diagnostic computation of the stress and velocity fields. We can therefore note the free surface as $h = h(x)$, the bedrock as $b = b(x)$, and the rate factor as $A = A(T'(x,z))$. In order to obtain the equations for ice flow in the first order plane strain approximation, the method described in Sect. 5.7.2 is applied. With these assumptions, Eqs. (5.62) and (5.69) collapse to the reduced set of equations

$$2\frac{\partial t^{\mathrm{D}}_{xx}}{\partial x} + \frac{\partial t_{xz}}{\partial z} = \rho g \frac{\mathrm{d}h}{\mathrm{d}x} \tag{7.36}$$

and

$$t^{\mathrm{D}}_{xx} = 2\eta\, \frac{\partial v_x}{\partial x}\,, \tag{7.37}$$

$$t_{xz} = \eta\, \frac{\partial v_x}{\partial z}\,. \tag{7.38}$$

Further, we assume the viscous rheology described by the regularised Glen flow law (4.24) with stress exponent $n = 3$.

By inserting Eqs. (7.37) and (7.38) in Eq. (7.36), we obtain the simplified form of the horizontal momentum balance (5.70),

$$\eta\left(4\,\frac{\partial^2 v_x}{\partial x^2} + \frac{\partial^2 v_x}{\partial z^2}\right) + 4\,\frac{\partial \eta}{\partial x}\,\frac{\partial v_x}{\partial x} + \frac{\partial \eta}{\partial z}\,\frac{\partial v_x}{\partial z} = \rho g \frac{\mathrm{d}h}{\mathrm{d}x}\,. \tag{7.39}$$

Equation (4.28) provides a cubic equation for the viscosity η,

$$1 - 2A\,\sigma_0^2\,\eta - 8A\,d_{\mathrm{e}}^2\,\eta^3 = 0\,, \tag{7.40}$$

with a simplified expression for the effective strain rate,

$$d_{\mathrm{e}}^2 = \frac{1}{4}\left(\frac{\partial v_x}{\partial z}\right)^2 + \left(\frac{\partial v_x}{\partial x}\right)^2\,. \tag{7.41}$$

As for the dynamic boundary condition at the free surface (stress-free condition), its z-component is given by Eq. (5.58), and the x-component follows from Eqs. (5.16), (5.17) and (5.23),

$$t_{zz}|_{z=h} = 0\,, \tag{7.42}$$

$$-t_{xx}|_{z=h}\frac{\mathrm{d}h}{\mathrm{d}x} + t_{xz}|_{z=h} = 0\,. \tag{7.43}$$

By splitting the normal stresses into the hydrostatic pressure and the respective deviatoric components, this reads

$$(-p + t^{\mathrm{D}}_{zz})|_{z=h} = 0\,, \tag{7.44}$$

$$-(-p + t^{\mathrm{D}}_{xx})|_{z=h}\frac{\mathrm{d}h}{\mathrm{d}x} + t_{xz}|_{z=h} = 0\,, \tag{7.45}$$

and we can use the former equation to replace the pressure in the latter,

$$-\,(t^{\mathrm{D}}_{xx} - t^{\mathrm{D}}_{zz})|_{z=h}\frac{\mathrm{d}h}{\mathrm{d}x} + t_{xz}|_{z=h} = 0\,. \tag{7.46}$$

Owing to $\operatorname{tr}\mathrm{t}^{\mathrm{D}} = 0$ and $t^{\mathrm{D}}_{yy} = 0$ (plane strain approximation!), $t^{\mathrm{D}}_{xx} + t^{\mathrm{D}}_{zz} = 0$ holds, thus Eq. (7.46) reduces to

$$-\,2t^{\mathrm{D}}_{xx}|_{z=h}\frac{\mathrm{d}h}{\mathrm{d}x} + t_{xz}|_{z=h} = 0\,. \tag{7.47}$$

By inserting the viscous rheology (7.37), (7.38), this can also be expressed in terms of the components of the velocity gradient,

$$\frac{1}{2}\left.\frac{\partial v_x}{\partial z}\right|_{z=h} - 2\left.\frac{\mathrm{d}h}{\mathrm{d}x}\frac{\partial v_x}{\partial x}\right|_{z=h} = 0\,. \tag{7.48}$$

Colinge and Rappaz (1999) demonstrated the well-posedness of the equations for the first order plane strain approximation for non-sliding basal boundary conditions.

7.5 Basal Sliding

7.5.1 General Remarks

In Sect. 5.1.2 the Weertman-type sliding law (5.35) was introduced, which is commonly used in ice sheet models when applying the shallow ice approximation. In the following section, the conditions at glacier beds and the justification of appropriate sliding parameterisations are presented.

It is a well established fact that the glacier base can move. Whether it is true sliding of the ice base over the glacier bed or a movement of the ice base on a deforming subglacial layer of some other material, or a combination of both, depends on local conditions at the glacier bed. Since observations of the glacier bed are difficult and expensive, and usually restricted to a few isolated points, basal motion of glaciers remains difficult to constrain. Furthermore, the strong dependence of basal motion on the local basal hydraulics makes it highly variable, temporally and spatially. Although variations in basal motion are reflected in variations of the ice velocity at the glacier surface, the transmission of basal movement to the surface is non-local and is strongly filtered. This is especially true for variations occurring on spatial scales smaller than the local ice thickness. Therefore, the determination of basal sliding from surface observations is limited to resolutions coarser than local ice thickness.

The sliding speed is essentially defined by the local basal conditions. A major distinction is given by so-called “hard” and “soft” beds.

A *hard bed* is a rigid rock bed on which the ice glides. The sliding of the ice base over the hard surface is sometimes assumed to be friction free due to a thin film of liquid water lubricating the interface perfectly. This type of sliding is thus essentially limited to temperate basal conditions. The basal motion of the ice is then resisted by the roughness of the bed that forces the ice to deform around and across the roughness elements. High pressure in the local basal hydraulic system may lead to partial decoupling of the ice from the bed and to formation of water filled cavities. As a consequence, the effective roughness of the bed, and thus, the resistance to basal motion is reduced.

A *soft bed* is given by a possibly flat and smooth layer of fine grained rocky material such as sand, gravel or till. This layer may deform due to the stress exerted by the ice flow across it. In this case, the thickness and the rheological properties of the layer define the local basal motion of the glacier. High water pressure may soften the material thus making the basal velocity of the ice dependent on the pressure in the local hydraulic system. In the extreme case, part of the layer may become almost liquid and slide almost friction free. In this limit the basal flow of the ice resembles the flow across basal lakes, such as occur beneath the Antarctic Ice Sheet, e.g., Lake Vostok.

7.5.2 Mean Sliding over Rough Hard Beds

Early attempts to parameterise basal motion assumed that sliding over a hard bed is locally friction free due to a thin water layer, but the mean resistance is caused by the flow of ice over some roughness elements at the glacier bed (Hutter 1983, p. 139). In this case, it can be assumed that the typical sliding velocity v_b is a function of the typical amplitude a and the mean wavelength λ of the bed undulation, the basal shear stress τ_b and the viscosity η of the ice. Under these assumptions, a sliding parameterisation must have the form

$$\mathcal{F}(\bar{v}_\mathrm{b}, \bar{\tau}_\mathrm{b}, \eta, a, \lambda) = 0\,, \tag{7.49}$$

where $\bar{v}_\mathrm{b}$ and $\bar{\tau}_\mathrm{b}$ are the mean basal velocity and basal shear stress, averaged over an appropriate area in consideration, η is the viscosity of ice, a and λ are the amplitude and wavelength of the bed undulations (Fig. 7.6).

Equation (7.49) can be written in a form such that it is dimensionally homogeneous. This means that the form of the equation does not change if any of the dimensional units used in the given quantities changes (Hutter and Jöhnk 2004). In this case, the dimensions of the quantities involved can be expressed as combinations of the dimensional quantities of length, L, time, T and mass, M,

physical quantity		dimension,
velocity	$[\bar{v}_\mathrm{b}]$	$\mathrm{L}\cdot\mathrm{T}^{-1}\,,$
stress	$[\bar{\tau}_\mathrm{b}]$	$\mathrm{L}^{-1}\cdot\mathrm{T}^{-2}\cdot\mathrm{M}\,,$
viscosity	$[\eta]$	$\mathrm{L}^{-1}\cdot\mathrm{T}^{-1}\cdot\mathrm{M}\,,$
amplitude	$[a]$	$\mathrm{L}\,,$
wavelength	$[\lambda]$	$\mathrm{L}\,.$

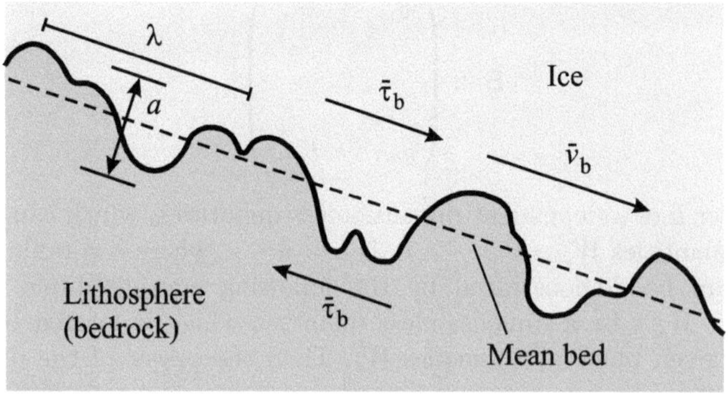

Fig. 7.6. Illustration of the basal roughness: mean amplitude a, mean wavelength λ, mean basal shear traction $\bar{\tau}_b$ and mean sliding velocity $\bar{v}_b$.

A more convenient representation of this information is given by the dimensional matrix of relation (7.49):

	$\bar{v}_b$	$\bar{\tau}_b$	η	a	λ
L	1	−1	1	1	1
T	−1	−2	−1	0	0
M	0	1	1	0	0

(7.50)

The dimensional matrix consists of the exponents of the corresponding dimensional quantities (rows) in the dimension of the corresponding physical quantities (columns).

With the five physical quantities in Eq. (7.49), an infinite number of dimensionless products of their powers can be formed, some of which are $a\lambda^{-1}$, $\bar{\tau}_b a \bar{v}_b^{-1}\eta^{-1}$, $\bar{\tau}_b \lambda \bar{v}_b^{-1}\eta^{-1}$, $\bar{\tau}_b \lambda^2 \bar{v}_b^{-1} a^{-1}\eta^{-1}$. It can easily be verified that for example, the third product can be obtained by dividing the second product by the first. Thus, the four examples given are not independent of each other. The question arises: Given a physical process described by a number of physical quantities involving a number of dimensional quantities, what is the maximum number of dimensionless quantities that are independent?

Let us consider a physical system described by n physical quantities W_i, $i = 1, \cdots n$. The physical quantities have dimensions L_k, $k = 1, \cdots m$, and there are m different dimensional quantities necessary, such as time, distance, mass, temperature or electric charge. Now, B is the dimensional matrix of the set $\{W_i | i = 1, \cdots, n\}$ of physical quantities,

$$\mathsf{B} \equiv \begin{pmatrix} b_{11} & \cdots & b_{1n} \\ \cdot & & \cdot \\ \cdot & & \cdot \\ \cdot & & \cdot \\ b_{m1} & \cdots & b_{mn} \end{pmatrix} . \tag{7.51}$$

The number k of independent dimensionless quantities, which can be formed with the quantities W_i, $i = 1, \cdots, n$, is $k = n - r$, where $r = \text{rank}(\mathsf{B})$.

This can be demonstrated by the following considerations. Let $A = W_1^{x_1} W_2^{x_2} \cdots W_n^{x_n}$ be a dimensionless quantity, which is formed by some or all of the given physical quantities W_i. Then the power of the dimensional unit L_i of A is

$$(L_i^{x_1})^{b_{i1}} (L_i^{x_2})^{b_{i2}} \cdots (L_i^{x_n})^{b_{in}} = (L_i)^0 = 1 , \tag{7.52}$$

which gives us an equation

$$b_{i1}x_1 + b_{i2}x_2 + \cdots + b_{in}x_n = 0 , \; i = 1, \cdots, m \tag{7.53}$$

for the x_i. This set of linear equations, $\mathsf{B} \cdot \mathbf{x} = 0$, has exactly $k = n - r$ linearly independent solutions, where r is the rank of the matrix B (Buckingham 1924).

For the example of Eq. (7.49), a dimensionless product D is given by

$$D = v_\mathrm{b}^{x_1} \tau_\mathrm{b}^{x_2} \eta^{x_3} a^{x_4} \lambda^{x_5} , \tag{7.54}$$

and consequently

$$\begin{aligned} 1 \cdot x_1 - 1 \cdot x_2 - 1 \cdot x_3 + 1 \cdot x_4 + 1 \cdot x_5 &= 0 , \\ -1 \cdot x_1 - 2 \cdot x_2 - 1 \cdot x_3 + 0 \cdot x_4 + 0 \cdot x_5 &= 0 , \\ 0 \cdot x_1 + 1 \cdot x_2 + 1 \cdot x_3 + 0 \cdot x_4 + 0 \cdot x_5 &= 0 . \end{aligned} \tag{7.55}$$

This is a set of three homogeneous linear equations for five unknowns. The number k of independent solutions of this set of equations is the number n of unknowns minus the rank r of the rectangular coefficient (dimensional) matrix, $k = n - r$. In the given case, $r = 3$ and thus, $k = 2$.

With these quantities, a set with a maximum of two independent dimensionless products can be formed, e.g.

$$\tilde{r} \equiv \frac{a}{\lambda} , \quad \tilde{\tau}_\mathrm{b} \equiv \frac{\tau_\mathrm{b}\, a}{v_\mathrm{b}\, \eta} , \tag{7.56}$$

where $\tilde{r}$ is called the bed roughness. With this, the sliding parameterisation (7.49) takes the form

$$\mathcal{F}(\tilde{r}, \tilde{\tau}_\mathrm{b}) = 0 . \tag{7.57}$$

If the mean sliding is determined by the drag inferred by deforming the ice over the roughness elements of the bed, the sliding parameterisation must reflect the stress-strain-rate relation of the ice, which for Glen's flow parameterisation is a power of the effective stress,

$$\eta = \frac{1}{c\,\tau^{n-1}}\,, \tag{7.58}$$

and the sliding parameterisation (7.57) takes the form

$$\mathcal{F}(\tilde{r}, a\,\frac{\tau_{\rm b}^n}{v_{\rm b}}) = 0\,. \tag{7.59}$$

The dimensional analysis alone does not give any information on the form of the functional $\mathcal{F}$. One of the simplest forms of this functional is a product of a function $f = f(\tilde{r})$ of the roughness and in proportion to $\tilde{\tau}_{\rm b}$,

$$v_{\rm b} = f(\tilde{r})\,a\,\tau_{\rm b}^n = C_1\,\tau_{\rm b}^n\,, \tag{7.60}$$

where the coefficient C_1 may depend on position and time. This form of sliding parameterisation has been confirmed by numerical modelling of ice flow over a sinusoidal bed (Gudmundsson 1994, 1997a,b).

The set of quantities chosen in Eq. (7.49) may be incomplete for some situations. The viability of the conclusions from a dimensional analysis depends on the completeness of the chosen set of relevant quantities. In the presented example, the possible influence of the ice thickness, basal hydrostatic pressure or basal hydraulic conditions are not yet considered.

For ice sheet modelling, the Weertman-type sliding parameterisation as given in Eq. (5.35) is commonly applied. The local ice thickness is a necessary quantity for modelling, and thus, basal hydrostatic pressure is known as well. The inclusion of the basal hydrostatic pressure into the sliding parameter,

$$C_1 \equiv C_{\rm b}\,\frac{1}{p_{\rm b}^q}\,, \tag{7.61}$$

may serve to remove some of the spatial variability from the coefficient. For the purpose of tuning an ice sheet model for a given application, it is justified to assume $C_{\rm b}$ to be spatially invariant, rather than to apply a sliding parameterisation in Eq. (7.60) with the coefficient C_1.

It is a well established fact that basal hydraulics plays an important role in the basal motion of glaciers. High basal water pressure reduces the effective pressure of the ice on the bed through buoyancy. If water pressure reaches the overburden pressure of the ice, the glacier locally becomes afloat. This leads to possible uplift of the ice base and formation of cavities. Cavities reduce the effective roughness of the bed which, in turn, leads to enhanced basal motion. These considerations suggest a dependency of basal motion on the effective pressure $p_{\rm eff} = p_{\rm b} - p_{\rm w}$, where p is the hydrostatic pressure of the ice and $p_{\rm w}$ is the water pressure. This allows the formation of one more dimensionless quantity

$$\tilde{p}_{\rm eff} \equiv \frac{p_{\rm eff}}{p}\,, \tag{7.62}$$

on which the functional $\mathcal{F}$ may depend:

$$\mathcal{F}(\tilde{r}, \tilde{p}_{\text{eff}}, \tilde{\tau}_{\text{b}}) = 0\,. \tag{7.63}$$

A proposed version of such a sliding parameterisation is

$$v_{\text{b}} = \frac{a\, f(\tilde{r})}{p_{\text{eff}}}\, \tau_{\text{b}}^{n}\,, \tag{7.64}$$

which for practical applications yields a sliding parameterisation of the form

$$v_{\text{b}} = C_2\, \frac{\tau_{\text{b}}^{n}}{p_{\text{eff}}}\,. \tag{7.65}$$

The coefficient C_2, sometimes called the sliding parameter, can serve as a tuning parameter. However, C_2 may depend on the local conditions at the bed, which in most cases are not well constrained, and it may strongly vary temporally and spatially.

7.5.3 Soft Beds on Sediment Layers

It has been determined by observations that glaciers may rest at least partially on layers of deformable granular material, such as till, sand or gravel. In such cases, not all of the basal motion of the glacier sole may be attributed to true sliding of the ice over the bed. Part of the basal motion then occurs due to the deformation of the sediment layer caused by the shear stress exerted on the layer by the motion of the ice.

The coupling of the motion of the ice to the motion of the deforming sediment depends on many factors that are difficult to observe. The layer may be inhomogeneous in its vertical and horizontal extents. The vertical sub-layering may be due to fractionation of coarse and fine grained material and to water content. The rheological properties of the layer may depend on the distribution of grain size, but may also strongly vary with varying water pressure. Fine grained material is known to liquify at high water pressure. In such cases, a thin layer may perfectly lubricate the interface between the ice and the bed, and the ice sole may glide almost friction free. Such a situation may occur in the fast flowing ice streams of the Siple Coast in Antarctica. The velocities of these streams seems to be limited essentially by the occurrence of distributed bounded areas of high sliding resistance, so-called 'sticky spots'. As a consequence, the local basal movement may be determined not only by the local basal conditions but by the conditions over a wider region.

The idea of a subglacial layer of deformable material can be used to implement basal sliding for the commonly assumed sliding parameterisations (Vieli et al. 2000). Let us assume that the layer is homogeneous in the vertical extent and over a large enough region in the horizontal extent. The thickness of the layer is assumed to be sufficiently small that the shear rate and shear stress are constant over its vertical extent. The base of the layer is kept at zero velocity and the layer undergoes a simple shearing resulting in a velocity

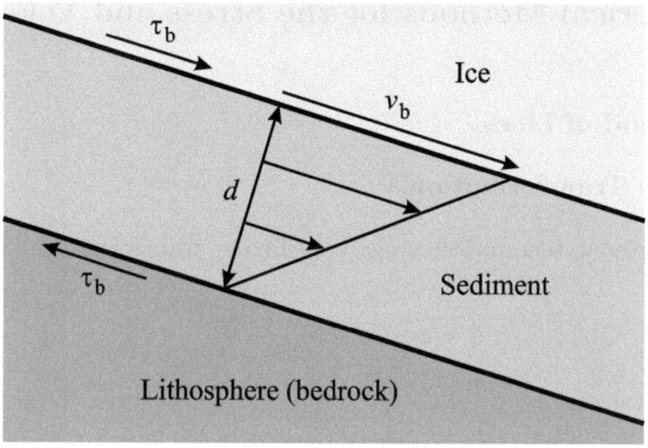

Fig. 7.7. Illustration of the shear flow of a basal layer of deformable sediment.

v_t at its upper surface equal to the velocity v_b of the ice base of the glacier (Fig. 7.7). Applying Eq. (3.135) for incompressible Newtonian fluids to this simple situation yields

$$\frac{\partial v}{\partial z} = \frac{1}{\eta_\mathrm{sed}}\,\tau_\mathrm{b}\,, \tag{7.66}$$

where η_sed is the constant viscosity of the sediment. Integration over the thickness of the layer yields

$$v_\mathrm{b} = \frac{d}{\eta_\mathrm{sed}}\,\tau_\mathrm{b}\,. \tag{7.67}$$

This establishes a parameterisation for the basal velocity of the ice on a deformable layer. If the layer deforms like a Newtonian fluid, the corresponding relation between the basal velocity v_b and the basal shear stress τ_b corresponds to Eq. (5.35) with $p = 1$ and $q = 0$. If the rheology of the layer corresponds to that of ice, such as given in Eq. (4.22), then the relation between basal velocity and stress also becomes a power-law-type relation,

$$v_\mathrm{b} = \frac{d}{B_\mathrm{sed}^n}\,\tau_\mathrm{b}^n \equiv C_\mathrm{sed}\,\tau_\mathrm{b}^n\,. \tag{7.68}$$

This result is the basis for a possible implementation of a sliding parameterisation in a numerical glacier model. The implemented subglacial layer, however, may be a feature only mimicking the true physical meaning. Depending on the type of numerics of the flow model, direct implementation of the sliding with adequate discretisation schemes may be more stable.

Basal motion parameterisations of the forms (7.60) and (7.64) fail for the case of free slip conditions over large areas of the bed, such as floating ice shelves, subglacial lakes or decoupled flat beds by high water pressure and extremely softened bed substrates. In such cases, the basal shear traction vanishes, giving a well defined basal boundary condition.

7.6 Numerical Methods for the Stress and Velocity Fields

7.6.1 Method of Lines

Coordinate Transformation

We now apply a terrain-following coordinate transformation according to Eq. (5.127),

$$\xi = x \,, \quad \zeta = \frac{z - b(x)}{h(x) - b(x)} \,. \tag{7.69}$$

With the notation

$$\tau = t_{xz} - c\, t^{\mathrm{D}}_{xx} \,, \quad \sigma = t^{\mathrm{D}}_{xx} \,, \tag{7.70}$$

where τ is the shear traction parallel to surfaces defined by ζ = const, and the abbreviations

$$c \equiv (1-\zeta)\frac{\mathrm{d}b}{\mathrm{d}x} + \zeta \frac{\mathrm{d}h}{\mathrm{d}x} \,, \tag{7.71}$$

$$a \equiv \frac{\partial c}{\partial \zeta} \,, \tag{7.72}$$

we obtain from Eqs. (7.36), (7.37) and (7.38)

$$\frac{\partial \tau}{\partial \zeta} = -2a\sigma - (h-b)\left(2\frac{\partial \sigma}{\partial \xi} - \rho g \frac{\mathrm{d}h}{\mathrm{d}\xi}\right) \,, \tag{7.73}$$

$$\frac{\partial u}{\partial \zeta} = 2(h-b)\, Af\, (\tau + 2c\sigma) \,, \tag{7.74}$$

$$0 = \frac{\partial u}{\partial \xi} - Af\, (2c\tau + \sigma) \,, \tag{7.75}$$

where f is defined in Eqs. (4.10) and (4.23) for the regularised Glen flow law. This is a set of three equations for the variables u, τ and σ. Solutions to it are sought in the domain bounded by the free and basal surfaces, for which boundary conditions need to be imposed. In the transformed terrain following coordinates, the locations of their imposition are $\zeta = 0$ and $\zeta = 1$.

The boundary condition at the free surface consists of a vanishing shear traction, Eq. (7.47), which in the transformed coordinates reduces to

$$\tau_{\mathrm{s}} = 0 \,. \tag{7.76}$$

For the boundary condition at the ice base, see Sect. 7.5, e.g., Eq. (7.60) or (7.65).

The field equations (7.73) to (7.75) may be regarded not as a system of partial differential equations, but rather as if they were ordinary differential equations to be integrated in the ζ-direction. This will formally be achieved by replacing all derivatives with respect to ξ by finite differences. We have made this interpretation implicitly apparent in (7.73) to (7.75) by writing the terms involving the ζ-derivatives on the left hand side and all other terms on the right hand side. Viewed this way, Eqs. (7.73) to (7.75) are two first order ordinary differential equations (ODEs) in ζ for τ and u and one algebraic equation for σ, of which the ζ-integration requires three boundary or initial conditions as defined above. Because one condition is imposed at $\zeta = 1$ and two are imposed at $\zeta = 0$, i.e., at the two end points of the interval $\zeta \in [0, 1]$, the problem is referred to as a two point boundary value problem, which is a well-posed mathematical problem (Colinge and Rappaz 1999).

Line Integration

The idea of discretizing the partial differential equations in all dimensions except one to obtain a system of ordinary differential equations is called the *method of lines*. By introducing a discrete grid on the ξ-axis and approximating the ξ-derivatives by finite differences, Eqs. (7.73) and (7.74) can be rewritten as ODEs, and Eq. (7.75) becomes algebraic.

A regular rectangular grid is defined on the rectangular domain of the mapped geometry. The grid consists of M and N grid points in the ξ and ζ directions, and grid points are numbered from $i = 1, \ldots, M$ and $j = 1, \ldots, N$, respectively, see Fig. 5.14 in Sect. 5.7.3. We assume that the ice thickness and all velocity and stress components are zero at the end points $i = 1$ and $i = M$. Thus, for $i = 2, \ldots, M-1$, substituting the differentials with respect to ξ by centered finite differences of order two in Eqs. (7.73) to (7.75) yields

$$\begin{aligned} \frac{\partial \tau}{\partial \zeta} &= (h_i - b_i)\, \rho g \frac{h_{i+1} - h_{i+1}}{2\,\Delta\xi} \\ &\quad - 2\,(h_i - b_i)\, \frac{\sigma_{i+1,j} - \sigma_{i-1,j}}{2\,\Delta\xi} - 2a_i\, \sigma_{i,j}\,, \end{aligned} \tag{7.77}$$

$$\frac{\partial u}{\partial \zeta} = (h_i - b_i)\, A f_{i,j}\, (2\,\tau_{i,j} + 4c_{i,j}\, \sigma_{i,j})\,, \tag{7.78}$$

$$0 = -\frac{u_{i+1,j} - u_{i-1,j}}{2\,\Delta\xi} + A f_{i,j}\, (2c_{i,j}\, \tau_{i,j} + \sigma_{i,j})\,. \tag{7.79}$$

For each vertical grid line with given $i = 2, \ldots, M-1$, this establishes a set of two ordinary differential equations of order one for the two unknowns $\tau_{i,j}$, and $u_{i,j}$, and one algebraic equation for $\sigma_{i,j}$. This large set of ODEs can be integrated by using a standard numerical integrator, for example, a forward Euler predictor-corrector scheme or a Runge-Kutta scheme. The integration begins at the base ($\zeta = 0$), with starting values for $\tau_{i,1}$, $\sigma_{i,1}$ and $u_{i,1}$ at each basal grid point. At each step of the numerical integration, the algebraic equation is solved explicitly or with a numerical root finder. It is important that the algebraic equation always has a unique real solution. This is the case for small aspect ratios generally relevant for glaciers and for the flow law (4.10) with a flow law exponent $n = 3$ (Colinge and Rappaz 1999).

To arrive at a proper finite difference scheme several points must be taken into account. The shear stress component τ is computed from Eq. (7.73) and depends on the stress components σ and its derivative with respect to ξ. Furthermore, the stress component σ is computed with Eq. (7.75) and depends on the derivative of the velocity component u with respect to ξ. This cascade of dependence on ξ-derivatives reduces the order of the whole difference scheme to $p-1$, even if all difference schemes for single derivatives are of the order p. Thus, to obtain a consistent difference scheme of a desired order p, the derivatives of the velocity components in Eqs. (7.74) and (7.75) must be discretised to order $p+1$ (Colinge and Blatter 1998).

The integration from $\zeta = 0$ to $\zeta = 1$ of Eqs. (7.73) and (7.74) with the chosen starting values for basal shear stress and velocity components, i.e., three conditions, does not automatically satisfy the surface boundary conditions (7.76). In order to solve the boundary value problem, the proper basal values for $\tau_{i,1}$ can be found iteratively. A good initial choice is the shallow ice approximation of the basal shear stress, which in many cases is already close to the solution. However, if basal conditions have a large variability, then an initial $\tau_{i,1}^0 = 0$ is often the better choice to start with. With the basal shear stress and the values for the basal velocity components, Eq. (7.75) is first used to calculate the basal values for $\sigma_{i,1}$. Integrating upwards from the base yields surface values $\tau_{i,M}^0 \neq 0$. By using this result, a correction to $\tau_{i,1}^0$ must be found to obtain $\tau_{i,M}^1$ closer to the required boundary condition. This method is called *shooting*, or *single shooting*. The type of basal boundary conditions that must be met is defined by the sliding parameterisation. In the proposed numerical scheme, prescribing basal velocity is the simplest possibility. Mixed boundary conditions, vanishing basal velocity at non-sliding locations and basal friction represented by basal shear traction at sliding locations may be useful for specific studies on basal conditions. The application of a sliding law which relates basal shear traction to basal velocity, however, requires a nested iteration of this sliding law within the shooting procedure.

Single Shooting Fixed Point Iteration

A simple way to define a single shooting procedure is to write the problem, Eqs. (7.73) to (7.75), as a fixed point problem. We introduce the following notation, where M is the number of grid points for the method of lines:

$$T_\mathrm{b} = \left(\tau_{\mathrm{b},2}, \cdots, \tau_{\mathrm{b,M}-1}\right)^\mathrm{T}, \tag{7.80}$$

$$T_\mathrm{s} = \left(\tau_{\mathrm{s},2}, \cdots, \tau_{\mathrm{s,M}-1}\right)^\mathrm{T}, \tag{7.81}$$

$$G(T_\mathrm{b}) = T_\mathrm{b} - \alpha\, T_\mathrm{s}(T_\mathrm{b}), \tag{7.82}$$

where $\alpha > 0$ is an underrelaxation factor. A converging fixed point iteration is equivalent to a constructive proof of Banach's fixed point theorem (Debnath and Mikusinski 1993, Colinge and Blatter 1998). Thus, for α sufficiently small, the fixed point T_b^* of G, $G(T_b^*) = T_b^*$, is the correct solution.

In the simplified case of an infinite slab with parallel flat surface and base, a simple criterion

$$\alpha < 4e^{-8/\Delta\xi} \tag{7.83}$$

can be rigorously proven to ensure the existence and uniqueness of the desired fixed point (Colinge and Rappaz 1999). This exact criterion for the slab is approximately valid for the case of realistic glacier geometries, and one can expect to be forced to decrease α either when $\Delta\xi$ decreases or the aspect ratio ε increases (Blatter 1995, Colinge and Blatter 1998).

The criterion is very restrictive and explains the impossibility of applying the method to small and steep glaciers. Moreover, it uses global quantities of ice geometry, and local conditions may require a smaller α to achieve convergence. This is especially true if the surface slope locally displays large longitudinal variations, such as in ice-falls or at steep glacier snouts. The criterion provides information regarding the stability of the numerical process. An unstable process is one which assigns largely different results to pairs of very close starting conditions. Since the idea of single shooting is to use the final results to learn something about the starting conditions, it is essential to avoid instability.

The restrictive criterion (7.83) also indicates the impossibility to apply the line integration in a single shot inverse mode. One may be tempted to start the integration at the surface with zero shear traction and measured velocity components to compute basal shear traction and basal velocity. Although the mathematical problem is well-posed, it is ill-conditioned, and thus basal conditions obtained by this integration are extremely sensitive to the input at the surface, and may be far away from the correct solution.

7.6.2 Global Discretisation Schemes

Transformed Equations

Equations (7.39) and (7.40) give the basis for an iterative solution procedure to solve the non-linear problem (Colinge and Rappaz 1999). For a given guess of the initial velocity field v_0, the algebraic equation (7.40) can be solved for the viscosity η, and with the result, the differential equation (7.39) can be solved for the velocity field v_x. If the starting field lies within the attractive domain for the solution, this fixed point iteration converges towards that solution. Since the solution of this equation is unique, we obtain the correct solution for the glacier flow.

The iteration must be started with some initial guess either for the fluidity or for the velocity field $v_0(x, z)$. One possibility of a velocity field may be the shallow ice approximation. However, this approximation is not always a good approximation and some other simpler initial field may yield a better start.

The system of equations is suitable for both finite element and finite difference discretisations. In this work, a finite difference scheme combined with a coordinate transformation is used, which maps the longitudinal section of a glacier onto a rectangular domain

The elliptic differential equation (7.39) and the algebraic equation (7.40) transform to

$$L_1 \frac{\partial^2 u}{\partial \xi^2} + L_2 \frac{\partial^2 u}{\partial \zeta^2} + L_3 \frac{\partial^2 u}{\partial \xi\, \partial \zeta} + L_4 \frac{\partial u}{\partial \xi} + L_5 \frac{\partial u}{\partial \zeta} = \rho g \frac{\mathrm{d}h}{\mathrm{d}x} \tag{7.84}$$

and

$$1 - 2\,A \sigma_0^2\, \eta - 8\,A \left[\left(\frac{\partial u}{\partial \xi} + \frac{\partial \zeta}{\partial x}\frac{\partial u}{\partial \zeta} \right)^2 + \frac{1}{4} \left(\frac{\partial \zeta}{\partial z}\frac{\partial u}{\partial \zeta} \right)^2 \right] \eta^3 = 0\,, \tag{7.85}$$

and the transformed surface boundary condition (7.48) becomes

$$\left(\frac{1}{2}\frac{\partial \zeta}{\partial z} - 2\frac{\mathrm{d}h}{\mathrm{d}x}\frac{\partial \zeta}{\partial x} \right) \frac{\partial u}{\partial \zeta} - 2\frac{\mathrm{d}h}{\mathrm{d}x}\frac{\partial u}{\partial \xi} = 0\,, \tag{7.86}$$

with

$$\begin{aligned}
L_1 &= 4\eta\,, \\
L_2 &= \eta \left(\frac{\partial \zeta}{\partial z} \right)^2 + 4\eta \left(\frac{\partial \zeta}{\partial x} \right)^2 , \\
L_3 &= 8\eta \frac{\partial \zeta}{\partial x}\,, \\
L_4 &= 4 \frac{\partial \eta}{\partial \xi} + 4 \frac{\partial \eta}{\partial \zeta}\frac{\partial \zeta}{\partial x}\,, \\
L_5 &= \frac{\partial \eta}{\partial x} \left[\left(\frac{\partial \zeta}{\partial z} \right)^2 + 4 \left(\frac{\partial \zeta}{\partial x} \right)^2 \right] \\
&\quad + 4 \frac{\partial \eta}{\partial \xi}\frac{\partial \zeta}{\partial x} + 4\eta \left[\left(\frac{\partial}{\partial \xi}\frac{\partial \zeta}{\partial x} \right) + \frac{\partial \zeta}{\partial z} \left(\frac{\partial}{\partial \zeta}\frac{\partial \zeta}{\partial x} \right) \right] .
\end{aligned} \tag{7.87}$$

At the surface, vanishing shear traction, Eq. (7.48), yields

$$L_\mathrm{h} \frac{\partial u}{\partial \zeta} - 2 \frac{\mathrm{d}h}{\mathrm{d}x}\frac{\partial u}{\partial \xi} = 0\,, \tag{7.88}$$

with

$$L_\mathrm{h} \equiv \left(\frac{1}{2}\frac{\partial \zeta}{\partial z} - 2\frac{\mathrm{d}h}{\mathrm{d}x}\frac{\partial \zeta}{\partial x} \right) . \tag{7.89}$$

At the base, a no slip condition, $u_\mathrm{b} = 0$, a sliding velocity, $u_\mathrm{b} = u_\mathrm{b}(\xi)$, or a basal traction, $\tau_\mathrm{b} = \tau_\mathrm{b}(\xi)$ may be prescribed,

$$L_{\rm b}\frac{\partial u}{\partial \zeta} - 2\,\frac{{\rm d}b}{{\rm d}x}\,\frac{\partial u}{\partial \xi} = \tau_{\rm b}\,, \tag{7.90}$$

with

$$L_{\rm b} \equiv \left(\frac{1}{2}\,\frac{\partial \zeta}{\partial z} - 2\,\frac{{\rm d}b}{{\rm d}x}\,\frac{\partial \zeta}{\partial x}\right), \tag{7.91}$$

or a functional relation between basal sliding and basal shear traction, as described in Sect. 7.5.

The resulting linear system of $n_x \times n_z$ unknowns, where n_x and n_z are the number of grid points in x- and z-directions, can be solved with an adequate linear solver for sparse systems.

Discretised Equations

To solve Eq. (7.84), a second order centered finite difference scheme is used on the same rectangular grid as used for the method of lines in Sect. 7.6.1. For the interior grid points, $i = 3, \ldots, M-2$, $j = 2, \ldots, N-1$, this yields

$$\begin{aligned}
\left(L_1\,\frac{\partial^2 u}{\partial \xi^2}\right)_{i,j} &= L_1|_{i,j}\,\frac{u_{i+1,j} - 2u_{i,j} + u_{i-1,j}}{\Delta\xi^2}\,,\\
\left(L_2\,\frac{\partial^2 u}{\partial \zeta^2}\right)_{i,j} &= L_2|_{i,j}\,\frac{u_{i,j+1} - 2u_{i,j} + u_{i,j-1}}{\Delta\zeta^2}\,,\\
\left(L_3\,\frac{\partial^2 u}{\partial \xi \partial \zeta}\right)_{i,j} &= L_3|_{i,j}\,\frac{u_{i+1,j+1} + u_{i-1,j-1}}{4\Delta\xi\Delta\zeta}\\
&\quad - L_3|_{i,j}\,\frac{u_{i+1,j-1} + u_{i-1,j+1}}{4\Delta\xi\Delta\zeta}\,,\\
\left(L_4\,\frac{\partial u}{\partial \xi}\right)_{i,j} &= L_4|_{i,j}\,\frac{u_{i+1,j} - u_{i-1,j}}{2\Delta\xi}\,,\\
\left(L_5\,\frac{\partial u}{\partial \zeta}\right)_{i,j} &= L_5|_{i,j}\,\frac{u_{i,j+1} - u_{i,j-1}}{2\Delta\zeta}\,,
\end{aligned} \tag{7.92}$$

and by substituting these terms in Eq. (7.84) and rearranging the terms with respect to the velocities at the different grid points we obtain

$$\begin{aligned}
&u_{i-1,j-1}\,\frac{L_3|_{i,j}}{4\Delta\xi\,\Delta\zeta}\\
&+ u_{i-1,j}\left(\frac{L_1|_{i,j}}{\Delta\xi^2} - \frac{L_4|_{i,j}}{2\Delta\xi}\right)\\
&+ u_{i-1,j+1}\,\frac{-L_3|_{i,j}}{4\Delta\xi\,\Delta\zeta}
\end{aligned}$$

$$
\begin{aligned}
&+ u_{i,j-1} \left(\frac{L_2|_{i,j}}{\Delta\zeta^2} - \frac{L_5|_{i,j}}{2\,\Delta\zeta} \right) \\
&+ u_{i,j} \left(-\frac{2\,L_1|_{i,j}}{\Delta\xi^2} - \frac{2\,L_2|_{i,j}}{\Delta\zeta^2} \right) \\
&+ u_{i,j+1} \left(\frac{L_2|_{i,j}}{\Delta\zeta^2} + \frac{L_5|_{i,j}}{2\,\Delta\zeta} \right) \\
&+ u_{i+1,j-1} \frac{-L_3|_{i,j}}{4\Delta\xi\,\Delta\zeta} \\
&+ u_{i+1,j} \left(\frac{L_1|_{i,j}}{\Delta\xi^2} + \frac{L_4|_{i,j}}{2\Delta\xi} \right) \\
&+ u_{i+1,j+1} \frac{L_3|_{i,j}}{4\Delta\xi\,\Delta\zeta} = R_i\,.
\end{aligned}
\tag{7.93}
$$

When $i = 2$ and $i = M - 1$, the same equation holds, however the terms containing the velocities at $i = 1$ and $i = M$ are zero and can be omitted.

To implement the surface boundary conditions, a fictitious external grid point with $j = N + 1$ above the surface was introduced and both the field equation (7.84) and the boundary condition (7.88) were discretised with second order centered differences on the boundary grid points. Subsequently, the values of the requested field on the fictitious grid point in the two equations are eliminated to obtain one finite difference equation for the boundary points. This procedure maintains second order discretisation accuracy and allows us to consider simultaneously both conditions, the field equation (7.84) and the boundary condition (7.88).

Applying second order centered finite differences to the boundary condition (7.88) for the surface grid points $i = 2, \ldots, M - 1$ and $j = N$ yields

$$
\begin{aligned}
u_{i-1,j+1} - u_{i-1,j-1} &= [u_{i,j} - u_{i-2,j}]\, L_{\mathrm{s},i-1}\,, \\
u_{i,j+1} - u_{i,j-1} &= [u_{i+1,j} - u_{i-1,j}]\, L_{\mathrm{s},i}\,, \\
u_{i+1,j+1} - u_{i+,1,j-1} &= [u_{i+2,j} - u_{i,j}]\, L_{\mathrm{s},i+1}\,,
\end{aligned}
\tag{7.94}
$$

and to the field equation (7.84) for $j = N$,

$$
\begin{aligned}
& u_{i-2,j}\, L_{i-1} \frac{L_3|_{i,j}}{4\Delta\xi\,\Delta\zeta} \\
&+ u_{i-1,j} \left[\frac{L_1|_{i,j}}{\Delta\xi^2} - \frac{L_4|_{i,j}}{2\Delta\xi} - L_{\mathrm{h},i} \left(\frac{L_2|_{i,j}}{\Delta\zeta^2} + \frac{L_5|_{i,j}}{2\,\Delta\zeta} \right) \right] \\
&+ u_{i,j-1} \frac{2\,L_2|_{i,j}}{\Delta\zeta^2} \\
&+ u_{i,j} \left[-\frac{2\,L_1|_{i,j}}{\Delta\xi^2} - \frac{2\,L_2|_{i,j}}{\Delta\zeta^2} - \frac{L_3|_{i,j}}{4\Delta\xi\,\Delta\zeta} (L_{\mathrm{s},i-1} + L_{\mathrm{s},i+1}) \right]
\end{aligned}
$$

$$+ \, u_{i+1,j} \left[\frac{L_1|_{i,j}}{\Delta\xi^2} + \frac{L_4|_{i,j}}{2\Delta\xi} + L_{\mathrm{h},i} \left(\frac{L_2|_{i,j}}{\Delta\zeta^2} + \frac{L_5|_{i,j}}{2\,\Delta\zeta} \right) \right]$$

$$+ \, u_{i+2,j} \; L_{i+1} \, \frac{L_3|_{i,j}}{4\Delta\xi\,\Delta\zeta} = R_i \,. \tag{7.95}$$

Note again that for $i = 2$, $i = 3$, $i = M - 1$ and $i = M - 2$ the vanishing velocities at the grid points $1, j$ and M, j occur in Eqs. (7.94) and (7.95).

At the base, $j = 1$, again the sliding velocity, $u_{i,1} = u_{\mathrm{b},i}$, or the basal shear traction, $\tau_{\mathrm{b},i}$, or a sliding relation according to Eq. (7.65) $u_{i,1} = u_{\mathrm{b},i} = f(\tau_{\mathrm{b},i})$ may be given.

This numerical scheme is of first order accuracy in the overall discretisation scheme. This is a consequence of the fact that for the algebraic equation, the available velocity field is of second order accuracy, and a second order discretisation scheme to obtain the derivatives is one order less than the given field. To obtain second order accuracy for the overall scheme, higher order difference schemes have to be used, which of course makes the linear system less sparse and its solution more time consuming.

To solve the large linear system for the unknown velocities $u_{i,j}$, $i = 2, \ldots, M-1$, $j = 1, \ldots, N$, we need to reformulate the equations in the standard form

$$\mathsf{A} \cdot \mathbf{x} = \mathbf{d} \,, \tag{7.96}$$

where A, $\mathbf{x}$ and $\mathbf{d}$ are the coefficient matrix, the n-tuple of the unknowns and the n-tuple of constants in the system of linear equations, respectively. To this end, the elements of the matrix of the unknown velocity field, $u_{i,j}$, must be renumbered to obtain a vector $\mathbf{v}$. For $i = 1, \ldots, M$ and $j = 1 \ldots, N$, the mapping

$$u_{i,j} \to v_k \quad \text{with} \quad k = (i-1)\,N + j \tag{7.97}$$

is applied to Eqs. (7.93) and (7.95) to obtain a system of linear equations,

$$\begin{aligned} & A_{k,k-N-1} \; v_{k-N-1} \\ + \; & A_{k,k-N} \; v_{k-N} \\ + \; & A_{k,k-N+1} \; v_{k-N+1} \\ + \; & A_{k,k-1} \; v_{k-1} \\ + \; & A_{k,k} \; v_k \\ + \; & A_{k,k+1} \; v_{k+1} \\ + \; & A_{k,k+N-1} \; v_{k+N-1} \\ + \; & A_{k,k+N} \; v_{k+N} \\ + \; & A_{k,k+N+1} \; v_{k+N+1} = -1 \,, \end{aligned} \tag{7.98}$$

with the coefficients

$$A_{k,k-N-1} = \frac{L_3|_{i,j}}{4\Delta x\,\Delta z} \,,$$

$$
\begin{aligned}
A_{k,k-N} &= \frac{L_1|_{i,j}}{\Delta x^2} - \frac{L_4|_{i,j}}{2\Delta x}, \\
A_{k,k-N+1} &= \frac{-L_3|_{i,j}}{4\Delta x\,\Delta z}, \\
A_{k,k-1} &= \frac{L_2|_{i,j}}{\Delta z^2} - \frac{L_5|_{i,j}}{2\,\Delta z}, \\
A_{k,k} &= -\frac{2\,L_1|_{i,j}}{\Delta x^2} - \frac{2\,L_2|_{i,j}}{\Delta z^2}, \\
A_{k,k+1} &= \frac{L_2|_{i,j}}{\Delta z^2} + \frac{L_5|_{i,j}}{2\,\Delta z}, \\
A_{k,k+N-1} &= \frac{-L_3|_{i,j}}{4\Delta x\,\Delta z}, \\
A_{k,k+N} &= \frac{L_1|_{i,j}}{\Delta x^2} + \frac{L_4|_{i,j}}{2\Delta x}, \\
A_{k,k+N+1} &= \frac{L_3|_{i,j}}{4\Delta x\,\Delta z},
\end{aligned}
\tag{7.99}
$$

and correspondingly at the base depending on the chosen boundary conditions.

The coefficient matrix A is a $(M+N)\times(M+N)$ matrix in which the lines or columns contain about nine non-vanishing elements. The non-zero elements are arranged in bands parallel to the main diagonal $i=j$.

To start the computation, an initial field, either of the velocity or the viscosity, must be assumed. If a constant viscosity field is assumed, Eq. (7.84) can be solved first for the velocity field, which in turn can be used in Eq. (7.85) to obtain a new viscosity field. Repetition of this process defines a fixed point iteration which converges to the correct solution, if it converges.

Figure 7.8 shows the computed horizontal component for a longitudinal section of Storglaciären, Sweden (Aschwanden and Blatter 2005). The non-linear algebraic equation (7.40) was solved with a Newton-Raphson iteration scheme. For this purpose, the derivatives of the velocity field were obtained with a centered second order difference scheme in the interior of the domain and with one-sided 3-point discretisation for boundary points.

The sliding velocities shown in Fig. 7.9 were chosen as basal boundary conditions. The values of the basal velocities were obtained by trial and error matching the observed surface velocities for the given viscosity of temperate ice. Although Storglaciären is not temperate throughout the entire ice body, the cold surface layer in the ablation zone does not influence the flow field substantially.

7.6.3 Vertical Velocity Component

In the first order approximation, the vertical velocity component does not influence the horizontal velocity component. Thus, once the horizontal com-

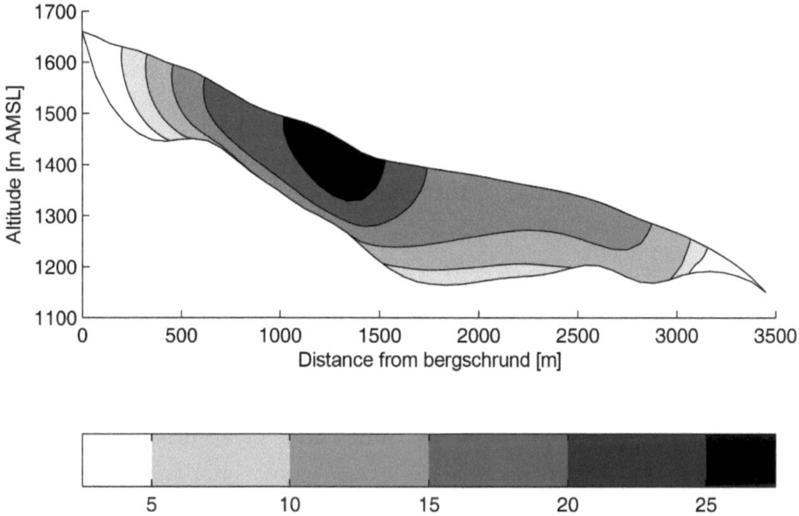

Fig. 7.8. Contour plot of the horizontal velocity component in $\mathrm{m\,a^{-1}}$ in a longitudinal section of Storglaciären, Sweden. The sliding velocities are taken from the scenario used in Fig. 7.9 to fit the observed annual mean surface velocity in 2001-2002. Adapted from Aschwanden and Blatter (2005), © American Geophysical Union.

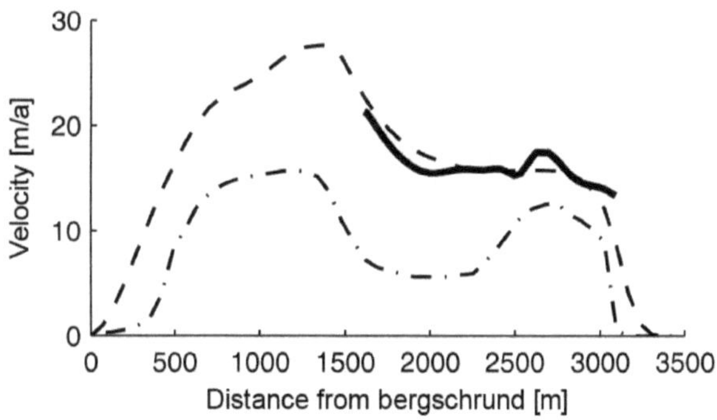

Fig. 7.9. Measured annual mean surface velocities along the central flow line on Storglaciären, Sweden, for 2001-2002 (solid line). The dashed and the dash-dotted lines correspond to the surface and basal velocities computed with the first order approximation, respectively. Adapted from Aschwanden and Blatter (2005), © American Geophysical Union.

ponents are known, the vertical components can be computed by solving the mass continuity equation, which in the two-dimensional form are

$$\frac{\partial v_z}{\partial z} + \frac{\partial v_x}{\partial x} = 0\,. \tag{7.100}$$

Equation (7.100) can be solved by a quadrature,

$$v_z - v_{z,\mathrm{b}} = -\int_b^z \frac{\partial v_x}{\partial x} \mathrm{d}z' \,, \tag{7.101}$$

where $v_{z,\mathrm{b}}$ is the vertical velocity at the base. Applying the coordinate transformation, Eq. (7.69), on Eq. (7.101) yields

$$w - w_\mathrm{b} = -(h - b) \int_0^\zeta \left(\frac{\partial u}{\partial \xi} + \frac{\partial \zeta'}{\partial x} \frac{\partial u}{\partial \zeta'} \right) \mathrm{d}\zeta' \,, \tag{7.102}$$

where w and w_b are the vertical components of the velocity in the transformed coordinates. In general, the integration must be performed numerically. Assuming that the given field of the horizontal velocity is exact, the resulting vertical velocity component is of the same order of accuracy as the chosen method for the quadrature.

7.6.4 Trajectories

Let us consider a given velocity field $\mathbf{v} = \mathbf{v}(\mathbf{x})$, where $\mathbf{x} = \overrightarrow{OP}$ points to a given position P. The starting point of a particle is P_0 with the position vector $\mathbf{x}_0 = \overrightarrow{OP_0}$ at the time t_0. The trajectory of the particle follows the differential equation

$$\frac{\mathrm{d}\mathbf{x}}{\mathrm{d}t} = \mathbf{v} \,. \tag{7.103}$$

A numerical solution of the differential equation (7.103) uses a forward Euler iteration scheme, sometimes called Petterssen iteration (Seibert 1993) with an initial step

$$\mathbf{x}_1 = \mathbf{x}_0 + \Delta t \, \mathbf{v}(\mathbf{x}_0) \tag{7.104}$$

and subsequent iteration steps

$$\mathbf{x}_{i+1} = \mathbf{x}_0 + \frac{\Delta t}{2} \left[\mathbf{v}(\mathbf{x}_0) + \mathbf{v}(\mathbf{x}_i) \right] \,, \tag{7.105}$$

for $i = 1, \ldots, N$. If $N = 1$, this scheme is called a predictor corrector scheme. The Petterssen iteration scheme is a fixed point iteration scheme of second order accuracy, which converges towards a fixed point $\mathbf{x}_\mathrm{f} = \overrightarrow{OP}_\mathrm{f}$,

$$\mathbf{x}_\mathrm{f} = \mathbf{x}_0 + \frac{\Delta t}{2} \left[\mathbf{v}_0 + \mathbf{v}(\mathbf{x}_\mathrm{f}) \right] \,. \tag{7.106}$$

Flow fields of glaciers are generally smooth and trajectories can be computed as accurately as required. Figure 7.10 shows a set of trajectories computed for the velocity field of Fig. 7.8 for Storglaciären. The trajectories were calculated with a time step of one day with a fixed number of five iteration steps. To test the accuracy, trajectories were computed starting at given points in the accumulation area and backward trajectories in the reversed velocity

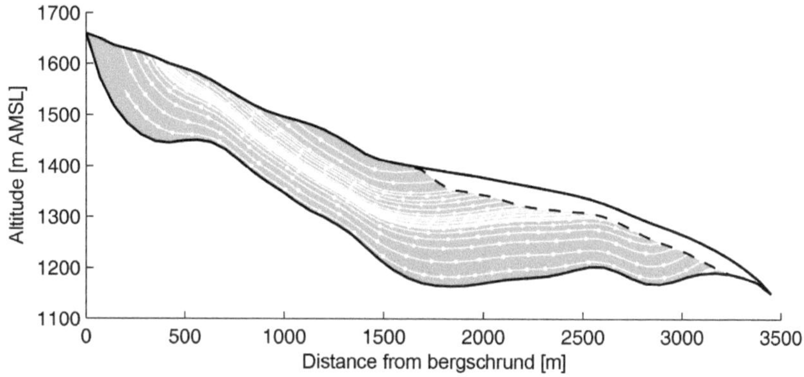

Fig. 7.10. Trajectories of ice particles in Storglaciären, Sweden, computed with the velocity field corresponding to scenario H2, Fig. 7.9. The distances between dots on the trajectories correspond to 10 years travelling time. Adapted from Aschwanden and Blatter (2005), © American Geophysical Union.

field starting at the end points of the previous trajectories. The distances between the starting points of the trajectories and the end points of the backward trajectories were generally smaller than 10^{-5} times the lengths of the trajectories, thus only a few millimetres.

7.6.5 Transverse First Order Flow Profiles

Often, limited information on the geometry or the flow field of a glacier is available. If only part of a glacier bed is mapped, model calculations are only possible for this part, and additional boundary conditions at the upper and lower end of the surveyed domain are required. In two dimensional flowline models, only shallow ice approximations can be inferred for the velocity profile at the upper end of the limited domain. In three dimensional models, an entire transverse profile must be prescribed. One way to impose a boundary condition is based on the assumption that the glacier is a uniform channel of given cross sectional shape at the boundary (Fig. 7.11). The velocity profile in a cross section of such a channel can be computed with either the shallow ice approximation, or better, with a first order approximation. This first order approximation takes into account the horizontal shearing due to differential motion across the channel.

To arrive at the field equation for the flow field in a cross section, the following assumptions are made (Sugiyama et al. 2007):

- The channel is inclined in the x-direction only.
- The channel is homogeneous in the flow (x-) direction, thus $\partial/\partial x = 0$ for all velocity components.
- No lateral component of the velocity, $v_y = 0$.

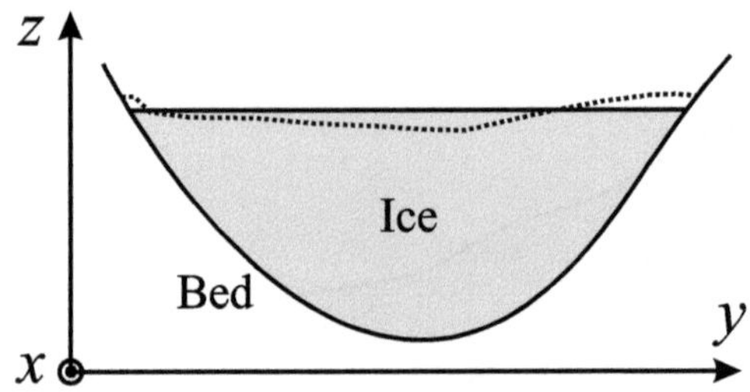

Fig. 7.11. Schematic of a cross section of a glacier. The dotted line indicates the glacier surface, the shaded area is the model domain.

With these assumptions, Eqs. (5.62) and (5.63) simplify to equations with a similar structure as the force balance, Eqs. (7.36), (7.37) and (7.38), for the longitudinal plane strain approximation

$$\frac{\partial t_{xy}}{\partial y} + \frac{\partial t_{xz}}{\partial z} = \rho g \frac{\mathrm{d}h}{\mathrm{d}x} \tag{7.107}$$

and the stress-strain-rate relations

$$t_{xz} = \eta \, \frac{\partial v_x}{\partial z} \, , \tag{7.108}$$

$$t_{xy} = \eta \, \frac{\partial v_x}{\partial y} \, . \tag{7.109}$$

The form of these equations is similar to the corresponding equations for the longitudinal plane strain equations. Thus, the same numerical solution procedure as described in Sect. 7.6.1 can be applied. A second set of equations suitable for the global discretisation scheme given in Sect. 7.6.2 can be obtained by eliminating the stress components in Eqs. (7.107), (7.108) and (7.109), we obtain an equation with the same structure as Eq. (7.39) for the plane flow approximation,

$$\eta \left(\frac{\partial^2 v_x}{\partial y^2} + \frac{\partial^2 v_x}{\partial z^2} \right) + \frac{\partial \eta}{\partial y} \frac{\partial v_x}{\partial y} + \frac{\partial \eta}{\partial z} \frac{\partial v_x}{\partial z} = \rho g \frac{\mathrm{d}h}{\mathrm{d}x} \, , \tag{7.110}$$

and

$$\frac{\partial h}{\partial y} = 0 \, . \tag{7.111}$$

For the viscosity η, Eq. (7.40) also holds in this case,

$$1 - 2A \, \sigma_0^2 \, \eta - 8A \, d_\mathrm{e}^2 \, \eta^3 = 0 \, , \tag{7.112}$$

with a simplified expression for the effective strain rate, Eq. (5.66),

$$d_\mathrm{e}^2 = \frac{1}{4} \left[\left(\frac{\partial v_x}{\partial z} \right)^2 + \left(\frac{\partial v_x}{\partial y} \right)^2 \right] . \tag{7.113}$$

Equations (7.41) hold, and for the boundary conditions at the ice surface

$$\frac{\partial v_x}{\partial z} - \frac{\partial h}{\partial x}\frac{\partial v_x}{\partial y} = 0\,. \tag{7.114}$$

The structure of these equations is basically the same as for the set of equations for the two-dimensional longitudinal plane strain flow field, Eqs. (5.70), (7.40) and (7.41). Numerical solutions for a cross section of Rhone Glacier, Swiss Alps, are shown in Fig. 7.12. In these simulations, the basal sliding velocity has been used as a fitting parameter in order to match modelled with measured surface velocities.

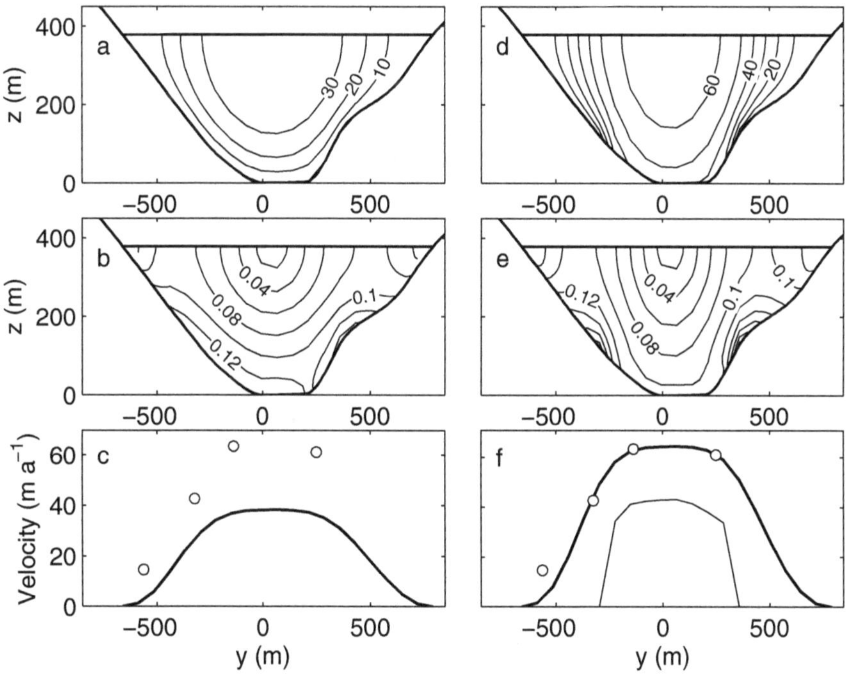

Fig. 7.12. Modelled longitudinal velocity component v_x (panels a and d, in m a^{-1}) and shear stress t_{xz} (panels b and e, in MPa) in a cross section of Rhone Glacier, Swiss Alps, about 3 km downglacier from the top of the glacier at the end of the 19th century. Panels c and f show the cross profiles of the longitudinal velocity components at the surface and at the base, together with measured values of surface velocities (open circles). The left panels show the non-sliding case, the right panels the sliding scenario for which good agreement between modelled and measured surface velocities is achieved. (Credit: S. Sugiyama, Hokkaido University, Sapporo, Japan.)

7.7 Applications and Limitations of Glacier Models

7.7.1 Information on Glaciers

Numerical models of glaciers require input data on the geometry, climatic boundary conditions at the ice surface and basal conditions. It is obvious that information on surface conditions are easier to obtain than on basal conditions. Furthermore, additional information on englacial temperature and/or velocity fields are required for the validation of model results.

The surface topography of glaciers can be determined by terrestrial or aerial photogrammetry to a high accuracy. Some limitation may be caused by low optical contrasts in firn areas. The bed topography can be obtained with radio echo soundings, either from surface based or air borne radar systems. The accuracy is limited by the possible resolution due to the wavelength of the radar beam, uncertainties in the propagation speed of microwaves in ice and firn and the uncertainties in the reflection points of the received signals. Other methods, such as seismic and gravimetric sounding seem to be less accurate. Direct control of sounding depths may be achieved by drilling to the bed. However, contrary to the accurate and highly resolved surface topography, the bed topography can only be sounded with lower accuracy and with lower spatial resolution.

The velocity field at the surface of the ice can be measured by means of repeated surveys of markers with theodolites, global positioning systems (GPS) or by time lapse photogrammetry of identifiable objects on the ice. Englacial movement and strain rates can only be measured at a limited number of sites in boreholes, either by inclinometry for shear strain rates or depth monitoring of markers in the borehole for vertical normal strain.

The same restrictions apply to the measurement of basal sliding velocities. Boreholes and subglacial tunnels or caves may give access to the glacier bed and allow us to measure the local sliding velocities. However, even if some local sliding velocities can be determined, their representativeness is limited by the spatial and temporal variations of the sliding patterns. On the other hand, basal sliding is a necessary boundary condition for reliable modelling of the thermomechanical behaviour of glaciers. Even if the details of sliding and its variations are unavailable, knowledge on the annual mean sliding pattern averaged over regions of a size comparable to the ice thickness is useful for modelling the surface evolution.

The ice thickness is never in equilibrium with the mass balance on the surface of a glacier. The rate of change of the surface topography does not influence the velocity field at any given moment, although the geometry itself determines the velocity. The adjustment of the velocities is instantaneous in terms of the time scales of possible observations. Thus, the velocity field is treated as quasi-stationary, and the acceleration term in the momentum equation can be omitted in all cases. The rate of change of the velocity field is strictly defined by the rate of change of the surface geometry, and it does not correspond to the Euler acceleration term. The sliding pattern, on the other hand, is highly variable in time and space. Although the surface reacts instantaneously to changes in the velocity field, and thus to changes in the sliding, variations of the latter do not contribute significantly to the long term changes in the surface. Due to the high variability in the sliding velocities, the representability of the resulting sliding pattern may be questionable.

To model the evolution of the ice surface topography in a given climate, a reliable distribution of the surface mass balance is required, in addition to a reliable velocity field. Ablation and accumulation rates can be determined in situ, however, it is laborious and expensive to maintain a network of mass balance sites. This is especially true for remote and inaccessible glaciers where a remotely sensed mass balance would be desirable. By repeated mapping of the glacier surface with sufficiently high accuracy, the total mass change over a given period can be determined by means of remote sensing techniques. To obtain the distributed mass balance, the vertical component of the surface velocity needs to be mapped in addition. However, no corresponding sounding method is available at present.

7.7.2 Inverse Problems

Except for tunnels or caves beneath the ice and boreholes to the base, the glacier bed is not accessible to observations. Indirect methods are a possibility to generate information on basal conditions by interpretation of information obtained on the surface of the ice. One common problem of this type of inverse method is the fact that the interpretation of basal conditions is not unique for a given information on the surface. A second problem is frequently that the mathematical problem relating the known surface conditions to the unknown basal conditions is unstable or ill-conditioned, thus, small errors in surface information result in large errors in the basal conditions.

An inverse problem may be formulated in the following way: For a given surface and bed topography and a given set of measured surface velocities, find an optimised basal velocity field consistent with the given rheological properties of the ice. The problem may be extended with the assumption that unknown bed undulations with short wavelengths also contribute to the variabilities in the surface velocity field. A difficulty lies in the fact that some types of bed undulations may produce similar signals at the surface as variation in the sliding pattern may produce. In this section, only the reduced problem stated above is considered in more details.

The simplest way to solve the inverse problem uses a forward approach. If for some other reasons, a sliding distribution can be assumed, this sliding can be used as a basal boundary condition in a numerical model. If the resulting surface velocities deviate from corresponding observed velocities, the assumptions for sliding may be modified to obtain a better match. These modifications may be performed by trial and error, or may be carried out by a systematic procedure, which defines an iterative process.

The system of equations for ice flow, Eqs. (7.77) to (7.79), requires boundary conditions on both sides, on the surface and at the base. Starting the line integration with a first guess for basal velocity or basal stress yields a surface shear traction generally different from the required vanishing shear traction. With the single shooting fixed point iteration, the correct solution may be obtained, provided the iteration converges, see Sect. 7.6.1. This defines a well-posed mathematical problem. If on the other hand, only information on the surface is available, the same set of equations may be used with prescribed vanishing surface shear traction together with given measured surface velocities, and the basal conditions can be found in a one-shot line integration starting at the surface. However, this integration is unstable and the resulting basal velocities may be totally wrong.

The problem becomes more serious the higher the chosen horizontal resolution of the model grid is. For grid sizes Δx smaller than about twice the local ice thickness, the resulting sliding velocity displays large numerically induced variations along the base. The numerical error becomes smaller for larger Δx, which indicates that only mean sliding variations over larger areas can be resolved with inverse methods. The glacier acts as an efficient low pass filter for variations in the basal velocity field, such that at the surface, the information on short wavelength basal variation is mostly diffused away (Bahr et al. 1994, Truffer 2004).

7.7.3 The Shallowness of Glaciers

The above considerations motivate the attempts to determine basal sliding and surface mass balance distributions not only by remote sensing techniques, but also by combinations of them with numerical modelling to complete the necessary information. In principle, both quantities can be obtained with defined procedures. However, the shallowness of glaciers and the ill-condition of the equations for ice flow seriously limits the achievable accuracy (Saito et al. 2006).

According to Eq. (5.47), the volume flux Q_x is

$$Q_x = H\,\bar{v}_x = \int_b^h v_x(z)\,\mathrm{d}z\,. \tag{7.115}$$

For an isothermal parallel sided slab, Q_x can be analytically integrated. With Eq. (7.13) and by setting $b = 0$ and $h = H$,

$$\begin{aligned} Q_x &= \frac{2A\,(\rho g \sin\alpha)^n}{n+1} \int_b^h \left[H^{n+1} - (H-z)^{n+1}\right]\,\mathrm{d}z \\ &= \frac{2A\,(\rho g \sin\alpha)^n}{n+2}\,H^{n+2}\,. \end{aligned} \tag{7.116}$$

Relative errors (perturbations) in the thickness, H, and in the inclination, $\sin\alpha$, are transformed to a relative error $\delta\bar{v}_x$ in the mass flux,

$$\frac{\delta\bar{v}_x}{\bar{v}_x} = (n+2)\,\frac{\delta H}{H} + n\,\frac{\delta(\sin\alpha)}{\sin\alpha}\,. \tag{7.117}$$

Thus a relative error in H results in a $(n+2)-$fold and a relative error in the inclination of the slab results in a $n-$fold relative error of the computed volume flux. The errors may be smaller in the first order approximation, since the velocity vectors are not defined solely by the local ice thickness and surface inclination, but also by their values within a domain up- and downglacier of about five to ten times the ice thickness.

The method generally used to compute the vertical component of the ice velocity at the surface is derived by integrating the continuity equation of mass, in plane flow,

$$v_z\big|_{z=h} = -\frac{\partial}{\partial x}\int_b^h v_x\,\mathrm{d}z' \equiv -\frac{\partial(H\bar{v}_x)}{\partial x}\,. \tag{7.118}$$

For a given horizontal component of the velocity field, v_x, and the surface geometry, h, the accuracy of the integration scheme is second order in the discrete element Δz, if it is assumed that the numerical quadrature is performed with a second order scheme. In a numerical scheme using finite differences, this yields

$$v_z\big|_{z=h} = -\frac{\partial(H\bar{v}_x)}{\partial x} \approx -\frac{\Delta H}{\Delta x}\,\bar{v}_x - H\,\frac{\Delta\bar{v}_x}{\Delta x}\,. \tag{7.119}$$

This computation turns out to be very sensitive to perturbations or errors in u as well as $\Delta H/\Delta x$. The term $\Delta H/\Delta x$ is influenced by numerical error due to the marginal slope singularities. If we assume that the error δv_z only stems from errors $\delta(\Delta\bar{u})$ in the difference $\Delta\bar{v}_x$ of horizontal components of the velocity, v_x, then

$$\delta v_z\big|_{z=h} = \frac{\partial v_z}{\partial(\Delta\bar{v}_x)}\Big|_{z=h}\,\delta(\Delta\bar{v}_x) = \frac{H}{\Delta x}\,\delta(\Delta\bar{v}_x)\,. \tag{7.120}$$

To estimate the magnitude of δv_z, we introduce the aspect ratio $\epsilon = [H]/[L] = [W]/[U]$. The length of the glacier is discretised with N grid cells of length $\Delta x = [L]/N$, and thus from Eq. (7.120) we get

$$\frac{\delta v_z}{[W]}\Big|_{z=h} \propto N\,\frac{\delta(\Delta\bar{v}_x)}{[U]}\,. \tag{7.121}$$

Since $\Delta\bar{v}_x$ can be relatively small, the accuracy of this difference may suffer from digit elimination, especially if $\Delta\bar{v}_x$ is of the same order of magnitude as the individual errors $\delta\bar{v}_x$ of the mean velocities $\bar{v}_x$. In other words, we need to differentiate v_x numerically in order to compute v_z, and numerical differentiation is not well-posed in the limit of taking small divided differences.

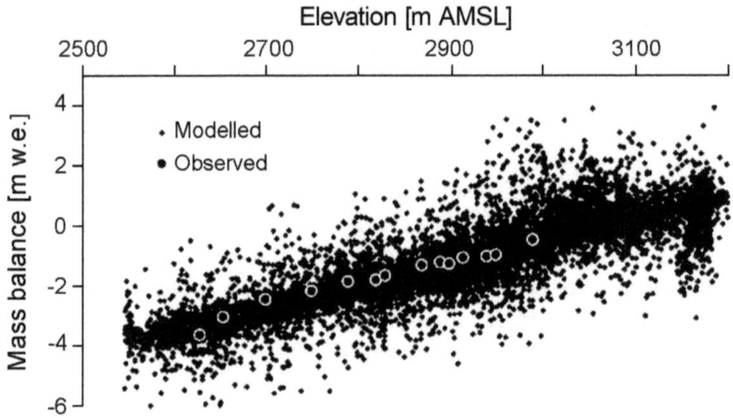

Fig. 7.13. Scatter plot of mass balance values versus altitude for Haut Glacier d'Arolla, Switzerland. The mass balance values were determined by a combination of remote sensing and modelling. Adapted from Hubbard et al. (2000), © International Glaciological Society.

This result is rather discouraging. Higher accuracy cannot be achieved by using higher resolution Δx. The best accuracy lies somewhere between too high a resolution, where digit elimination becomes damaging, and too low a resolution, where the discretisation error becomes too large. In any case, the relative error in the computed vertical velocity component is at least an order of magnitude larger than the relative error in the computed or modelled horizontal velocity component.

This result seriously limits the possibility of determining the distributed mass balance with remote sensing and modelling. In principle, the surface changes can be measured with repeated altimetry from airplanes or satellites. The horizontal component of the velocity field can be obtained with numerical modelling, provided that the bed topography is known, and it can be validated for the surface by measurements using time lapse photogrammetry. Based on this information, the vertical component of the velocity field at the surface could in principle be calculated. However, here the above described error amplification reduces the achievable accuracy for the local values of vertical velocities, and thus for the local mass balance determination. The overall pattern of the mass balance distribution may still be reasonably well reproduced. Figure 7.13 shows a scatter plot of mass balance values determined for all model grid points on the surface of Haut Glacier d'Arolla, Switzerland, together with a few points with measured mass balances. Although the scatter is quite large, the measured points lie well centered in the general pattern of the cloud of points.

7.7.4 Discontinuities

The dynamics of glaciers and ice sheets is not only controlled by continuous processes, such as creep flow and heat flux, but may be dominated by discontinuities, such as sliding and calving. Here, we consider discontinuities from a macroscopic point of view. On the much smaller scales of molecules, creep flow and fracturing lie much closer together, and the transition from continuous to discontinuous becomes very gradual, if not irrelevant.

We define a discontinuity in a field variable by the magnitude of its numerical change compared with the spatial or temporal scales over which it occurs. Sometimes, a discontinuous surface may be accompanied by additional processes that may occur on one side but not on the other side, or only along the surface. In the following, we give a list, possibly incomplete, of discontinuities that occur in glaciers and ice sheets, some of them are already described in more detail in earlier sections of this book.

- *Ice surface*: The ice surface is treated as a non-material surface where the density jumps. "Non-material" indicates that material flows through the surface, in contrast to material surfaces, where invariant individual material particles form the surface.
- *Ice base*: The ice base can be considered as material in the case of a cold base, and as non-material in the case of a temperate ice base with melt. Across the base, fields of mechanical qualities of the materials jump. Of the prognostic fields, velocity may jump, stress and temperature fields are continuous, however, stress gradients and heat flux may be discontinuous. The sliding problem is treated in Sect. 7.5, however, melt is a process which may occur only on one side of the base or at the very base alone.
- *Cold-temperate transition surface*: The cold-temperate transition surface (CTS) is characterised by the change between dry cold and wet temperate ice (see Sects. 7.1 and 9.3.5). Velocity, stress and temperature fields are continuous, but heat flux may be discontinuous. For heat flux, two different situations must be distinguished, (i) ice flows from the cold to the temperate part, (ii) ice flows from the temperate to the cold part. In the first case, no local heat source can produce a jump in the heat flux, and all fields, including moisture content, are continuous. In the second case, the liquid water contained in the temperate ice freezes at the CTS, thus moisture content and heat flux jump.
- *Englacial debris layers*: Debris layers are observed in several glaciers. However, it is not clear if these layers lead to gliding and thrusting in the ice. The debris-ice mixture exhibits different mechanical and thermal properties than pure ice, and on scales of the ice grains and the pieces of debris, mechanical fields may be discontinuous.
- *Hydraulic system*: Water storage and flow occurs on a wide range of scales on the surface, within and at the base of the ice. Subglacial water strongly influences sliding, thus plays a crucial role in the overall dynamics of the

dynamics of the ice masses, together with the capacity of the englacial conduit system and the basal drainage system. Although the hydraulic system works on spatial width scales on the order of only decimetres to metres, it connects through the entire glacier.

- *Fracturing and calving*: Fracturing occurs locally at very short time scales and produces local variation on short spacial scales in the mechanical fields. If fracturing leads to mass loss by calving over cliffs or into water, it contributes to the dynamics of the glacier through mechanisms, which may be independent of or only weekly coupled to climatic forcings. Calving involves many different processes, such as frontal melt, buoyancy, crevassing, but may be different for grounded and floating ice tongues.

8

Glacial Isostasy

8.1 Background

The ice sheets on Earth have undergone very large changes over the glacial-interglacial cycles in the past. Today, ice sheets of significant size occur only in Antarctica and Greenland, whereas during the Last Glacial Maximum (LGM), 21,000 years ago, extended ice sheets also covered large parts of North America, northern Europe, etc. (see Chapter 1). These ice sheets, with typical thicknesses of several kilometres, impose therefore large, time-dependent loads on the crust of the Earth, to which the body of the Earth as a visco-elastic, multi-layer system reacts with a delayed, essentially vertical displacement.

At present, this effect is most conspicuous in Scandinavia and North America, where a land uplift of the order of millimetres per year can be detected as a consequence of the unloading from the glacial ice sheets. Figure 8.1 displays the measured postglacial Fennoscandian land uplift, and it becomes evident that the pattern reflects the topography of the Fennoscandian glaciation at the LGM, with maximum values of more than 8 $\mathrm{mm\,a^{-1}}$ in the northern Gulf of Bothnia.

The first known documentation of postglacial land uplift in Scandinavia dates back to the year 1491, when the inhabitants of the Swedish town of Östhammar located at the southern Gulf of Bothnia reported that their town could no longer be reached by fishing-boats due to a growth of the land at the sea [for the historical outline given here cf. Ekman (1991), and references therein]. First attempts to explain this phenomenon from the 18th century blamed a decrease of the sea level for the water retreat. So in 1706 Hiärne proposed that the water of the Baltic Sea runs off into the deeper Atlantic, and in 1719 Swedenborg claimed that a gradual deceleration of the rotation of the Earth leads to a general lowering of the sea level in higher latitudes. In 1743, Celsius calculated the water decrease by reconstructing the sea level for more than 100 years for a rock at the coast of the small island Iggön, and published the value 13 $\mathrm{mm\,a^{-1}}$, which is, according to modern knowledge, about 60% too large, yet of the right order of magnitude. The remarkable

R. Greve, H. Blatter, *Dynamics of Ice Sheets and Glaciers*, Advances in Geophysical and Environmental Mechanics and Mathematics,
DOI 10.1007/978-3-642-03415-2_8,

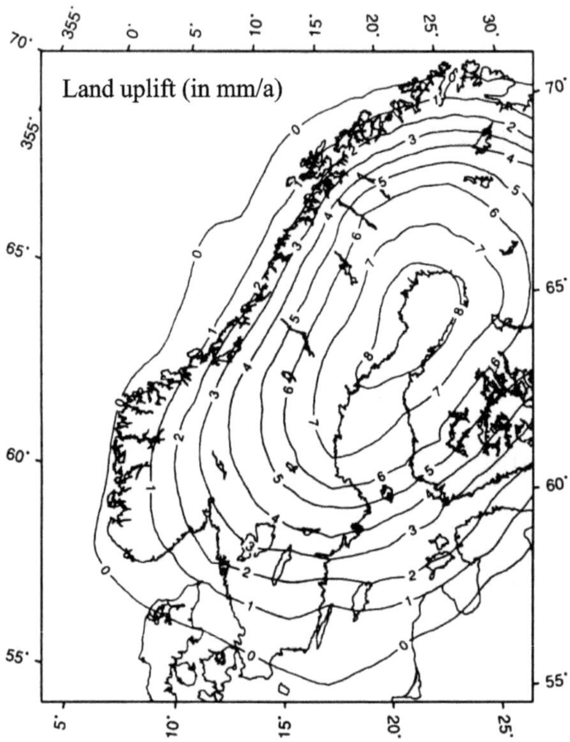

Fig. 8.1. Measured present land uplift in Scandinavia. Adapted from Thoma and Wolf (1999), © M. Thoma and D. Wolf.

explanation Celsius had in mind is that the water runs off through a hole at the sea bottom, just like water in a bath-tub.

An essential prerequisite for understanding the sea-water retreat in the Gulf of Bothnia, as a consequence of land uplift, was provided in the year 1837, when Agassiz discovered that the Earth had experienced ice ages in the past, when large parts of the polar and temperate zones of the northern hemisphere were glaciated. It took another three decades before Jamieson, in 1865, formulated the idea of glacial isostasy, that is, that the crust of the Earth rises due to the unloading from the glacial ice coverage. This idea was finally confirmed by De Geer in 1888 and 1890, who mapped the land uplift in Scandinavia and eastern North America and concluded that only Jamieson's explanation agrees with the spatial uplift pattern.

In 1935, Haskell determined the viscosity of the mantle of the Earth by inverting land uplift data. He found the value $\eta_\mathrm{m} = 10^{21}$ Pa s, which is still a reference for this important quantity. More recently, sophisticated self-gravitating, spherical, visco-elastic multi-layer (SGVE) models of the Earth have been developed which are able to describe the process of glacial isostasy in great detail, made possible only by modern computer performance.

8.2 Structure of the Earth

From seismic studies it can be inferred that the interior of the Earth reveals a layered structure, the different layers being mutually separated by changes in the chemical composition and/or phase transitions (Fig. 8.2, left). The uppermost layer, the crust, has a mean thickness of 20 km (which, however, varies significantly from place to place; it is thickest below mountains and thinnest below oceans), a mineral composition and a mean density of 2900 kg m^{-3}. The crust is underlain by a mineral mantle of 2900 km in thickness and a density varying from 3300 kg m^{-3} at the top to 5700 kg m^{-3} at the bottom. Below, the 3500 km thick, metallic core is found, which can be subdivided into an outer, liquid, and an inner, solid part. The density of the core increases from approximately 9400 kg m^{-3} at the top to 13500 kg m^{-3} at the center of the Earth.

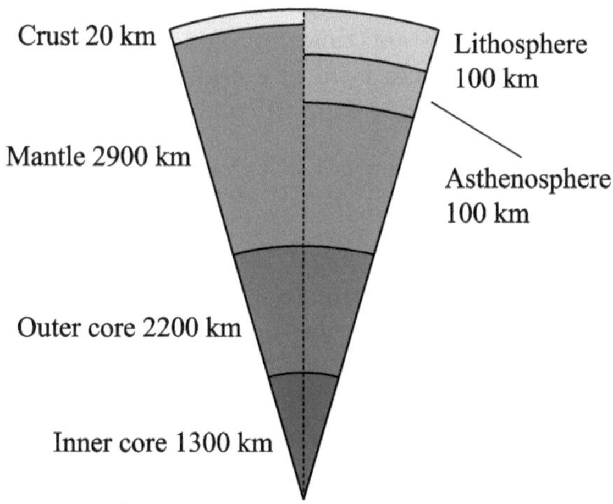

Fig. 8.2. Layered structure of the Earth's interior. Left: Classification by chemical composition. Right: Classification by rheological properties.

However, from the point of view of rheological properties, this classification is not satisfactory for the upper regions. The crust and the uppermost part of the mantle behave mainly as an elastic solid; however, with increasing depth, viscous fluid properties become more dominant over geological periods due to the increasing temperature of the mantle material. Further down, the viscosity increases again due to the changing composition. Because of this, the crust and the uppermost part of the mantle comprise the $\sim$ 100 km thick elastic lithosphere, underlain by a further $\sim$ 100 km thick layer of low viscosity, known as the asthenosphere (Fig. 8.2, right). By contrast, the remaining part of the mantle down to the outer core is characterised by a less pronounced, but nonetheless relevant, fluid behaviour.

Within this rheological classification, the well-known process of plate tectonics can be interpreted as the motion of solid lithosphere plates on the viscous asthenosphere layer, driven by internal convection currents. As we will see below, for the dynamics of glacial isostasy the rheological classification is also favourable and usually applied.

8.3 Simple Isostasy Models

In this section, simple models for the problem of glacial isostasy, based on a plane, two-layer, lithosphere/asthenosphere system, will be presented. Hereby, the lithosphere layer is treated in two different ways, either as a locally deforming plate (local lithosphere, "LL"), or as a thin elastic plate (elastic lithosphere, "EL"). For the asthenosphere, two different approaches are also employed, namely the relaxing asthenosphere ("RA") where the viscous effects are parameterised by a constant time lag, and the diffusive asthenosphere ("DA") which consists of a thin layer of a horizontally flowing viscous fluid. The possible combinations lead to four different models, referred to as LLRA, ELRA, LLDA and ELDA.

8.3.1 LLRA Model

The most simple isostasy model combines the local lithosphere with the relaxing asthenosphere. The idea of the local lithosphere is that an ice load $q(x,y) = \rho g H(x,y)$, at a given position $(x,\ y)$, causes a steady-state displacement of the lithosphere, w_{ss}, in the vertical, z, direction at the position $(x,\ y)$ only (Fig. 8.3). Note that, in this chapter, we take z as positive *downward*, and identify the undisturbed position of the lithosphere base (in the absence of any ice loading) with $z = 0$. The value of w_{ss} (taken positive downward as well) is determined by the balance between the ice load and the buoyancy force which the lithosphere experiences in the underlying viscous asthenosphere,

$$\rho_{\mathrm{a}} g w_{\mathrm{ss}} = \rho g H \quad \Rightarrow \quad w_{\mathrm{ss}} = \frac{\rho}{\rho_{\mathrm{a}}} H\,, \tag{8.1}$$

where $\rho_{\mathrm{a}} = 3300\,\mathrm{kg\,m^{-3}}$ is the density of the asthenosphere.

Due to the viscous properties of the asthenosphere, for the case of time-dependent ice loads $q(x,y,t)$, the lithosphere cannot assume the steady-state displacement w_{ss} immediately. Instead, the response will have a time lag, and the relaxing asthenosphere parameterises this lag by a single time constant,

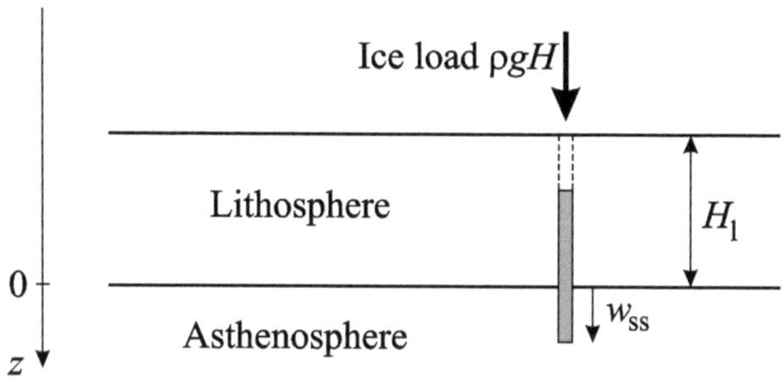

Fig. 8.3. Local lithosphere: A localised ice load $q = \rho g H$ causes a vertical steady-state displacement of the lithosphere, w_{ss}, at the position of the load only. H_l denotes the lithosphere thickness.

τ_a, of the order of 1000s of years (a widely used value is $\tau_a = 3\,\text{ka}$). The evolution of the actual displacement, w, with time, t, is then determined by

$$\frac{\partial w}{\partial t} = -\frac{1}{\tau_a}(w - w_{ss}), \tag{8.2}$$

which relates the vertical velocity of the lithosphere, $\partial w/\partial t$, linearly to the deviation of the displacement from the steady state, $w - w_{ss}$. Provided the ice thickness H, and therefore the steady-state displacement w_{ss}, do not change with time, the solution of (8.2) is

$$w(x, y, t) = w_{ss}(x, y) + [w_0(x, y) - w_{ss}(x, y)]\, e^{-t/\tau_a}, \tag{8.3}$$

so that any non-steady-state initial displacement w_0 relaxes exponentially into its steady-state value with the time constant τ_a (Fig. 8.4).

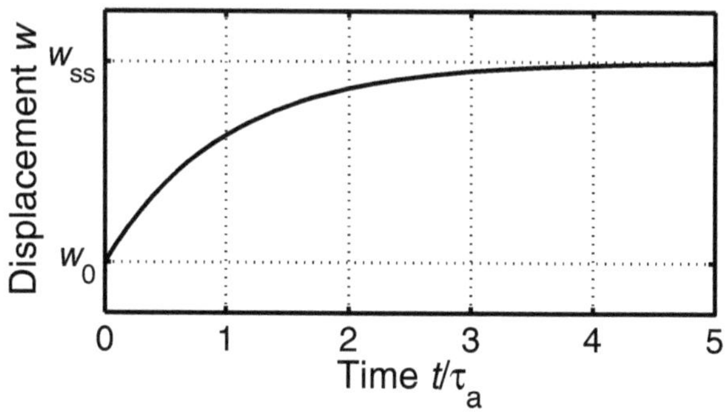

Fig. 8.4. Vertical displacement of the lithosphere as a function of time for the relaxing asthenosphere, with an initial displacement w_0 and a steady-state displacement w_{ss} (assumed to be constant over time). Time is shown in units of the time lag τ_a.

The great strength of the LLRA model is its simplicity (only a single parameter, the time-lag constant τ_{a}, is needed) and easy implementation in numerical models. It feeds back to the ice sheet equations of Chapter 5 as follows. Let $z_{\mathrm{l},0}(x,y)$ be the reference position of the lithosphere surface in the coordinate system of Chapter 5 (vertical coordinate positive upward, see Fig. 5.1) for $w=0$ (no displacement). The actual position of the lithosphere surface is then

$$z_{\mathrm{l}}(x,y,t) = z_{\mathrm{l},0}(x,y) - w(x,y,t)\,, \tag{8.4}$$

where the displacement w is governed by Eq. (8.3). Since the ice base b and the lithosphere surface z_{l} fall together under a grounded ice sheet, the time derivative reads

$$\frac{\partial z_{\mathrm{l}}}{\partial t} = \frac{\partial b}{\partial t} = -\frac{\partial w}{\partial t}\,. \tag{8.5}$$

This result enters the ice surface equation, for instance in its shallow ice form (5.100).

8.3.2 ELRA Model

The local lithosphere assumption, that the response to an ice load is a purely local displacement of the lithosphere, is certainly unrealistic. Instead, it is clear that the elasticity of the lithosphere layer must result in a non-local response, so that positions at some distance from an imposed ice load still suffer a displacement (Fig. 8.5).

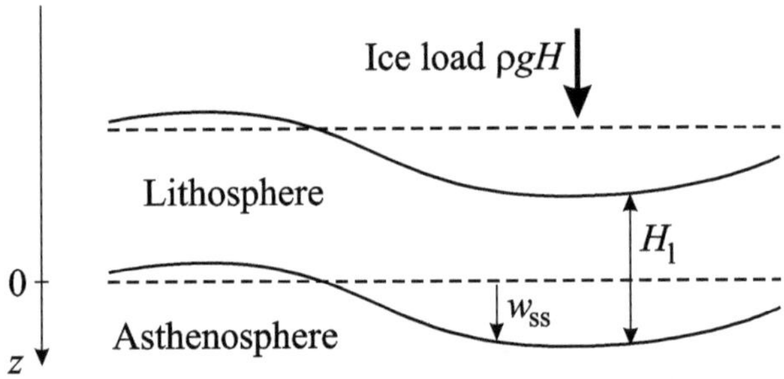

Fig. 8.5. Elastic lithosphere: A localised ice load $q=\rho g H$ causes a horizontally distributed (non-local) vertical steady-state displacement of the lithosphere, w_{ss}. H_{l} denotes the lithosphere thickness.

A rather easy way to describe this behaviour more adequately is the thin elastic plate, governed by the biharmonic equation

$$K_{\mathrm{l}}\,\Delta^2 w_{\mathrm{ss}} + \rho_{\mathrm{a}} g w_{\mathrm{ss}} = \rho g H \tag{8.6}$$

[see Sect. 3.4.2, Eq. (3.122)]. The parameter K_{l} is the flexural stiffness of the lithosphere,

$$K_\mathrm{l} = \frac{E_\mathrm{l} H_\mathrm{l}^3}{12(1-\nu_\mathrm{l}^2)} = \frac{\mu_\mathrm{l} H_\mathrm{l}^3}{6(1-\nu_\mathrm{l})}\,, \tag{8.7}$$

where H_l is the thickness, E_l Young's modulus, ν_l Poisson's ratio and μ_l the shear modulus of the lithosphere (indicated by the subscript "l").

Compared to the local lithosphere equation (8.1), the thin plate equation (8.6) contains an additional term of fourth order in the spatial derivative of the displacement which represents the flexural rigidity. Note that this term depends, apart from the elastic coefficients, on the lithosphere thickness H_l, whereas the simpler equation (8.1) is independent of H_l. The Green's function of (8.6) is known and can be given analytically. Assume that the ice load is localised according to

$$q(x,y) = \rho g H(x,y) = F_0\,\delta(x-\breve{x})\,\delta(y-\breve{y})\,, \tag{8.8}$$

which describes a single force F_0 imposed on the lithosphere surface at the position $(\breve{x},\ \breve{y})$ [$\delta(\cdot)$ denotes Dirac's δ function], then the solution of (8.6) is

$$w_\mathrm{ss}(r) = -\frac{F_0 L_\mathrm{r}^2}{2\pi K_\mathrm{l}}\,\mathrm{kei}\left(\frac{r}{L_\mathrm{r}}\right)\,, \tag{8.9}$$

with

$$r = \sqrt{(x-\breve{x})^2 + (y-\breve{y})^2}\,, \quad L_\mathrm{r} = \left(\frac{K_\mathrm{l}}{\rho_\mathrm{a} g}\right)^{1/4} \tag{8.10}$$

(Brotchie and Silvester 1969). Here, kei(·) is a Kelvin function of zero order, which can be derived from the general Bessel function, and whose values are tabulated in mathematical handbooks [e.g., Abramowitz and Stegun (1970)]. It is plotted in Fig. 8.6. Evidently, the radius of relative stiffness, L_r, determines the non-locality of the lithosphere displacement. The Kelvin function takes the value zero at approximately $r = 4L_\mathrm{r}$, followed by a slight forebulge (upward displacement) further away from the load.

The Green's function G of the thin plate equation (8.6) is given by Eq. (8.9) with the normalised force $F_0 = 1$,

$$G(r) = G(x,\breve{x},y,\breve{y}) = -\frac{L_\mathrm{r}^2}{2\pi K_\mathrm{l}}\,\mathrm{kei}\left(\frac{r}{L_\mathrm{r}}\right)\,. \tag{8.11}$$

Since the differential equation (8.6) is linear, the Green's function can be used to construct its solution for a general distribution of the ice load $q(x,y) = \rho g H(x,y)$ by superposition,

$$w_\mathrm{ss}(x,y) = \int_{A_\mathrm{ice}} \rho g H(x,y)\, G(x,\breve{x},y,\breve{y})\,\mathrm{d}\breve{x}\,\mathrm{d}\breve{y}\,, \tag{8.12}$$

where A_ice denotes the ice-covered area. Note that the Green's function method is not restricted to the thin plate equation; however, it can be used

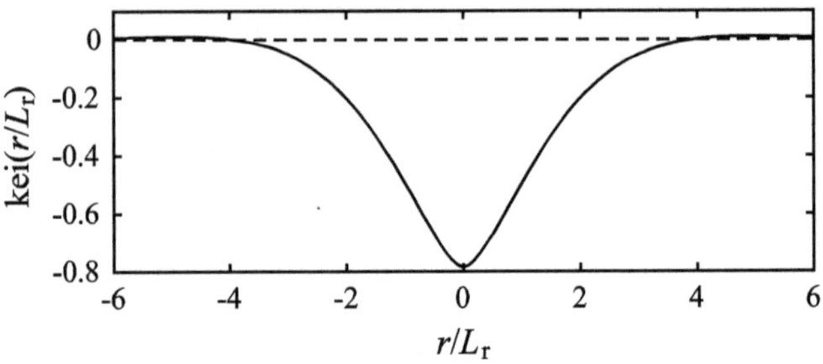

Fig. 8.6. Normalised displacement of the elastic lithosphere under a point load. r/L_r is the normalised distance from the load, $\mathrm{kei}(\cdot)$ a zero-order Kelvin function.

to solve linear, inhomogeneous differential equations in general [see e.g. Bronshtein et al. (2004)].

The asthenosphere component in the ELRA model is the same as in the LLRA model described above. So once the steady-state displacement w_ss is computed by solving the thin plate equation (8.6), the actual displacement w follows from the relaxing asthenosphere evolution equation (8.2) as before. Also, the feedback to the ice sheet equations of Chapter 5 is as described by Eqs. (8.4) and (8.5).

For terrestrial conditions, a typical value for the radius of relative stiffness L_r is $\sim$ 100 km. It is much smaller than the typical horizontal extent $[L] = 1000$ km of a large ice sheet like Antarctica or Greenland [see Eq. (5.5)], and consequently the effect of a non-local lithosphere displacement is not too pronounced. For that reason, the simpler LLRA model still provides reasonable results. The differences between the two models will be most significant in regions with large ice thickness gradients, which occur, in particular, close to the ice margins. By contrast, the extent of ice caps like Vatnajökull or Austfonna is similar to L_r, so that the non-locality of the elastic lithosphere approach is essential and must be accounted for. Glaciers are typically much smaller compared to L_r, which leads to an extreme spreading of the isostatic displacement and renders it negligible. Thus isostasy need not be considered for glaciers.

8.3.3 LLDA Model

The LLDA model employs, for the lithosphere component, the local lithosphere description, which uses the simple force balance (8.1) in order to obtain the steady-state lithosphere displacement, w_ss.

However, for the asthenosphere, the simple, parametric treatment with the relaxing asthenosphere equation (8.2) is now replaced by a more physical approach. To this end, the asthenosphere material is described as an incompressible, linear viscous (Newtonian) fluid of viscosity η_a which flows horizontally in a thin channel of constant asthenosphere thickness H_a (Fig. 8.7). As

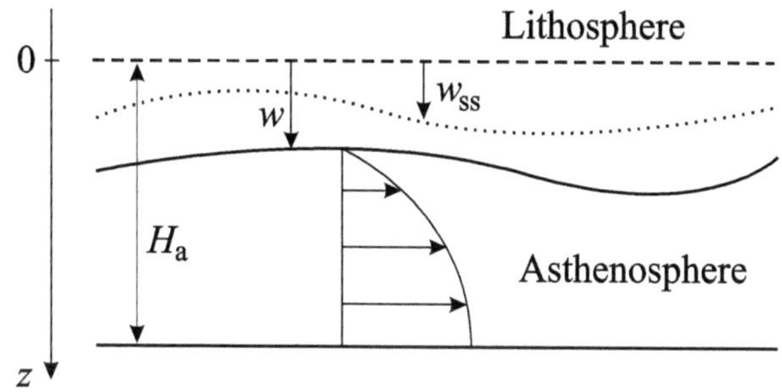

Fig. 8.7. Diffusive asthenosphere: Thin channel horizontal flow in the viscous asthenosphere of thickness H_a, driven by the deviation of the actual vertical lithosphere displacement, w, from its steady state, w_{ss}.

for the ice flow problems we have discussed in Chapters 5, 6 and 7, the acceleration and Coriolis terms in the momentum balance are neglected (Stokes flow). Therefore, the Navier-Stokes equation (3.138) reads

$$-\operatorname{grad} p + \eta_a \,\Delta \mathbf{v}_h = \mathbf{0}\,, \tag{8.13}$$

where we have considered the horizontal components only, so that $\mathbf{v}_h = (v_x, v_y)$ is the horizontal velocity, and grad $= (\partial/\partial x, \partial/\partial y)$. It is reasonable to assume that for steady-state conditions, that is, $w = w_{ss}$, the adjusting flow in the asthenosphere vanishes ($\mathbf{v}_{h,ss} = \mathbf{0}$), so that for the corresponding pressure, p_{ss}, the relation

$$-\operatorname{grad} p_{ss} + \eta_a \,\Delta \mathbf{v}_{h,ss} = -\operatorname{grad} p_{ss} = \mathbf{0} \tag{8.14}$$

holds. By subtracting (8.13) and (8.14),

$$-\operatorname{grad}(p - p_{ss}) + \eta_a \,\Delta \mathbf{v}_h = \mathbf{0}\,, \tag{8.15}$$

which tells us that the driving force for asthenospheric flow is the non-steady-state pressure, $p - p_{ss}$. Under the assumption of hydrostatic conditions, it can be related to the non-steady-state lithosphere displacement, $w - w_{ss}$, via

$$p - p_{ss} = -\rho_a g(w - w_{ss}) \tag{8.16}$$

(the minus sign is due to the fact that the displacements are counted positive downward). Inserting this result in (8.15) and assuming that velocity changes occur mainly in the vertical direction, that is

$$\frac{\partial \mathbf{v}_h}{\partial x}, \frac{\partial \mathbf{v}_h}{\partial y} \ll \frac{\partial \mathbf{v}_h}{\partial z}, \qquad \frac{\partial^2 \mathbf{v}_h}{\partial x^2}, \frac{\partial^2 \mathbf{v}_h}{\partial y^2} \ll \frac{\partial^2 \mathbf{v}_h}{\partial z^2}, \tag{8.17}$$

yields

$$\frac{\partial^2 \mathbf{v}_\mathrm{h}}{\partial z^2} = -\frac{\rho_\mathrm{a} g}{\eta_\mathrm{a}} \operatorname{grad}(w - w_\mathrm{ss}) \,. \tag{8.18}$$

Since the right-hand side is independent of z, this can easily be integrated. We neglect the lithosphere displacements compared to the asthenosphere thickness,

$$w,\, w_\mathrm{ss} \ll H_\mathrm{a} \,, \tag{8.19}$$

so that the top of the asthenosphere is always approximately at $z = 0$ and the bottom at $z = H_\mathrm{a}$. Further, we assume rather arbitrarily that the velocity vanishes at the top of the asthenosphere, and that the shear stress (and therefore the vertical derivative of the velocity) vanishes at the bottom:

$$\mathbf{v}_\mathrm{h}\big|_{z=0} = \mathbf{0} \,, \qquad \left.\frac{\partial \mathbf{v}_\mathrm{h}}{\partial z}\right|_{z=H_\mathrm{a}} = \mathbf{0} \,. \tag{8.20}$$

With these boundary conditions, double integration of (8.18) yields the parabolic velocity field

$$\mathbf{v}_\mathrm{h} = -\frac{\rho_\mathrm{a} g}{2\eta_\mathrm{a}} (z^2 - 2H_\mathrm{a} z) \operatorname{grad}(w - w_\mathrm{ss}) \,, \tag{8.21}$$

the shape of which is indicated in Fig. 8.7.

By integrating the mass balance $\operatorname{div} \mathbf{v} = 0$ vertically from $z = 0$ to $z = H_\mathrm{a}$, one finds, analogous to the derivation of the ice thickness equation (5.55),

$$\frac{\partial H_\mathrm{a}}{\partial t} = -\operatorname{div} \int_0^{H_\mathrm{a}} \mathbf{v}_\mathrm{h} \, \mathrm{d}z \tag{8.22}$$

(note that any mass exchange at the top and the bottom has been neglected). If we further assume that the bottom of the asthenosphere is rigid, then $\partial H_\mathrm{a}/\partial t = -\partial w/\partial t$, so that

$$\frac{\partial w}{\partial t} = \operatorname{div} \int_0^{H_\mathrm{a}} \mathbf{v}_\mathrm{h} \, dz \,. \tag{8.23}$$

Inserting the velocity profile (8.21) in this relation yields

$$\begin{aligned}
\frac{\partial w}{\partial t} &= -\frac{\rho_\mathrm{a} g}{2\eta_\mathrm{a}} \operatorname{div}\operatorname{grad}(w - w_\mathrm{ss}) \int_0^{H_\mathrm{a}} (z^2 - 2H_\mathrm{a} z) \, \mathrm{d}z \\
&= -\frac{\rho_\mathrm{a} g}{2\eta_\mathrm{a}} \Delta(w - w_\mathrm{ss}) \left[\frac{z^3}{3} - 2H_\mathrm{a} \frac{z^2}{2} \right]_0^{H_\mathrm{a}} \\
&= \frac{\rho_\mathrm{a} g H_\mathrm{a}^3}{3\eta_\mathrm{a}} \Delta(w - w_\mathrm{ss}) \,,
\end{aligned} \tag{8.24}$$

and, by introducing the diffusivity

$$D_{\mathrm{a}} = \frac{\rho_{\mathrm{a}} g H_{\mathrm{a}}^3}{3\eta_{\mathrm{a}}}, \tag{8.25}$$

one finally obtains the diffusive thin channel equation

$$\frac{\partial w}{\partial t} = D_{\mathrm{a}} \Delta (w - w_{\mathrm{ss}}). \tag{8.26}$$

With the scale $[L]$ for the horizontal extent of the ice load [see Eq. (5.5)], the associated time scale for the diffusive asthenosphere motion is

$$\tau_{\mathrm{a}} = \frac{[L]^2}{D_{\mathrm{a}}} = \frac{3\eta_{\mathrm{a}}}{\rho_{\mathrm{a}} g H_{\mathrm{a}}^3} [L]^2. \tag{8.27}$$

Note that here τ_{a} is proportional to the square of the extent of the ice load, whereas it is a constant parameter in the relaxing asthenosphere approach.

8.3.4 ELDA Model

Finally, the ELDA model combines the elastic lithosphere and the diffusive asthenosphere. Therefore, the steady-state lithosphere displacement is computed by solving the elastic thin plate equation (8.6), and the actual displacement then follows from the diffusive thin channel equation (8.26).

8.4 Analytical Solution for the Local Lithosphere

We come back to the simplified problem formulated in Sect. 5.6.1, and the corresponding ice thickness equation (5.107) with the volume flux (5.108). Let us now drop the assumption of a flat, rigid bed, assume $z_{1,0}(x) = 0$ instead (flat lithosphere surface in the absence of an ice load) and allow isostatic displacement by the local lithosphere mechanism. Since the steady-state assumption is kept, the displacement of the lithosphere is given by $w = w_{\mathrm{ss}}$, and it is not necessary to specify an asthenosphere model.

With Eqs. (8.1) and (8.4), we find for the ice base

$$b(x) = z_1(x) = -w(x) = -w_{\mathrm{ss}}(x) = -\frac{\rho}{\rho_{\mathrm{a}}} H(x), \tag{8.28}$$

or, by using $H = h - b$ and rearranging terms,

$$\begin{aligned} b(x) &= -\lambda\, h(x), \\ H(x) &= (1+\lambda)\, h(x), \end{aligned} \qquad \text{with } \lambda = \frac{\rho}{\rho_{\mathrm{a}} - \rho}. \tag{8.29}$$

Inserting Eq. $(8.29)_2$ into Eq. (5.108) shows that the steady-state ice thickness equation (5.111) now reads

$$\frac{\mathrm{d}Q}{\mathrm{d}x} = -\frac{\mathrm{d}}{\mathrm{d}x}\left(A_0\,(1+\lambda)^{n+2}\,h^{n+2}\left|\frac{\partial h}{\partial x}\right|^{n-1}\frac{\partial h}{\partial x}\right) = a_\mathrm{s}\,. \tag{8.30}$$

By introducing the transformation

$$h(x) = \tilde{h}(x)\left(\frac{1}{1+\lambda}\right)^{(n+2)/(2n+2)}, \tag{8.31}$$

Eq. (8.30) changes to

$$\frac{\mathrm{d}Q}{\mathrm{d}x} = -\frac{\mathrm{d}}{\mathrm{d}x}\left(A_0\,\tilde{h}^{n+2}\left|\frac{\partial \tilde{h}}{\partial x}\right|^{n-1}\frac{\partial \tilde{h}}{\partial x}\right) = a_\mathrm{s}\,, \tag{8.32}$$

which has the same form as the original equation (5.111) for a flat, rigid bed. Hence, the transformed quantity $\tilde{h}(x)$ is identical to the surface topography for a flat, rigid bed (e.g., the Vialov or Bueler profile derived in Sects. 5.6.2 and 5.6.3, respectively), and the surface topography with isostasy results from Eq. (8.31). The basal topography and ice thickness can subsequently be computed by Eq. (8.29).

For the parameter values $\lambda = 0.3$ and $n = 3$, the surface elevation of the ice sheet with isostasy is $\sim$ 15% lower and the thickness $\sim$ 10% larger than for the flat-based ice sheet. Figure 8.8 shows a comparison for the Bueler profile.

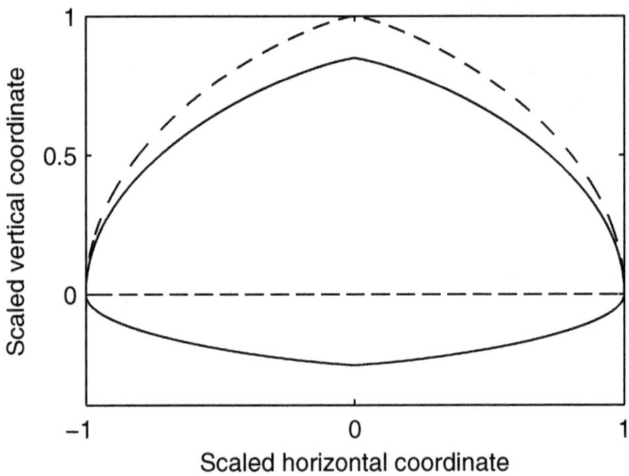

Fig. 8.8. Bueler solution for the ice surface and base with (solid lines) and without (dashed lines) isostatic displacement of the lithosphere (in scaled coordinates).

8.5 Numerical Methods

In the finite difference scheme for the plane strain shallow ice equations described in Sect. 5.7, a rigid ice base was assumed. We are now going to extend the scheme by including isostasy, modelled by the four simple models introduced in Sect. 8.3.

8.5.1 Local Lithosphere

The simple, algebraic equation (8.1) for the local lithosphere can readily be discretised. It yields

$$(w_{\mathrm{ss}})_{i,n} = \frac{\rho}{\rho_{\mathrm{a}}} H_{i,n} \,. \tag{8.33}$$

8.5.2 Elastic Lithosphere

In the case of the elastic lithosphere, things are more complicated. We have seen above that the Green's function of the thin plate equation (8.6) is given by Eq. (8.11). However, under two-dimensional, plane strain conditions, it changes to

$$G(r) = G(x, \check{x}) = -\frac{(\sqrt{2}L_{\mathrm{r}})^3}{8K_{\mathrm{l}}} f(\alpha) \,, \tag{8.34}$$

with

$$r = |x - \check{x}| \,, \quad \alpha = \frac{r}{\sqrt{2}L_{\mathrm{r}}} \,, \tag{8.35}$$

and

$$f(\alpha) = \mathrm{e}^{-\alpha}(\cos\alpha + \sin\alpha) \tag{8.36}$$

[see e.g. Turcotte and Schubert (2002)].

Analogous to the continuous superposition (8.12), we can construct the discretised steady-state displacement $(w_{\mathrm{ss}})_{i,n}$ at the point ξ_i by superposing the effects of the discretised ice loads at all positions $i' = 0 \ldots I$. The imposed load at the position i' is equal to $\rho g H_{i',n}$, and so we obtain

$$(w_{\mathrm{ss}})_{i,n} = -\frac{(\sqrt{2}L_{\mathrm{r}})^3}{8K_{\mathrm{l}}} \sum_{i'=0}^{I} \rho g H_{i',n} \, f(\alpha_{ii'}) \, \Delta\xi \,, \tag{8.37}$$

where

$$\alpha_{ii'} = \frac{|\xi_i - \xi_{i'}|}{\sqrt{2}L_{\mathrm{r}}} = \frac{|i - i'| \, \Delta\xi}{\sqrt{2}L_{\mathrm{r}}} \,. \tag{8.38}$$

8.5.3 Relaxing Asthenosphere

The discretisation of the evolution equation for the displacement of the relaxing asthenosphere model, Eq. (8.2), is straightforward. By employing Euler forward stepping for the time derivative [see e.g. Morton and Mayers (1994), Hundsdorfer and Verwer (2003)], we obtain the explicit scheme

$$\frac{w_{i,n+1} - w_{i,n}}{\Delta\tau} = -\frac{1}{\tau_\mathrm{a}}\Big[w_{i,n} - (w_\mathrm{ss})_{i,n}\Big] \,. \tag{8.39}$$

Alternatively, the scheme can be made implicit by replacing $w_{i,n}$ on the right-hand side by the value at the new time, $w_{i,n+1}$. The update of the vertical position of the lithosphere surface follows from Eq. (8.4),

$$(z_\mathrm{l})_{i,n+1} = (z_{\mathrm{l},0})_i - w_{i,n+1} \,, \tag{8.40}$$

and, according to Eq. (8.5), its discretised time derivative is the negative of Eq. (8.39),

$$\begin{aligned}\frac{(z_\mathrm{l})_{i,n+1} - (z_\mathrm{l})_{i,n}}{\Delta\tau} &= \frac{b_{i,n+1} - b_{i,n}}{\Delta\tau} = -\frac{w_{i,n+1} - w_{i,n}}{\Delta\tau} \\ &= \frac{1}{\tau_\mathrm{a}}\Big[w_{i,n} - (w_\mathrm{ss})_{i,n}\Big] \,. \end{aligned} \tag{8.41}$$

8.5.4 Diffusive Asthenosphere

The plane strain version of the diffusive thin channel equation (8.26) in the transformed coordinates of Sect. 5.7 reads

$$\frac{\partial w}{\partial \tau} = D_\mathrm{a}\frac{\partial^2}{\partial \xi^2}(w - w_\mathrm{ss}) \,. \tag{8.42}$$

Employing Euler forward stepping for the time derivative and central differences for the diffusion term [see e.g. Morton and Mayers (1994), Hundsdorfer and Verwer (2003)] yields the discretised form

$$\begin{aligned}\frac{w_{i,n+1} - w_{i,n}}{\Delta\tau} = D_\mathrm{a}&\frac{w_{i+1,n} - 2w_{i,n} + w_{i-1,n}}{\Delta\xi^2} \\ -D_\mathrm{a}&\frac{(w_\mathrm{ss})_{i+1,n} - 2(w_\mathrm{ss})_{i,n} + (w_\mathrm{ss})_{i-1,n}}{\Delta\xi^2} \,. \end{aligned} \tag{8.43}$$

This explicit scheme can be made implicit by replacing all occurrences of $w_{i(\pm 1),n}$ on the right-hand side by the corresponding values at the new time, $w_{i(\pm 1),n+1}$. Analogous to Eqs. (8.40) and (8.41), the vertical position of the lithosphere surface is updated according to

$$(z_\mathrm{l})_{i,n+1} = (z_{\mathrm{l},0})_i - w_{i,n+1} \,, \tag{8.44}$$

and its discretised time derivative is

$$\frac{(z_\mathrm{l})_{i,n+1} - (z_\mathrm{l})_{i,n}}{\Delta\tau} = \frac{b_{i,n+1} - b_{i,n}}{\Delta\tau} = -\frac{w_{i,n+1} - w_{i,n}}{\Delta\tau}$$
$$= -D_\mathrm{a}\frac{w_{i+1,n} - 2w_{i,n} + w_{i-1,n}}{\Delta\xi^2}$$
$$+D_\mathrm{a}\frac{(w_\mathrm{ss})_{i+1,n} - 2(w_\mathrm{ss})_{i,n} + (w_\mathrm{ss})_{i-1,n}}{\Delta\xi^2}\,. \quad (8.45)$$

The discretisations of the four simple models are now given by the following equations:

- LLRA model: Eqs. (8.33) and (8.39)–(8.41).
- ELRA model: Eqs. (8.37)–(8.38) and (8.39)–(8.41).
- LLDA model: Eqs. (8.33) and (8.43)–(8.45).
- ELDA model: Eqs. (8.37)–(8.38) and (8.43)–(8.45).

8.6 Model Intercomparison

The simple isostasy models LLRA, ELRA, LLDA and ELDA described above will now be compared to the more sophisticated SGVE approach (see the end of Sect. 8.1). To this end, we follow the study by Le Meur and Huybrechts (1996), who coupled a three-dimensional dynamic/thermodynamic ice sheet model with these isostasy models, simulated the evolution of the Antarctic Ice Sheet from 126,000 years ago until today, and examined the computed present uplift rates at the top of the lithosphere. The main parameters of the applied isostasy models are:

- LL: densities $\rho = 910\ \mathrm{kg\,m^{-3}}$, $\rho_\mathrm{a} = 3300\ \mathrm{kg\,m^{-3}}$; ratio $\rho/\rho_\mathrm{a} = 0.276$.
- EL: flexural stiffness $K_\mathrm{l} = 10^{25}\ \mathrm{N\,m}$.
- RA: relaxation time $\tau_\mathrm{a} = 3000$ a.
- DA: diffusivity $D_\mathrm{a} = 50\ \mathrm{km^2\,a^{-1}}$.
- SGVE:
  - Lithosphere layer down to 100 km depth, linear elastic solid (Hookean body; see Sect. 3.4.2).
  - Upper-mantle layer from 100 km to 670 km depth, visco-elastic Maxwell fluid (Le Meur 1996), viscosity $\eta_\mathrm{um} = 5 \times 10^{20}$ Pa s.
  - Lower-mantle layer below 670 km depth, visco-elastic Maxwell fluid, viscosity $\eta_\mathrm{lm} = 10^{21}$ Pa s.

In Fig. 8.9 the computed uplift rates are shown for the five models. Evidently, the uplift is most pronounced in the region between the East and the West Antarctic Ice Sheet, and reaches maximum values of more than 10 $\mathrm{cm\,a^{-1}}$ in the vicinity of the grounding line of the Filchner-Rønne Ice Shelf. This is the case because, during the last ice age, the large Filchner-Rønne and Ross Ice Shelves were partly grounded, which increased the ice

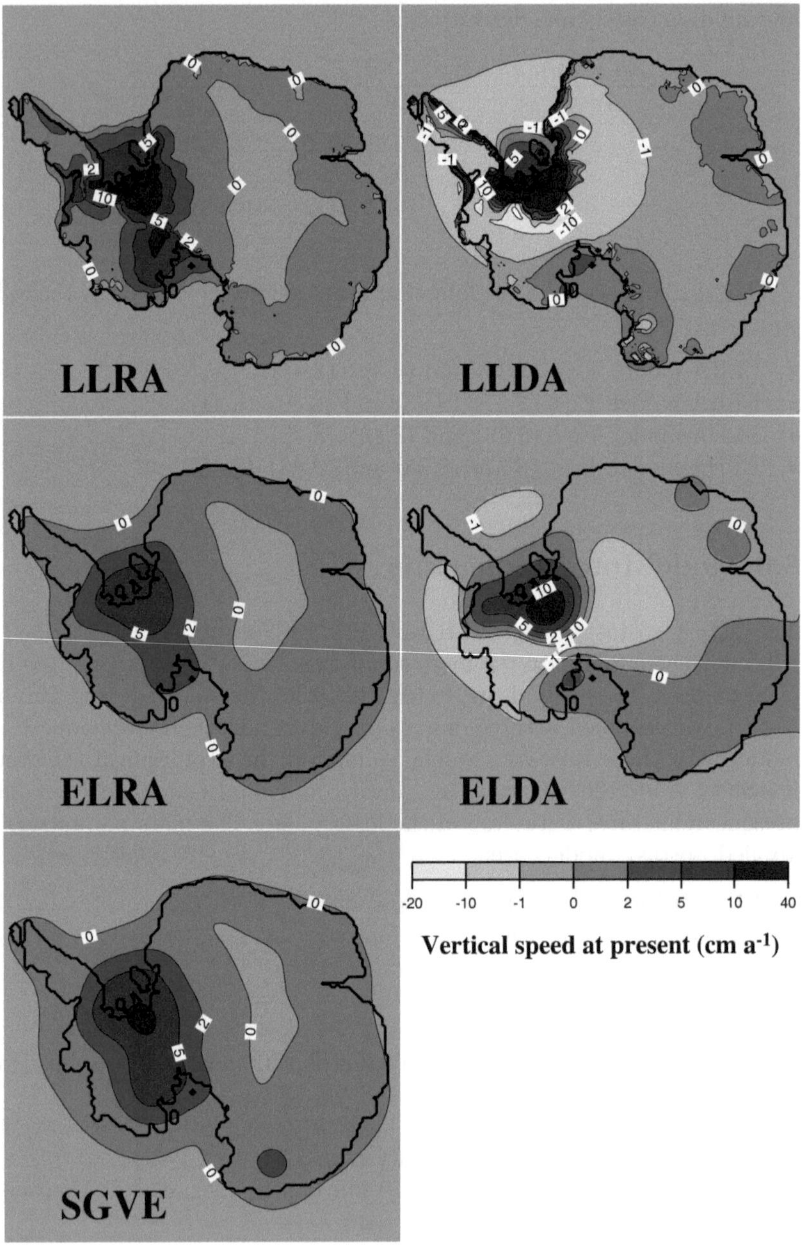

Fig. 8.9. Lithosphere uplift rates for the modern Antarctic Ice Sheet, computed with the five isostasy models LLRA, ELRA, LLDA, ELDA and SGVE coupled to a dynamic/thermodynamic ice sheet model. Reproduced from Le Meur and Huybrechts (1996), © International Glaciological Society.

thickness between these shelves significantly. Consequently, this region has experienced a large decrease of the ice thickness since then, which leads to the large uplift rates. By contrast, the other parts of the Antarctic Ice Sheet did not suffer similar changes, and therefore the uplift is much smaller.

If we take the result of the SGVE model as a reference, the performance of the simple models decreases in the order ELRA > LLRA > ELDA > LLDA. As far as the lithosphere parameterisations LL and EL are concerned, this order confirms the expectation that the more elaborate EL description should provide better results. The LL models show distinctly more non-physical fine structure in the uplift pattern due to the missing low-pass filter of the elastic lithosphere. By contrast, for the asthenosphere parameterisations RA and DA, the behaviour is opposite. The apparently more realistic DA models show an exaggerated concentration of positive uplift rates around the grounding line of the Filchner-Rønne Ice Shelf, surrounded by a narrow band of negative uplift rates which does not have any correspondence in the SGVE result. On the other hand, both RA models reproduce the uplift distribution of the SGVE model rather well. A likely explanation for this counter-intuitive finding is that the motion of the mantle induced by the varying ice load reaches far deeper than to the bottom of the asthenosphere, so that the thin channel assumption of the diffusive asthenosphere approach is not adequate.

Consequently, as long as the high accuracy of a SGVE model is dispensable (for instance, in a typical ice-sheet-modelling application), use of the ELRA model is favourable. The even simpler LLRA model is still a reasonable, computationally faster alternative (at the cost of some artificial fine structure in the computed uplift patterns), whereas the models with the diffusive asthenosphere parameterisation, LLDA and ELDA, should be discarded.

9

Advanced Topics

While in the previous chapters relatively well-established aspects of ice dynamics have been treated, we now turn to some more advanced and demanding topics at the forefront of current research. The selection of the topics (induced anisotropy, compressible firn, polythermal glaciers) is strongly influenced by the authors' own research interests and makes no claim to be complete. Other issues, such as subglacial hydrology, ice stream dynamics or calving mechanics, deserve equal attention, and we explicitly encourage the interested reader to follow these paths as well.

9.1 Induced Anisotropy

9.1.1 Background

Let us come back to the basic creep properties of polycrystalline ice. At the beginning of our discussion in Sect. 4.2, we made the assumption that the ice crystallites (grains) in a polycrystalline aggregate are essentially randomly oriented, so that the macroscopic behaviour is isotropic. However, observations show that this is not always the case. While the assumption holds well at the surface of an ice sheet or glacier where the ice has formed only recently out of accumulated snowfall, deeper down into the ice, different types of anisotropic fabrics with preferred orientations of the c-axes tend to develop (e.g., Paterson 1994).

Many models have been proposed to include anisotropy in the flow law of polycrystalline ice. On one end of the range in complexity, a simple flow enhancement factor E is introduced in an *ad-hoc* fashion as a multiplier of the isotropic ice fluidity in order to account for anisotropy and/or impurities. This can be conveniently achieved by replacing the rate factor $A(T')$ in the several versions of isotropic flow laws by $EA(T')$ (see Sect. 4.3.4), and it is done in most current large-scale ice sheet models, often without explicitly mentioning anisotropy (e.g., Saito and Abe-Ouchi 2004, Greve 2005, Huybrechts et al.

R. Greve, H. Blatter, *Dynamics of Ice Sheets and Glaciers*, Advances in Geophysical and Environmental Mechanics and Mathematics,
DOI 10.1007/978-3-642-03415-2_9, © Springer-Verlag Berlin Heidelberg 2009

2007). In macroscopic, phenomenological models, an anisotropic macroscopic formulation for the flow law of the polycrystal is postulated. To be usable, the rheological parameters that enter this law must be evaluated as functions of the anisotropic fabric. The concept of homogenisation models, also called micro-macro models, is to derive the polycrystalline behaviour at the level of individual crystallites and the fabric. As for the "high-end" complexity, full-field models solve the Stokes equation for ice flow properly by decomposing the polycrystal into many elements, which makes it possible to infer the stress and strain-rate heterogeneities at the microscopic scale. A very comprehensive overview of these different types of models and many references are given by Gagliardini et al. (2009). However, the more sophisticated models are usually too complex and computationally time-consuming to be included readily in a model of macroscopic ice flow.

Here, the Continuum-mechanical, Anisotropic Flow model, based on an anisotropic Flow Enhancement factor ["CAFFE model"; see Placidi et al. (2009), Greve et al. (2009), and references therein], will be described as an example. The CAFFE model belongs to the class of macroscopic models. The flow enhancement factor is taken as a function of a newly introduced scalar quantity called *deformability*, which is essentially a non-dimensional invariant related to the shear stress acting on the basal plane of a crystallite, weighted by the orientation-distribution function which describes the anisotropic fabric of the polycrystal. Fabric evolution is modelled by an orientation mass balance which accounts for grain rotation and recrystallisation processes. The CAFFE model is a good compromise between necessary simplicity on the one hand, and consideration of the major effects of anisotropy on the other.

9.1.2 Anisotropic Generalisation of Glen's Flow Law

Deformation of a Crystallite

In order to derive a generalisation of Glen's flow law (4.16) which accounts for general, anisotropic fabrics of the ice polycrystal, we first consider the deformation of a crystallite embedded in the polycrystalline aggregate. Only the dominant deformation along the basal plane is accounted for, and deformations along prismatic and pyramidal planes, which are at least 60 times more difficult to activate, are neglected (Fig. 4.2).

The tensor t is interpreted as the *macroscopic* stress tensor which describes the stress on a control volume spanning a sufficiently large number of individual crystallites. Therefore, t does not depend on the orientation $\mathbf{n}$ (unit normal vector of the basal plane, direction of the c-axis), but only on the position $\mathbf{x}$ and time t. Note that the CAFFE model does not deal with microscopic stresses on the level of the crystallites, and attempts in no way to relate microscopic and macroscopic stresses. The macroscopic stress vector on the basal plane is given by the expression $\mathsf{t}\cdot\mathbf{n}$ (Fig. 9.1). It is reasonable to

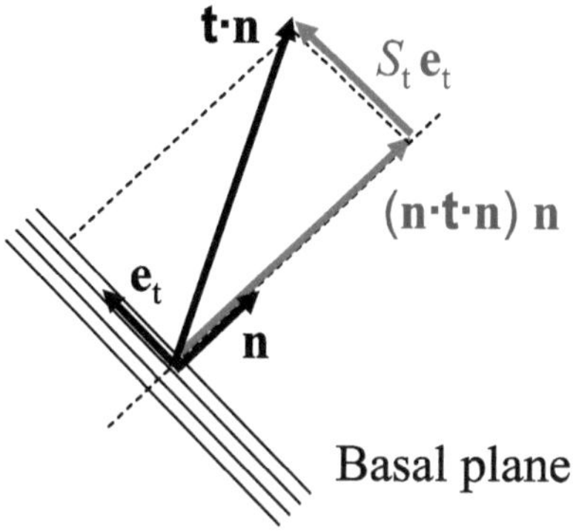

Fig. 9.1. Decomposition of the stress vector into a part normal and a part tangential to the basal plane of the ice crystallite.

assume that only the stress component S_t tangential to the basal plane contributes to its shear deformation, while the component normal to the basal plane has no effect.

According to Fig. 9.1, the decomposition of the stress vector reads

$$\mathsf{t} \cdot \mathbf{n} = (\mathbf{n} \cdot \mathsf{t} \cdot \mathbf{n})\mathbf{n} + S_t \mathbf{e}_t \,, \tag{9.1}$$

where $\mathbf{e}_t$ denotes the tangential unit vector. Inserting the decomposition (3.133) of the stress tensor t for incompressible fluids readily eliminates the pressure p and leaves

$$\mathsf{t}^D \cdot \mathbf{n} = (\mathbf{n} \cdot \mathsf{t}^D \cdot \mathbf{n})\mathbf{n} + S_t \mathbf{e}_t \,. \tag{9.2}$$

As mentioned above, deformation of the crystallite in the polycrystalline aggregate is attributed to the tangential component S_t only. Since we aim at a theory which describes the effects of anisotropy by a *scalar*, anisotropic flow enhancement factor, we define the scalar invariant

$$S_t^2 = (\mathsf{t}^D \cdot \mathbf{n})^2 - (\mathbf{n} \cdot \mathsf{t}^D \cdot \mathbf{n})^2 \,. \tag{9.3}$$

This quantity has the unit of a stress squared, and a natural way to non-dimensionalise it is by the square of the effective stress σ_e [Eq. (4.9)], which is also a scalar invariant. Thus, we introduce the *crystallite deformability*, which is loaded by the stress t, as

$$\mathcal{A}^\star(\mathbf{n}) = \frac{5}{2}\,\frac{S_t^2(\mathbf{n})}{\sigma_e^2} = 5\,\frac{S_t^2(\mathbf{n})}{\mathrm{tr}\,(\mathsf{t}^D)^2} \,. \tag{9.4}$$

The factor 5/2 has been introduced merely for reasons of convenience, as it will become clear below.

Flow Law for Polycrystalline Ice

In polycrystalline ice, the crystallites within a control volume (which is assumed to be large compared to the crystallite dimensions, but small compared to the macroscopic scale of ice flow) show a certain fabric. Extreme cases are on the one hand the single maximum fabric, for which all c-axes are perfectly aligned, and on the other hand the isotropic fabric with a random distribution of the c-axes. A general fabric, which is usually in between these cases, can be described by the *orientation mass density* (OMD) $\rho^\star(\mathbf{n})$. It is defined as the mass per volume and orientation, the latter being specified by the unit normal vector (direction of the c-axis) $\mathbf{n} \in S^2$ (S^2 is the unit sphere). When integrated over all orientations, the OMD must yield the ordinary mass density ρ, which leads to the normalisation condition

$$\int_{S^2} \rho^\star(\mathbf{n})\,\mathrm{d}^2 n = \rho\,. \tag{9.5}$$

Alternatively, an *orientation distribution function* (ODF) $f^\star(\mathbf{n})$ can be defined as

$$f^\star(\mathbf{n}) = \frac{\rho^\star(\mathbf{n})}{\rho}\,. \tag{9.6}$$

The ODF is normalised to unity when integrated over all orientations,

$$\int_{S^2} f^\star(\mathbf{n})\,\mathrm{d}^2 n = 1\,. \tag{9.7}$$

It is physically impossible to distinguish between the orientations $\mathbf{n}$ and $-\mathbf{n}$. This can be accounted for in the OMD and ODF either by allowing non-zero values on one half of S^2 (e.g., the "upper" hemisphere with $z \geq 0$) only, or by imposing the symmetry condition

$$\rho^\star(\mathbf{n}) = \rho^\star(-\mathbf{n})\,, \qquad f^\star(\mathbf{n}) = f^\star(-\mathbf{n})\,. \tag{9.8}$$

We use the ODF in order to define the *deformability* of polycrystalline ice by weighting the crystallite deformability (9.4),

$$\begin{aligned} \mathcal{A} &= \int_{S^2} \mathcal{A}^\star(\mathbf{n})\, f^\star(\mathbf{n})\,\mathrm{d}^2 n \\ &= \frac{5}{2}\int_{S^2} \frac{S_\mathrm{t}^2(\mathbf{n})}{\sigma_\mathrm{e}^2} f^\star(\mathbf{n})\,\mathrm{d}^2 n = 5\int_{S^2} \frac{S_\mathrm{t}^2(\mathbf{n})}{\mathrm{tr}\,(\mathbf{t}^\mathrm{D})^2} f^\star(\mathbf{n})\,\mathrm{d}^2 n\,. \end{aligned} \tag{9.9}$$

Note that, for the isotropic case, the ODF is $f^\star(\mathbf{n}) = 1/(4\pi)$, and a laborious, but straightforward calculation of the integral in Eq. (9.9) [writing the terms $S_\mathrm{t}^2(\mathbf{n})$ and $\mathrm{tr}\,(\mathbf{t}^\mathrm{D})^2$ in component form, expressing n_x, n_y, n_z and $\mathrm{d}^2 n$

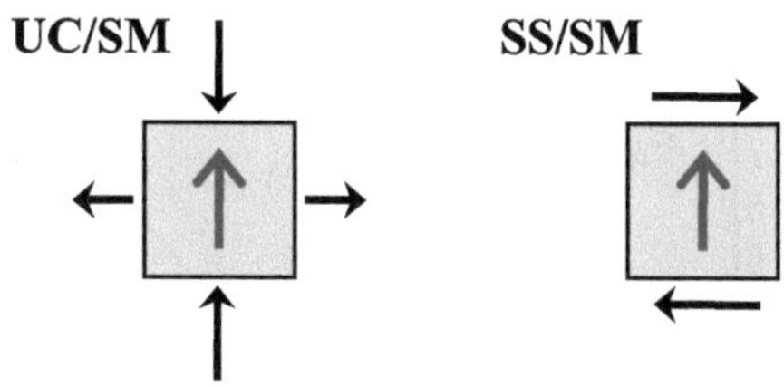

Fig. 9.2. Uniaxial compression on single maximum (UC/SM) and simple shear on single maximum (SS/SM) for a small sample of polycrystalline ice. Stresses are indicated as black arrows, and the single maximum fabric is marked by the dark-grey arrows within the ice sample.

in spherical coordinates, integrating over the zenith angle θ and the azimuth angle ϕ; best to be done with a computer algebra tool] yields a deformability of $\mathcal{A} = 1$ for the isotropic polycrystal. For that reason, the factor 5/2 has been introduced in Eqs. (9.4) and (9.9).

The CAFFE flow law for anisotropic polar ice can now be formulated. Essentially, we keep the form of Glen's flow law (4.16), but with a scalar, anisotropic enhancement factor $E(\mathcal{A})$,

$$\mathsf{D} = E(\mathcal{A})\, A(T')\, \sigma_\mathrm{e}^{n-1}\, \mathrm{t}^\mathrm{D}\,. \tag{9.10}$$

The function $E(\mathcal{A})$ is supposed to be strictly increasing with the deformability $\mathcal{A}$, and has the fixed points

$$\begin{aligned} E(0) &= E_\mathrm{min} \ \text{(uniaxial compression on single maximum)},\\ E(1) &= 1 \qquad \text{(arbitrary stress on isotropic fabric)},\\ E(\tfrac{5}{2}) &= E_\mathrm{max} \ \text{(simple shear on single maximum)}. \end{aligned} \tag{9.11}$$

The "hard" case $(9.11)_1$ and the "soft" case $(9.11)_3$ are illustrated in Fig. 9.2. Note also that the deformability cannot take values larger than $\mathcal{A} = 5/2$.

As for the detailed functional form of the anisotropic enhancement factor, experimental data suggest that the enhancement factor depends on the "Schmid factor" (which can be identified with the shear stress in the basal plane, S_t, of the CAFFE model) to the fourth power (Azuma 1995, Miyamoto 1999). Since the polycrystal deformability $\mathcal{A}$ contains a dependency of S_t^2 [see Eq. (9.9)], it is reasonable to assume a dependency of E on $\mathcal{A}^2$. However, this does not allow Eq. (9.11) for arbitrary choices of the parameters E_min and E_max to be fulfilled. Hence the function $E(\mathcal{A})$ is chosen to depend on $\mathcal{A}^2$ in the interval $[1, \frac{5}{2}]$ only, and for the interval $[0, 1]$ a dependency on $\mathcal{A}^t$ is introduced. The exponent t is adjusted such that the function is continuously differentiable at $\mathcal{A} = 1$. This yields

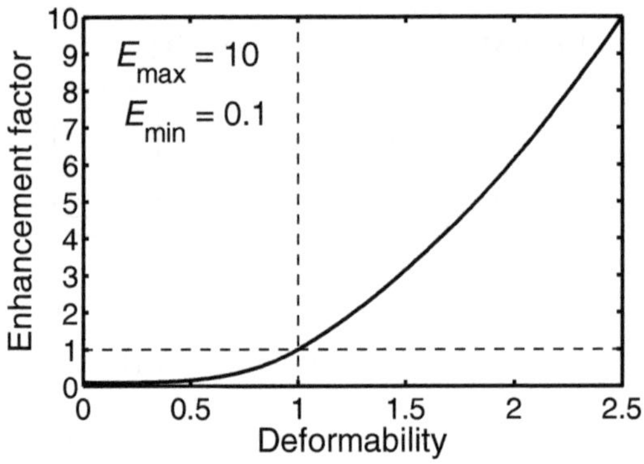

Fig. 9.3. Anisotropic enhancement factor $E(\mathcal{A})$ as a function of the deformability $\mathcal{A}$ according to Eq. (9.12), for $E_{\text{max}} = 10$ and $E_{\text{min}} = 0.1$.

$$E(\mathcal{A}) = \begin{cases} E_{\text{min}} + (1 - E_{\text{min}})\mathcal{A}^t\,, \;\; t = \dfrac{8}{21}\,\dfrac{E_{\text{max}} - 1}{1 - E_{\text{min}}}\,, & 0 \le \mathcal{A} \le 1\,, \\[2ex] \dfrac{4\mathcal{A}^2(E_{\text{max}} - 1) + 25 - 4E_{\text{max}}}{21}\,, & 1 \le \mathcal{A} \le \dfrac{5}{2}\,. \end{cases} \tag{9.12}$$

Several studies (e.g., Russell-Head and Budd 1979, Pimienta et al. 1987, Budd and Jacka 1989) indicate that the parameter E_{max} (maximum softening) is ~ 10. The parameter E_{min} (maximum hardening) can be realistically chosen between 0 and ~ 0.1, a non-zero value serving mainly the purpose of avoiding numerical problems. The function (9.12) is shown in Fig. 9.3.

Owing to the collinearity of the tensors t^{D} and D, the CAFFE flow law (9.10) can be written as

$$\mathsf{D} = \frac{1}{2\eta(T', \sigma_{\text{e}}, \mathcal{A})}\,\mathsf{t}^{\text{D}}\,, \tag{9.13}$$

with the shear viscosity

$$\eta(T', \sigma_{\text{e}}, \mathcal{A}) = \frac{1}{2E(\mathcal{A})\,A(T')\,\sigma_{\text{e}}^{n-1}}\,. \tag{9.14}$$

Inversion of the Flow Law

The anisotropic CAFFE flow law (9.10) can be inverted analytically in the same way as the isotropic Glen flow law; see Eqs. (4.19) – (4.22). Analogous to Eq. (4.20), the result is

$$\mathsf{t}^{\text{D}} = [E(\mathcal{A})]^{-1/n}\,B(T')\,d_{\text{e}}^{-(1-1/n)}\,\mathsf{D}\,. \tag{9.15}$$

The deformability $\mathcal{A}$ also needs to be expressed by strain rates instead of stresses [see Eq. (9.9)]. In analogy to Eq. (9.2), we consider the macroscopic strain-rate vector $\mathsf{D} \cdot \mathbf{n}$ in a crystallite, and decompose it according to

$$\mathsf{D}\cdot\mathbf{n} = (\mathbf{n}\cdot\mathsf{D}\cdot\mathbf{n})\mathbf{n} + D_{\mathrm{t}}\mathbf{e}_{\mathrm{t}}\,, \tag{9.16}$$

where D_{t} is the shear rate in the basal plane (see also Fig. 9.1). Analogous to Eq. (9.3), we define the scalar invariant

$$D_{\mathrm{t}}^2 = (\mathsf{D}\cdot\mathbf{n})^2 - (\mathbf{n}\cdot\mathsf{D}\cdot\mathbf{n})^2\,. \tag{9.17}$$

The crystallite deformability [Eq. (9.4)] can be readily expressed by D_{t} and the effective strain rate d_{e},

$$\begin{aligned}
\mathcal{A}^\star(\mathbf{n}) &= 5\,\frac{S_{\mathrm{t}}^2(\mathbf{n})}{\mathrm{tr}\,(\mathsf{t}^{\mathrm{D}})^2} \\
&\overset{(9.3)}{=} 5\,\frac{(\mathsf{t}^{\mathrm{D}}\cdot\mathbf{n})^2 - (\mathbf{n}\cdot\mathsf{t}^{\mathrm{D}}\cdot\mathbf{n})^2}{\mathrm{tr}\,(\mathsf{t}^{\mathrm{D}})^2} \\
&\overset{(9.13)}{=} 5\,\frac{(2\eta\mathsf{D}\cdot\mathbf{n})^2 - (\mathbf{n}\cdot 2\eta\mathsf{D}\cdot\mathbf{n})^2}{\mathrm{tr}\,(2\eta\mathsf{D})^2} \\
&= 5\,\frac{(\mathsf{D}\cdot\mathbf{n})^2 - (\mathbf{n}\cdot\mathsf{D}\cdot\mathbf{n})^2}{\mathrm{tr}\,\mathsf{D}^2} \\
&\overset{(9.17)}{=} 5\,\frac{D_{\mathrm{t}}^2(\mathbf{n})}{\mathrm{tr}\,(\mathsf{D}^2)} = \frac{5}{2}\,\frac{D_{\mathrm{t}}^2(\mathbf{n})}{d_{\mathrm{e}}^2}\,,
\end{aligned} \tag{9.18}$$

and the deformability of polycrystalline ice [Eq. (9.9)] yields

$$\begin{aligned}
\mathcal{A} &= \int_{S^2} \mathcal{A}^\star(\mathbf{n})\, f^\star(\mathbf{n})\,\mathrm{d}^2 n \\
&= \frac{5}{2}\int_{S^2} \frac{D_{\mathrm{t}}^2(\mathbf{n})}{d_{\mathrm{e}}^2} f^\star(\mathbf{n})\,\mathrm{d}^2 n = 5\int_{S^2} \frac{D_{\mathrm{t}}^2(\mathbf{n})}{\mathrm{tr}\,(\mathsf{D}^2)} f^\star(\mathbf{n})\,\mathrm{d}^2 n\,.
\end{aligned} \tag{9.19}$$

The inverse form of the CAFFE flow law (9.15) with the deformability in the form of (9.19) can also be written with the shear viscosity η,

$$\mathsf{t}^{\mathrm{D}} = 2\eta(T', d_{\mathrm{e}}, \mathcal{A})\,\mathsf{D}\,, \tag{9.20}$$

where

$$\eta(T', d_{\mathrm{e}}, \mathcal{A}) = \frac{1}{2}[E(\mathcal{A})]^{-1/n}\, B(T')\, d_{\mathrm{e}}^{-(1-1/n)}\,. \tag{9.21}$$

9.1.3 Proof of Anisotropy for the CAFFE Flow Law

We will now prove explicitly that the flow law (9.10) of the CAFFE model [or its equivalent, inverse form (9.15)] is anisotropic, despite the fact that the collinearity between the tensors t^{D} and D has been retained and the effects of anisotropy are expressed by a scalar enhancement factor only.

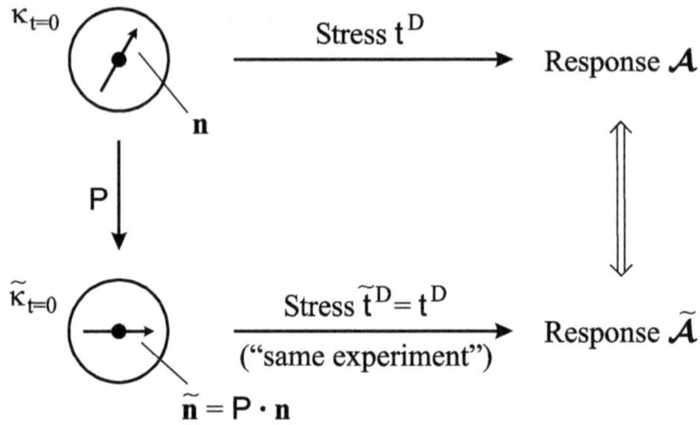

Fig. 9.4. Anisotropy of the CAFFE flow law: If the same stress ($\tilde{\mathsf{t}}^{\mathrm{D}} = \mathsf{t}^{\mathrm{D}}$) is applied to two rotated initial configurations ($\kappa_{t=0}$, $\tilde{\kappa}_{t=0}$), the responses $\mathcal{A}$ and $\tilde{\mathcal{A}}$ are different in general.

In the context of the theory of constitutive equations, the definition of isotropy states that any rotation of the body in question does not alter its material response. Mathematically speaking, this means invariance of the material functions (or functionals) to arbitrary orthogonal transformations P of an undistorted configuration κ (e.g., Liu 2002, p. 86). Anisotropy is the logical opposite: for at least one orthogonal transformation P, the invariance does not hold.

By construction, the anisotropy of the flow law (9.10) is contained in the enhancement factor $E(\mathcal{A})$ via the polycrystal deformability $\mathcal{A}$. So let us assume that, at the time $t = 0$, the initial configuration $\kappa_{t=0}$ is given by an unloaded ice specimen with the ODF $f^\star(\mathbf{n})$. At $t = 0^+$, it is subjected to the stress t^{D}, and, according to Eqs. (9.3) and (9.9), the resulting deformability is

$$\mathcal{A} = 5 \int\limits_{S^2} \frac{(\mathsf{t}^{\mathrm{D}} \cdot \mathbf{n})^2 - (\mathbf{n} \cdot \mathsf{t}^{\mathrm{D}} \cdot \mathbf{n})^2}{\mathrm{tr}\,(\mathsf{t}^{\mathrm{D}})^2} \, f^\star(\mathbf{n}) \, \mathrm{d}^2 n \,. \tag{9.22}$$

Now let us consider a second initial configuration $\tilde{\kappa}_{t=0}$ rotated by an orthogonal transformation P with respect to $\kappa_{t=0}$. The rotated orientations are given by

$$\tilde{\mathbf{n}} = \mathsf{P} \cdot \mathbf{n} \tag{9.23}$$

(Fig. 9.4). The ODF follows the rotation, so that

$$\tilde{f}^\star(\tilde{\mathbf{n}}) = f^\star(\mathbf{n}) \quad \overset{(9.23)}{\Rightarrow} \quad \tilde{f}^\star(\tilde{\mathbf{n}}) = f^\star(\mathsf{P}^{\mathrm{T}} \cdot \tilde{\mathbf{n}}) \,. \tag{9.24}$$

At $t = 0^+$, the rotated configuration is subjected to the stress $\tilde{\mathsf{t}}^{\mathrm{D}}$, which is supposed to be the same as before,

$$\tilde{\mathsf{t}}^{\mathrm{D}} = \mathsf{t}^{\mathrm{D}} \tag{9.25}$$

(Fig. 9.4). The deformability with respect to the rotated configuration is then

$$\tilde{\mathcal{A}} \overset{(9.22)}{=} 5 \int_{S^2} \frac{(\tilde{\mathsf{t}}^{\mathrm{D}} \cdot \tilde{\mathbf{n}})^2 - (\tilde{\mathbf{n}} \cdot \tilde{\mathsf{t}}^{\mathrm{D}} \cdot \tilde{\mathbf{n}})^2}{\mathrm{tr}\,(\tilde{\mathsf{t}}^{\mathrm{D}})^2} \tilde{f}^\star(\tilde{\mathbf{n}})\,\mathrm{d}^2\tilde{n}$$

$$\overset{(9.24),(9.25)}{=} 5 \int_{S^2} \frac{(\mathsf{t}^{\mathrm{D}} \cdot \tilde{\mathbf{n}})^2 - (\tilde{\mathbf{n}} \cdot \mathsf{t}^{\mathrm{D}} \cdot \tilde{\mathbf{n}})^2}{\mathrm{tr}\,(\mathsf{t}^{\mathrm{D}})^2} f^\star(\mathsf{P}^{\mathrm{T}} \cdot \tilde{\mathbf{n}})\,\mathrm{d}^2\tilde{n}\,. \tag{9.26}$$

We change the name of the integration variable in the last integral of Eq. (9.26) from $\tilde{\mathbf{n}}$ to $\mathbf{n}$,

$$\tilde{\mathcal{A}} = 5 \int_{S^2} \frac{(\mathsf{t}^{\mathrm{D}} \cdot \mathbf{n})^2 - (\mathbf{n} \cdot \mathsf{t}^{\mathrm{D}} \cdot \mathbf{n})^2}{\mathrm{tr}\,(\mathsf{t}^{\mathrm{D}})^2} f^\star(\mathsf{P}^{\mathrm{T}} \cdot \mathbf{n})\,\mathrm{d}^2 n\,. \tag{9.27}$$

This is the same as the deformability with respect to $\kappa_{t=0}$ [Eq. (9.22)] for arbitrary orthogonal transformations P if and only if $f^\star(\mathbf{n}) = \mathrm{const} = 1/(4\pi)$. In this case, the flow law (9.10) is isotropic. For the general case of a non-constant ODF, the deformabilities (9.22) and (9.27) are not equal for arbitrary orthogonal transformations P, so that the flow law (9.10) is anisotropic, QED.

9.1.4 Some Examples

The properties of the CAFFE flow law will now be investigated for three examples, namely (i) simple shear on a rotated single maximum fabric, (ii) uniaxial compression on a horizontal girdle fabric and (iii) combined shear and compression on a vertical single maximum fabric.

Simple Shear on Rotated Single Maximum Fabric

We consider an ice sample with a single maximum fabric that has been rotated by the angle θ_0 in the x-z-plane away from the vertical direction,

$$f^\star(\mathbf{n}) = \delta(\mathbf{n} - \mathbf{n}_0)\,, \quad \text{with } \mathbf{n}_0 = \begin{pmatrix} \sin\theta_0 \\ 0 \\ \cos\theta_0 \end{pmatrix} \tag{9.28}$$

[$\delta(\cdot)$ denotes Dirac's δ function]. In order to visualise the ODF (9.28), we use *Schmidt diagrams.* In a Schmidt diagram, the upper hemisphere $S^2|_{z\geq 0}$ is projected on the horizontal plane by a Lambert azimuthal equal-area projection (Fig. 9.5). This provides for an easy and intuitive representation of ODFs, and therefore Schmidt diagrams are very commonly used for that purpose. Figure 9.6 shows the Schmidt diagrams of the ODF (9.28) for three different rotation angles θ_0.

The ice sample is subjected to simple shear in the x-z-plane. The according Cauchy stress tensor and deviator are

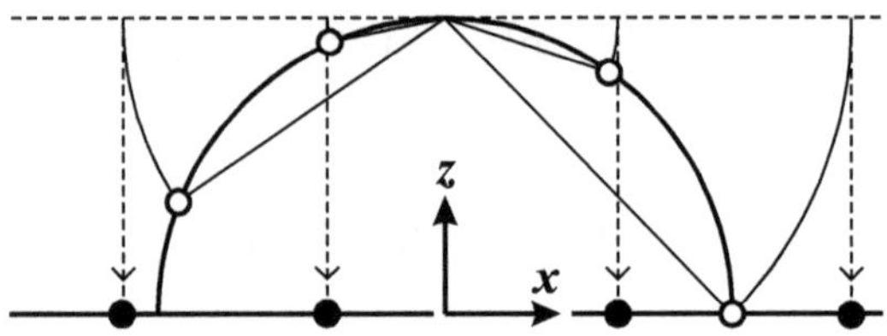

Fig. 9.5. Construction of a Schmidt diagram: Lambert azimuthal equal-area projection of points on the upper hemisphere (open circles) on the horizontal plane (solid circles).

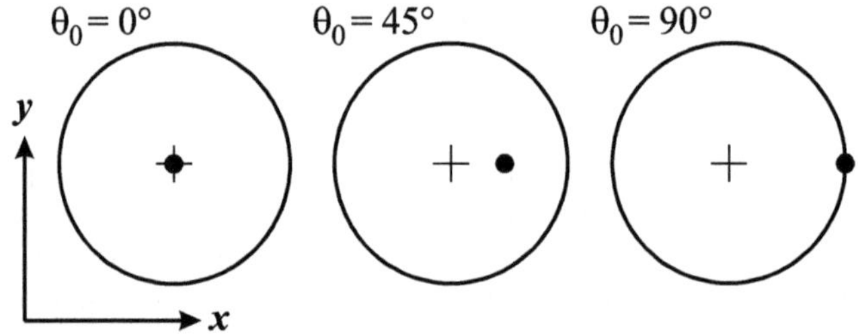

Fig. 9.6. Schmidt diagrams of the rotated single maximum fabric (9.28) for $\theta_0 = 0°$, 45° and 90°.

$$\mathsf{t} = \mathsf{t}^{\mathrm{D}} = \begin{pmatrix} 0 & 0 & \tau \\ 0 & 0 & 0 \\ \tau & 0 & 0 \end{pmatrix}, \tag{9.29}$$

where $\tau = t_{xz}$ is the shear stress.

Inserting Eq. (9.28) in Eq. (9.9) yields the polycrystal deformability

$$\begin{aligned} \mathcal{A} &= 5 \int_{S^2} \frac{S_{\mathrm{t}}^2(\mathbf{n})}{\mathrm{tr}\,(\mathsf{t}^{\mathrm{D}})^2}\, \delta(\mathbf{n} - \mathbf{n}_0)\, \mathrm{d}^2 n \\ &= 5 \frac{S_{\mathrm{t}}^2(\mathbf{n}_0)}{\mathrm{tr}\,(\mathsf{t}^{\mathrm{D}})^2} = 5 \frac{(\mathsf{t}^{\mathrm{D}} \cdot \mathbf{n}_0)^2 - (\mathbf{n}_0 \cdot \mathsf{t}^{\mathrm{D}} \cdot \mathbf{n}_0)^2}{\mathrm{tr}\,(\mathsf{t}^{\mathrm{D}})^2}, \end{aligned} \tag{9.30}$$

where, by using Eq. (9.29),

$$\begin{aligned} (\mathsf{t}^{\mathrm{D}} \cdot \mathbf{n}_0)^2 &= \tau^2 \cos^2\theta_0 + \tau^2 \sin^2\theta_0 = \tau^2, \\ (\mathbf{n}_0 \cdot \mathsf{t}^{\mathrm{D}} \cdot \mathbf{n}_0)^2 &= (2\tau \sin\theta_0 \cos\theta_0)^2 = \tau^2 \sin^2 2\theta_0, \\ \mathrm{tr}\,(\mathsf{t}^{\mathrm{D}})^2 &= 2\tau^2, \end{aligned} \tag{9.31}$$

so that the result is

$$\mathcal{A} = \frac{5}{2}(1 - \sin^2 2\theta_0) = \frac{5}{2} \cos^2 2\theta_0. \tag{9.32}$$

The corresponding enhancement factor [see Eq. (9.12)] is shown in Fig. 9.7. Maximum softness ($E = E_{\mathrm{max}} = 10$) occurs for a vertical ($\theta_0 = 0°$) and a

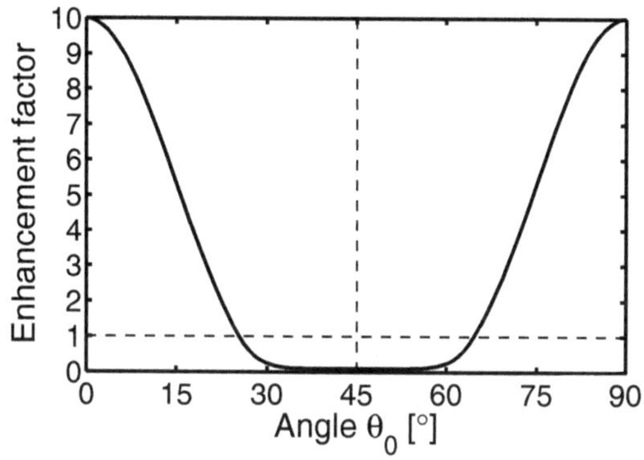

Fig. 9.7. Anisotropic enhancement factor $E(\mathcal{A})$ as a function of the rotation angle θ_0 for the configuration "simple shear on rotated single maximum fabric" [Eq. (9.32)]. Parameters: $E_{\max} = 10$, $E_{\min} = 0.1$.

horizontal ($\theta_0 = 90°$) single maximum, maximum hardness ($E = E_{\min} = 0.1$) for a single maximum rotated by $\theta_0 = 45°$. The ice is softer than isotropic ice ($E > 1$) for angles $\theta_0 < 25.4°$ and $\theta_0 > 64.6°$, and harder than isotropic ice ($E < 1$) for angles in between ($25.4° < \theta_0 < 64.6°$).

The result (9.32) is a nice illustration for the proof of anisotropy given above. Let us assume that the stress exponent n is equal to 3, the temperature is fixed so that $A(T') = A = \text{const}$, and the test is done for the extreme cases (1) $\theta_0 = 0°$ and (2) $\theta_0 = 45°$. Test (1) gives $\mathcal{A}_1 = \frac{5}{2}$, thus $E(\mathcal{A}_1) = E_{\max}$, and due to the flow law (9.10) the resulting shear rate in the x-z-plane is

$$\dot{\gamma}_1 = 2D_{xz} = 2AE_{\max}\tau^3. \tag{9.33}$$

Test (2) gives $\mathcal{A}_2 = 0$, thus $E(\mathcal{A}_2) = E_{\min}$ and

$$\dot{\gamma}_2 = 2D_{xz} = 2AE_{\min}\tau^3. \tag{9.34}$$

Since $E_{\min} \ll E_{\max}$, the shear rate of Eq. (9.34) is much smaller than that of Eq. (9.33). In other words, the material response of the ice specimen has changed considerably due to the 45° rotation of its initial configuration. This clearly fulfills the criterion for an anisotropic material.

Uniaxial Compression on Horizontal Girdle Fabric

As a second example, we consider an ice sample with a girdle fabric that is circularly symmetric with respect to the vertical direction (z-axis) and has the opening angle θ_0.

In order to formulate the ODF of the horizontal girdle fabric, we employ the spherical coordinates θ (zenith angle) and ϕ (azimuth angle) [see, e.g., Bronshtein et al. (2004)]. With respect to a fixed Cartesian coordinate system x, y, z, the orientation $\mathbf{n}$ reads

$$\mathbf{n} = \begin{pmatrix} \cos\phi\sin\theta \\ \sin\phi\sin\theta \\ \cos\theta \end{pmatrix}, \tag{9.35}$$

and the ODF is

$$f^\star(\mathbf{n}) = \frac{1}{2\pi\sin\theta_0}\delta(\theta-\theta_0)\,. \tag{9.36}$$

The factor $1/(2\pi\sin\theta_0)$ is required in order to fulfill the normalisation condition (9.7). Figure 9.8 shows the Schmidt diagrams of the ODF (9.36) for three different opening angles θ_0.

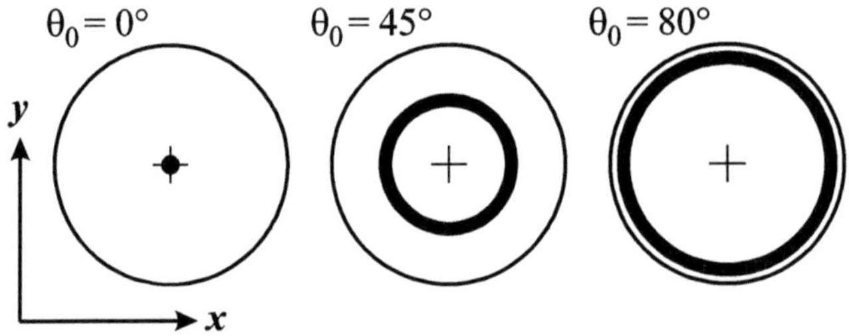

Fig. 9.8. Schmidt diagrams of the horizontal girdle fabric (9.36) for $\theta_0 = 0°$, $45°$ and $80°$.

The ice sample is subjected to uniaxial compression in the vertical direction. The Cauchy stress tensor and deviator read

$$\mathsf{t} = \begin{pmatrix} 0 & 0 & 0 \\ 0 & 0 & 0 \\ 0 & 0 & -\sigma \end{pmatrix} \quad\Rightarrow\quad \mathsf{t}^\mathrm{D} = \begin{pmatrix} \frac{1}{3}\sigma & 0 & 0 \\ 0 & \frac{1}{3}\sigma & 0 \\ 0 & 0 & -\frac{2}{3}\sigma \end{pmatrix}, \tag{9.37}$$

where $\sigma = -t_{zz}$ is the vertical normal stress, counted positive for compression.

The polycrystal deformability results from Eqs. (9.9) and (9.36),

$$\begin{aligned} \mathcal{A} &= \frac{5}{2\pi\sin\theta_0}\int\limits_{S^2} \frac{S_\mathrm{t}^2(\mathbf{n})}{\mathrm{tr}\,(\mathsf{t}^\mathrm{D})^2}\,\delta(\theta-\theta_0)\,\mathrm{d}^2n \\ &= \frac{5}{2\pi\sin\theta_0}\int\limits_{S^2} \frac{(\mathsf{t}^\mathrm{D}\cdot\mathbf{n})^2-(\mathbf{n}\cdot\mathsf{t}^\mathrm{D}\cdot\mathbf{n})^2}{\mathrm{tr}\,(\mathsf{t}^\mathrm{D})^2}\,\delta(\theta-\theta_0)\,\mathrm{d}^2n\,. \end{aligned} \tag{9.38}$$

By using Eqs. (9.35) and (9.37), we find for the several terms the expressions

$$
\begin{aligned}
(\mathrm{t}^\mathrm{D}\cdot\mathbf{n})^2 &= \tfrac{1}{9}\sigma^2(\cos^2\phi\sin^2\theta+\sin^2\phi\sin^2\theta+4\cos^2\theta)\\
&= \tfrac{1}{9}\sigma^2(\sin^2\theta+4\cos^2\theta)\,,\\
(\mathbf{n}\cdot\mathrm{t}^\mathrm{D}\cdot\mathbf{n})^2 &= (\tfrac{1}{3}\sigma\cos^2\phi\sin^2\theta+\tfrac{1}{3}\sigma\sin^2\phi\sin^2\theta-\tfrac{2}{3}\sigma\cos^2\theta)^2\\
&= (\tfrac{1}{3}\sigma\sin^2\theta-\tfrac{2}{3}\sigma\cos^2\theta)^2\\
&= \tfrac{1}{9}\sigma^2(\sin^4\theta-4\sin^2\theta\cos^2\theta+4\cos^4\theta)\,,\\
\mathrm{tr}\,(\mathrm{t}^\mathrm{D})^2 &= \tfrac{2}{3}\sigma^2\,,\\
\mathrm{d}^2n &= \sin\theta\,\mathrm{d}\theta\,\mathrm{d}\phi\,,
\end{aligned}
\tag{9.39}
$$

and Eq. (9.38) can be evaluated as

$$
\begin{aligned}
\mathcal{A} &= \frac{5}{6}(\sin^2\theta_0+4\cos^2\theta_0-\sin^4\theta_0+4\sin^2\theta_0\cos^2\theta_0-4\cos^4\theta_0)\\
&= \frac{5}{6}[\sin^2\theta_0(1-\sin^2\theta_0)+4\cos^2\theta_0(1-\cos^2\theta_0)+4\sin^2\theta_0\cos^2\theta_0]\\
&= \frac{15}{2}\sin^2\theta_0\cos^2\theta_0 \;=\frac{15}{8}(1-\cos^2 2\theta_0)\,.
\end{aligned}
\tag{9.40}
$$

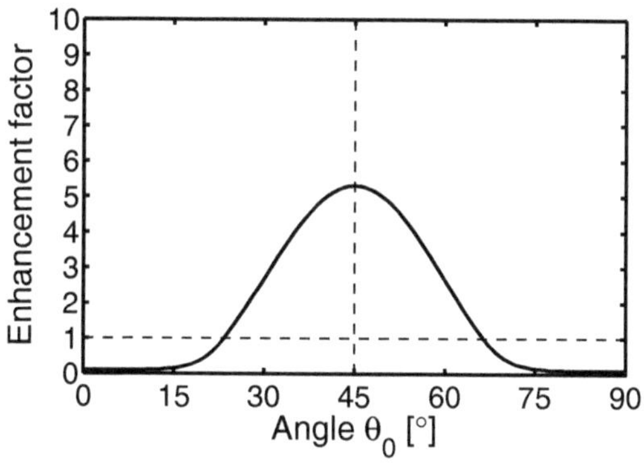

Fig. 9.9. Anisotropic enhancement factor $E(\mathcal{A})$ as a function of the opening angle θ_0 for the configuration "uniaxial compression on horizontal girdle fabric" [Eq. (9.40)]. Parameters: $E_{\max}=10$, $E_{\min}=0.1$.

Figure 9.9 shows the enhancement factor which results from the deformability (9.40). The ice sample is softest for a girdle fabric with opening angle $\theta_0=45°$; however, in contrast to the above case of simple shear on a rotated single maximum fabric, the enhancement factor does not reach the maximum value $E_{\max}=10$, but only $E(\theta_0\!=\!45°)=5.31$. Within the interval of opening angles $23.5°<\theta_0<66.5°$, the ice sample is softer than isotropic ice ($E>1$). Outside of this interval it is harder than isotropic ice ($E<1$), and maximum hardness ($E=E_{\min}=0.1$) occurs for $\theta_0=0°$ (vertical single maximum) and $\theta_0=90°$ ("equatorial" girdle).

Combined Shear and Compression on Vertical Single Maximum

As a last example, let us assume an ice sample with a vertical single maximum fabric,

$$f^\star(\mathbf{n}) = \delta(\mathbf{n} - \mathbf{e}_z)\,, \tag{9.41}$$

loaded by a combination of simple shear and uniaxial compression,

$$\mathsf{t} = \begin{pmatrix} 0 & 0 & \tau \\ 0 & 0 & 0 \\ \tau & 0 & -\sigma \end{pmatrix} \quad \Rightarrow \quad \mathsf{t}^{\mathrm{D}} = \begin{pmatrix} \frac{1}{3}\sigma & 0 & \tau \\ 0 & \frac{1}{3}\sigma & 0 \\ \tau & 0 & -\frac{2}{3}\sigma \end{pmatrix}. \tag{9.42}$$

Computation of the polycrystal deformability (9.9) is straightforward,

$$\begin{aligned} \mathcal{A} &= 5 \int_{S^2} \frac{S_{\mathsf{t}}^2(\mathbf{n})}{\mathrm{tr}\,(\mathsf{t}^{\mathrm{D}})^2}\, \delta(\mathbf{n} - \mathbf{e}_z)\, \mathrm{d}^2 n \\ &= 5 \frac{S_{\mathsf{t}}^2(\mathbf{e}_z)}{\mathrm{tr}\,(\mathsf{t}^{\mathrm{D}})^2} = 5 \frac{(\mathsf{t}^{\mathrm{D}} \cdot \mathbf{e}_z)^2 - (\mathbf{e}_z \cdot \mathsf{t}^{\mathrm{D}} \cdot \mathbf{e}_z)^2}{\mathrm{tr}\,(\mathsf{t}^{\mathrm{D}})^2}\,, \end{aligned} \tag{9.43}$$

and with the expressions

$$\begin{aligned} (\mathsf{t}^{\mathrm{D}} \cdot \mathbf{e}_z)^2 &= \tau^2 + \tfrac{4}{9}\sigma^2\,, \\ (\mathbf{e}_z \cdot \mathsf{t}^{\mathrm{D}} \cdot \mathbf{e}_z)^2 &= \tfrac{4}{9}\sigma^2\,, \\ \mathrm{tr}\,(\mathsf{t}^{\mathrm{D}})^2 &= 2\tau^2 + \tfrac{2}{3}\sigma^2 \end{aligned} \tag{9.44}$$

we obtain

$$\mathcal{A} = \frac{5}{2} \times \frac{\tau^2}{\tau^2 + \frac{1}{3}\sigma^2} = \frac{5}{2} \times \frac{1}{1 + \frac{1}{3}\left(\frac{\sigma}{\tau}\right)^2}\,. \tag{9.45}$$

Figure 9.10 shows the resulting enhancement factor as a function of the stress ratio σ/τ. The limits are as expected, namely $E \to E_{\mathrm{max}} = 10$ for $\sigma/\tau \to 0$ (shear dominates) and $E \to E_{\mathrm{min}} = 0.1$ for $\sigma/\tau \to \infty$ (compression dominates). In case of equal stresses the enhancement factor has the value $E(\sigma/\tau = 1) = 5.31$, so that softening due to shear outweighs hardening due to compression. Only for stress ratios $\sigma/\tau > 2.12$ is the enhancement factor less than unity, so that hardening is dominant.

In fact, the behaviour of the configuration for similar stresses ($\sigma/\tau \sim 1$) reveals a general weakness of the CAFFE flow law for complex states of stress or deformation. The modelled softening or hardening is necessarily the same for all directions, whereas in reality different directions will show different degrees of softening or hardening. Concretely, for the above example the ice sample will be soft for shear and hard for compression, while the CAFFE flow law predicts some average softening or hardening for both shear and compression, depending on the stress ratio. This shortcoming is a tribute to the simple formulation with a scalar enhancement factor, which allows the

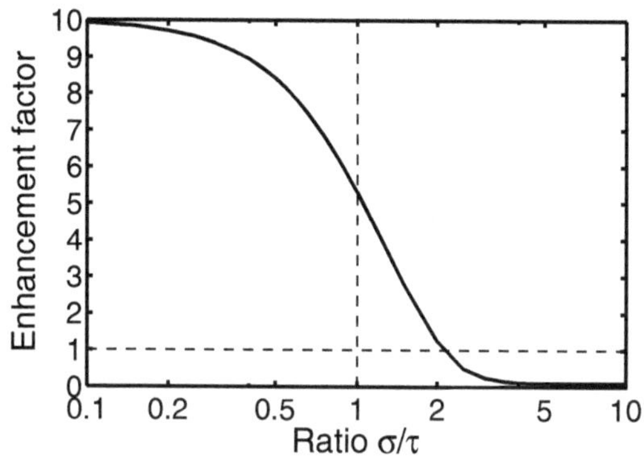

Fig. 9.10. Anisotropic enhancement factor $E(\mathcal{A})$ as a function of the stress ratio σ/τ for the configuration "combined shear and compression on vertical single maximum" [Eq. (9.45)]. Parameters: $E_{\text{max}} = 10$, $E_{\text{min}} = 0.1$.

flow law to be set up with only two well-known parameters (E_{max}, E_{min}). More complex anisotropic flow laws which give up the collinearity between the stress deviator t^{D} and the strain-rate tensor D have been formulated [see, e.g., the overview by Gagliardini et al. (2009)], but the price to pay is an increased number of parameters and greater computational demands.

9.1.5 Evolution of Anisotropy

Orientation Mass Balance

The anisotropic flow law in the form (9.10) or (9.15) needs to be complemented by an evolution equation for the anisotropic fabric. This is done by formulating an *orientation mass balance* for the OMD $\rho^\star(\mathbf{n})$.

We are not going to enter into the detailed formalism of orientation balance equations here [see, e.g., Faria (2003), Placidi (2004)]. Instead, we rather motivate the form of the orientation mass balance by generalizing the ordinary mass balance (3.58). The difference is that, in addition to the dependencies on the position vector $\mathbf{x} \in \mathcal{E}$ and the time t, the density and velocity fields also depend on the orientation vector $\mathbf{n} \in S^2$, which is indicated by the notation $\rho^\star(\mathbf{n})$ and $\mathbf{v}^\star(\mathbf{n})$. The velocity, which describes motions in the physical space $\mathcal{E}$, is complemented by an *orientation transition rate* $\mathbf{u}^\star(\mathbf{n})$, which describes motions on the unit sphere, that is, changes of the orientation due to grain rotation (Fig. 9.11). An *orientation flux* $\mathbf{q}^\star(\mathbf{n})$ is introduced, which allows redistributions of the OMD due to polygonisation (rotation recrystallisation). Consequently, the orientation mass balance reads

$$\frac{\partial \rho^\star}{\partial t} + \operatorname{div}(\rho^\star \mathbf{v}^\star) + \operatorname{div}_{S^2}(\rho^\star \mathbf{u}^\star + \mathbf{q}^\star) = \rho^\star \Gamma^\star \,. \qquad (9.46)$$

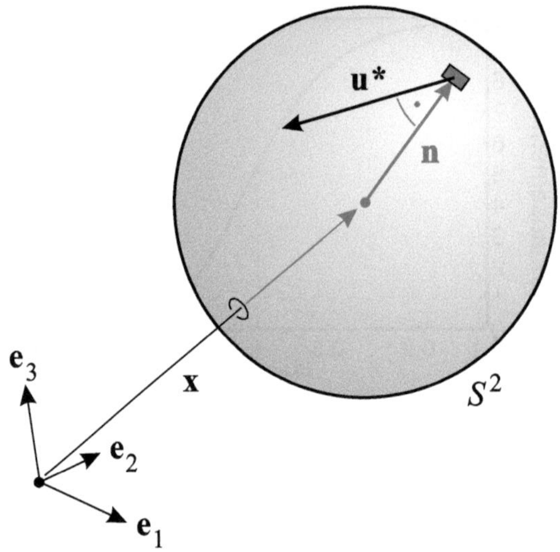

Fig. 9.11. Orientation transition rate $\mathbf{u}^{\star}(\mathbf{n})$ on the unit sphere S^2.

The first two terms on the left-hand side are straightforward generalisations of the terms in the ordinary mass balance (3.58). The third term on the left-hand side is the equivalent of the second term for the orientation transition rate $\mathbf{u}^{\star}(\mathbf{n})$ and the orientation flux $\mathbf{q}^{\star}(\mathbf{n})$, where div_{S^2} is the divergence operator on the unit sphere. On the right-hand side, a source term appears which allows certain orientations to be produced at the expense of others. The quantity $\Gamma^{\star}(\mathbf{n})$ is therefore called the *orientation production rate.* Physically, it describes dynamic recrystallisation (migration recrystallisation) and all other processes in which the transport of mass from one grain, having a certain orientation, to another grain, having a different orientation, cannot be neglected.

In the following, we will make the reasonable assumption that the spatial velocity does not depend on the orientation, that is, $\mathbf{v}^{\star}(\mathbf{n}) = \mathbf{v}$. Therefore, the orientation mass balance (9.46) simplifies to

$$\frac{\partial \rho^{\star}}{\partial t} + \mathrm{div}\,(\rho^{\star}\mathbf{v}) + \mathrm{div}_{S^2}(\rho^{\star}\mathbf{u}^{\star} + \mathbf{q}^{\star}) = \rho^{\star}\Gamma^{\star}\,. \tag{9.47}$$

By using the Gauss theorem and the mass-conservation requirement

$$\int\limits_{S^2} \rho^{\star}(\mathbf{n})\,\Gamma^{\star}(\mathbf{n})\,\mathrm{d}^2 n = 0\,, \tag{9.48}$$

integration of Eq. (9.47) over S^2 (all orientations) gives the ordinary mass balance

$$\frac{\partial \rho}{\partial t} + \mathrm{div}\,(\rho\mathbf{v}) = 0\,. \tag{9.49}$$

In order to solve the orientation mass balance (9.47), constitutive relations for the orientation transition rate $\mathbf{u}^\star(\mathbf{n})$, the orientation flux $\mathbf{q}^\star(\mathbf{n})$ and the orientation production rate $\Gamma^\star(\mathbf{n})$ must be provided as closure conditions.

Constitutive Relation for the Orientation Transition Rate

As mentioned above, the orientation transition rate corresponds physically to grain rotation. Since grain rotation is induced by shear deformation in the basal plane, it is reasonable to assume that it is controlled by the shear rate $D_\mathrm{t}\mathbf{e}_\mathrm{t}$ [Eq. (9.16)]. In the CAFFE model, a linear relation is applied,

$$\mathbf{u}^\star(\mathbf{n}) = -\iota\, D_\mathrm{t}\mathbf{e}_\mathrm{t} + \mathsf{W}\cdot\mathbf{n} = -\iota\cdot[\mathsf{D}\cdot\mathbf{n} - (\mathbf{n}\cdot\mathsf{D}\cdot\mathbf{n})\mathbf{n}] + \mathsf{W}\cdot\mathbf{n}\,. \tag{9.50}$$

The parameter ι is assumed to be a positive constant. The additional term $\mathsf{W}\cdot\mathbf{n}$ involving the spin tensor W describes the contribution of local rigid-body rotations; see Eqs. (3.29) and (3.36) – (3.39).

In the special case $\iota = 1$, the basal planes are material area elements, that is, they carry out an affine rotation. However, due to geometric incompatibilities of the deformation of individual crystallites in the polycrystalline aggregate, affine rotations are not plausible, and we expect realistic values of ι to be less than unity. In fact, Placidi (2004) showed that the fabrics in the upper 2000 m of the GRIP ice core in central Greenland can be best explained by the value $\iota \approx 0.4$. A study by Seddik et al. (2008) on the EDML ice core in Dronning Maud Land, East Antarctica, provided a best fit between modelled and measured fabrics for $\iota = 0.6$ (see below, Sect. 9.1.6).

Constitutive Relation for the Orientation Flux

The orientation flux is supposed to describe polygonisation (rotation recrystallisation). It is modelled as an isotropic diffusive process,

$$\mathbf{q}^\star(\mathbf{n}) = -\lambda\,\mathrm{grad}_{S^2}\rho^\star(\mathbf{n})\,, \tag{9.51}$$

where the parameter $\lambda > 0$ is the orientation diffusivity and grad_{S^2} is the gradient operator on the unit sphere. Equation (9.51) is equivalent to Fick's law of diffusion on the unit sphere.

A problem is that very few data exist which allow values of the parameter λ to be constrained. This requires further attention.

Constitutive Relation for the Orientation Production Rate

The driving force for the orientation production rate, which essentially models dynamic recrystallisation (migration recrystallisation), is macroscopic deformations of the polycrystal, which can be more easily followed on the microscopic scale by grains oriented favourably for the given deformation. Therefore,

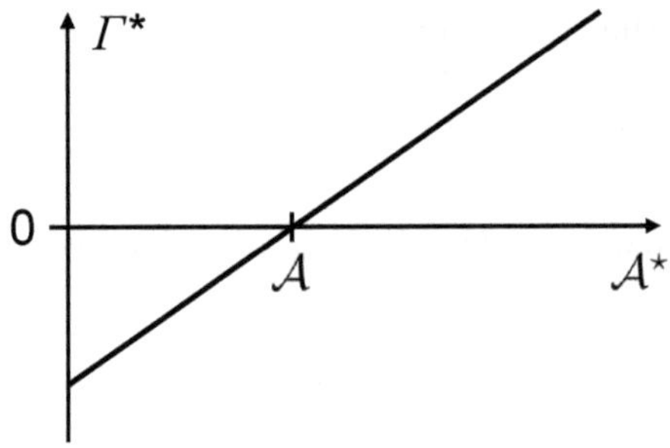

Fig. 9.12. Orientation production rate according to Eq. (9.52).

it is reasonable to assume that the orientation production rate for a certain orientation $\mathbf{n}$ is related to the crystallite deformability $\mathcal{A}^\star(\mathbf{n})$ [Eqs. (9.4), (9.18)]. In the CAFFE model, the linear relation

$$\Gamma^\star(\mathbf{n}) = \Gamma\,[\mathcal{A}^\star(\mathbf{n}) - \mathcal{A}] \tag{9.52}$$

is proposed. Subtraction of the polycrystal deformability $\mathcal{A}$ is required in order to fulfill the mass-conservation condition (9.48). The parameter Γ is assumed to be positive, which guarantees a positive mass production for favourably oriented grains, and a negative production for unfavourably oriented grains (Fig. 9.12). Since dynamic recrystallisation is expected to be strongly dependent on temperature, Γ should increase with increasing temperature. However, as for the case of the orientation diffusivity λ, it is not possible at the moment to constrain values of Γ in a reasonable fashion.

The formulation of the CAFFE model is now complete. Equation (9.10) is the actual flow law, which replaces its isotropic counterpart (4.16). Anisotropy enters via the enhancement factor $E(\mathcal{A})$ [Eq. (9.12)], which depends on the deformability $\mathcal{A}$ defined in Eq. (9.9). Computation of the deformability requires knowledge of the orientation mass density $\rho^\star$, which is governed by the evolution equation (9.47) and the constitutive relations (9.50), (9.51) and (9.52).

9.1.6 Application to the EDML Core, Antarctica

The fabric evolution equation (9.47) of the CAFFE model was solved numerically by Seddik et al. (2008) for the site of the EPICA (European Project for Ice Coring in Antarctica) ice core at Kohnen Station, Dronning Maud Land, Antarctica (referred to in short as "EDML core"). Kohnen Station is situated at 75°00′06″S, 00°04′04″E, at an altitude of 2892 m AMSL (above mean sea level), and the overall length of the EDML core is 2774 m (EPICA Community Members 2006). Preliminary fabric data are available from 50 m until 2570 m depth (I. Hamann, personal communication 2007, Eisen et al. 2007; shown in Fig. 9.13), against which the results of the CAFFE model can be tested. The measured fabrics show a gradual transition from randomly oriented c-axes in shallower depths (down to 600 m) to a broad vertical girdle fabric ($\sim$ 600 to 1000 m depth). A narrowing of the vertical girdle fabric follows between $\sim$ 1000 and 2000 m depth. A sudden change in the flow regime is indicated by a vertical alignment of the c-axes over only 10 m towards a single maximum ($\sim$ 2040 m depth). Near the bottom (Schmidt diagram for 2563 m depth), the single maximum fabric shows a slightly widened structure.

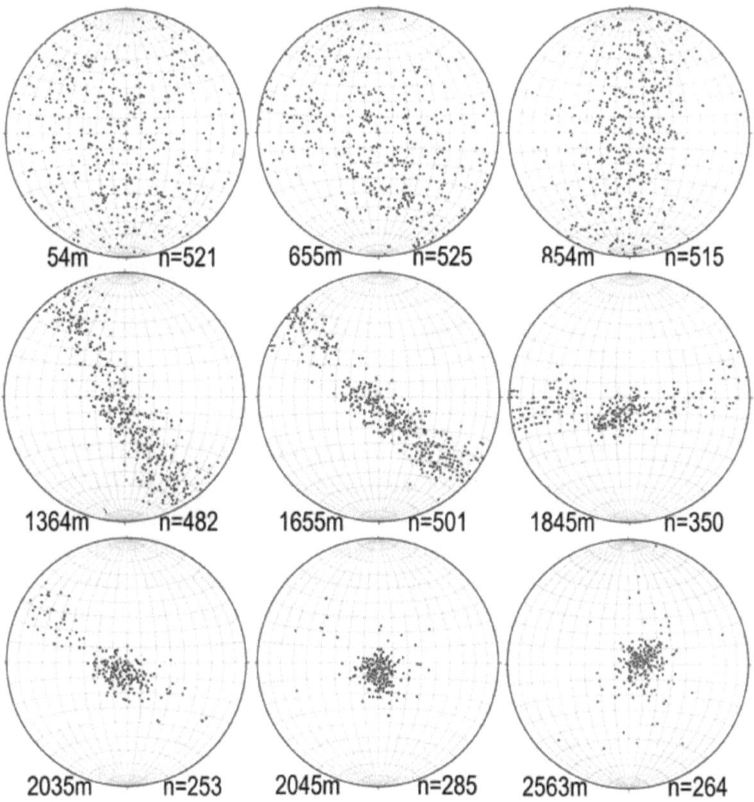

Fig. 9.13. Selected Schmidt diagrams for the observed fabrics of the EDML ice core between 54 m and 2563 m depth. Each dot represents the orientation of a single grain, and n denotes the number of grains included. Note that the orientations of the horizontal planes with respect to the ice flow direction are unknown. Reproduced from Seddik et al. (2008), based on I. Hamann (personal communication 2007) and Eisen et al. (2007), © International Glaciological Society.

Based on the fact that Kohnen Station is located on a flank of the Antarctic Ice Sheet (rather than a dome like most other deep ice cores), the local flow profile $\mathbf{v}(z)$ for the EDML core was reconstructed by Seddik et al. (2008) with the following settings and assumptions:

- Horizontal coordinates x, y defined such that Kohnen Station marks the origin, and the x-axis points in the downslope ($\sim 260°$, WSW) direction.
- Horizontal flow field governed by the shallow ice approximation , so that $v_x(z)$ is essentially governed by Eq. (5.84) [neglected basal sliding $v_{\mathrm{b}x}$, $A(T')$ replaced by $E(\mathcal{A})\,A(T')$], and $v_y(z) = 0$.
- Vertical velocity $v_z(z)$ results from a Dansgaard-Johnsen-type distribution (Dansgaard and Johnsen 1969) of the vertical strain rate $D_{zz} = \partial v_z/\partial z$ (constant negative D_{zz} from the surface down to two thirds of the ice thickness, linearly increasing D_{zz} below, almost zero at the base).
- Similar for the temperature profile $T(z)$: constant $T = T_{\mathrm{s}}$ (surface temperature) from the surface down to two thirds of the ice thickness, linearly increasing T below, pressure melting point at the base.
- Horizontal stretching in x-direction only: $D_{xx}(z) = -D_{zz}(z)$, $D_{yy}(z) = 0$.

Under the additional assumptions of steady-state conditions, that is, $\partial(\cdot)/\partial t = 0$, and an isotropic fabric at the ice surface, this allows a numerical solution of the fabric evolution equation (9.47), which yields the OMB $\rho^{\star}$ as a function of the vertical coordinate z [for details see Seddik et al. (2008)]. Recrystallisation processes were neglected ($\lambda = 0$, $\Gamma = 0$), and the remaining parameter ι was set to the value 0.6, which yielded the best fit between modelled and measured fabrics.

The computed Schmidt diagrams for the ODF $f^{\star}(z) = \rho^{\star}(z)/\rho$ are presented in Fig. 9.14. For the sake of easy comparison with the fabric data, the depths are the same as in Fig. 9.13. We see a rather good agreement between the data and the model results for the transition with depth from isotropy via broad and narrow vertical girdle fabrics to a single maximum at about 2040 m. The computed girdle fabrics are aligned with the y-axis (perpendicular to the flow direction), while the alignments of the observed fabrics appear erratic because the processing of the ice core did not preserve the orientation in the horizontal plane. Compared to the data, the computed girdle fabrics show a stronger concentration of c-axes towards the vertical. According to Paterson (1994, p. 198), this discrepancy indicates that some amount of compression occurs in transverse (y-) direction, which has not been accounted for in the above assumptions. At the depths of 2035 m and 2045 m, the single maximum fabrics obtained have a noticeable deviation from the vertical (center of the Schmidt diagrams). This is also in agreement with the data, and is an effect of the transition of the deformation regime from mainly pure shear (compression in the z-direction, stretching in the x-direction) to mainly bed-parallel simple shear. However, at 2563 m depth, the modelled fabric shows a widened and somewhat decayed structure, which is reflected in the data only to a limited extent. The spurious spatial oscillations in this fabric indicate that it is likely

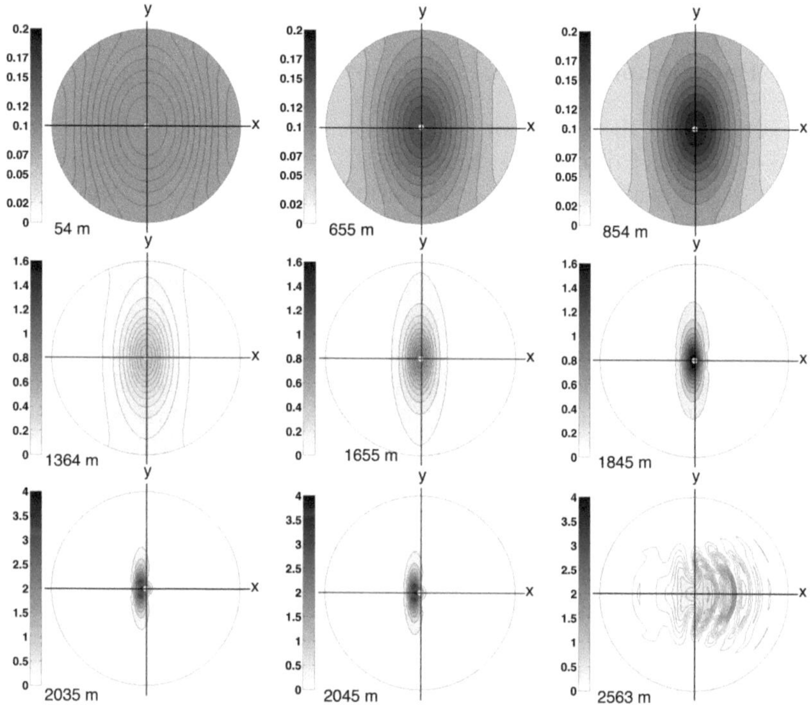

Fig. 9.14. Schmidt diagram representation of the EDML fabrics [ODF $f^{\star}(\mathbf{n})$] at depths between 54 m and 2563 m computed by the evolution equation (9.47) of the CAFFE model. Reproduced from Seddik et al. (2008), © International Glaciological Society.

influenced by a numerical instability of the applied finite volume scheme near the base. Further, the disregard of recrystallisation processes is critical in the warm, near-basal part of the ice core, where dynamic recrystallisation is expected to be important. It is thus understandable that the model results do not reproduce the observations so well below $\sim$ 2100 m depth. Future work will aim at constraining values for the parameters λ and $\varGamma$ in order to include recrystallisation processes in applications of the CAFFE model.

9.2 Compressible Firn

9.2.1 Background

Glaciers and ice sheets do not only consist of ice that is incompressible, as assumed in earlier discussions. In most glaciers and ice sheets we find a layer of *firn* at the surface in the accumulation area. Firn is sometimes defined as snow that is older than one year, thus having survived at least one melt season. A characteristic property of firn is its porosity; that is, it consists of a mixture of ice crystallites and air. Thus, the density of firn is smaller than that of ice.

The density of a firn layer increases as it moves deeper into the glacier. This is a consequence of the increasing pressure due to the subsequent deposition of additional firn layers every year. At some depth, the firn is transformed into ice at almost the density of pure polycrystalline ice. This depth defines the thickness of the firn layer.

The accumulation area, and thus the firn zone of glaciers, may extend over a large range of altitude and thus climatic conditions. Depending on the summer climate at the given altitudes, a variable fraction of the firn and snow at the surface may melt and the meltwater may percolate into the firn to variable depths. At very high altitudes the snow may stay dry throughout the melt season and melt never occurs.

The densification and transformation of firn into ice strongly depends on the amount of water percolating into the firn. The presence of water makes the densification process extremely complex due to melting, refreezing, recrystallisation, and the formation of ice lenses and ice layers within the firn. Generally, the densification occurs faster with higher water content. Consequently, the total thickness of the firn layer decreases with higher water content. Wet firn also loses more of the enclosed air and the resulting ice contains fewer bubbles than cold firn and ice.

Generally, firn only occurs in a surface layer in the higher parts of glaciers and ice sheets and the thickness of the layer is generally much smaller than the total ice thickness. A few exceptional cases exist, such as the highest parts of cold mountain glaciers (Lüthi and Funk 2001) and firn-filled volcanic craters (Zwinger et al. 2007).

9.2.2 Densification of Firn

Figure 9.15 shows a density profile in the uppermost part of the Fiescherhorn Glacier in the Swiss Alps, at about 3800 m AMSL (above mean sea level). The firn is not entirely dry and some ice layers are found in the firn. The density profile shows four typical and clearly distinguishable parts separated by kinks in the profile. In the layer between the surface and about 10 m depth the density increases from about 400 to 550 $\mathrm{kg\,m^{-3}}$. In the second layer down to 40 m the density increases nearly linearly with depth to about 800 $\mathrm{kg\,m^{-3}}$.

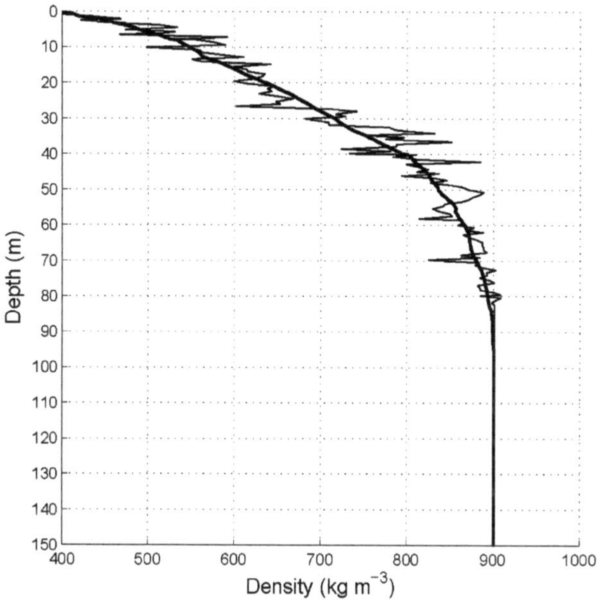

Fig. 9.15. Density profile measured on Fiescherhorn Glacier, Swiss Alps. Reproduced from Schwerzmann (2006), © A. A. Schwerzmann.

Further down, the density increases at a decreasing rate, and comes close to pure ice density at a depth of 90 m. Below this depth the ice has an almost constant density of $900\,\mathrm{kg\,m^{-3}}$, slightly less than the density of pure ice. This is a typical density profile for dry or nearly dry snow zones. The profile gives rise to a distinction of consecutive stages of different types of densification processes, sometimes called *initial, intermediate* and *final stages of densification.*

The initial stage of densification is complex and conceptually difficult to describe and quantify. In low density snow and firn, the ice forms a fragile grid prone to fracturing and rearrangement of the ice grains. When temperature varies, transport of vapour, evaporation and recrystallisation lead to grain growth and support the sintering process. Since low density snow and firn occur only in the uppermost few metres of the firn area, it may not contribute significantly to the overall glacier flow. However, since low density firn is much softer than ice, it may lead to substantial shearing in the surface layer even at low stresses (Fig. 9.16).

In the intermediate stage, the densification is determined by viscoplastic deformation of the ice at the contact points between ice grains. The air is still connected throughout this part of the firn and is at atmospheric pressure. To

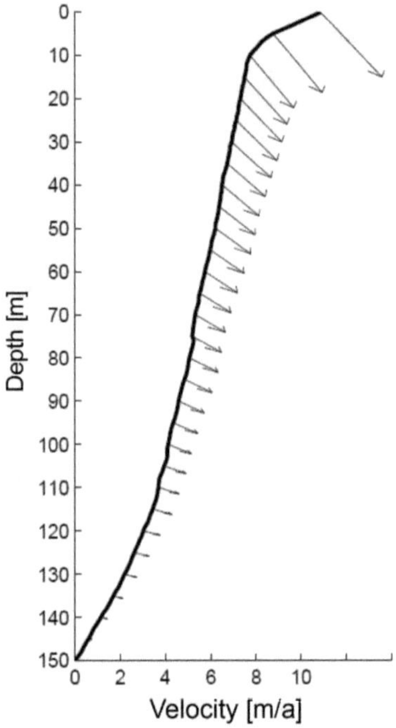

Fig. 9.16. Velocity profile measured on Fiescherhorn Glacier, Swiss Alps. Reproduced from Schwerzmann (2006), © A. A. Schwerzmann.

a good approximation, the intermediate stage is modelled as an assembly of closely packed spherical grains of ice.

The final stage is determined by the closure of air bubbles. Thus, the final stage firn consists of a solid matrix of ice with pressurised air bubbles completely enclosed by the ice. A conceptual model assumes a spherical void filled with air under pressure enclosed in a spherical grain of ice under isotropic pressure.

At even higher pressure the air in bubbles may start to immerse into the ice crystallites. Due to this air the ice formed by densification of dry or nearly dry firn cannot reach the density of pure ice. At high enough pressures and low enough temperatures, the enclosed gases may form clathrates, i.e., the gas molecules may become enclosed in polyhedral cavities of water molecules in the ice lattice. These clathrates may form macroscopically visible crystals enclosed in normal ice crystallites.

9.2.3 Constitutive Relation for Firn

Similar to ice, firn is assumed to be an isotropic, viscous, heat-conducting fluid. However, from the above considerations it has become clear that compressibility must be taken into account. Therefore, a material function of the form (3.127) is appropriate; however, the bulk and shear viscosities cannot be taken as constants.

Since hydrostatic equilibrium in the presence of gravity does not exist for firn, the thermodynamic pressure is negligible. Thus, the total pressure is equal to the viscous pressure,

$$p_{\text{tot}} = p_{\text{visc}} = -\zeta \,\text{tr}\,\mathsf{D} = -\zeta \,\text{div}\,\mathbf{v}\,, \tag{9.53}$$

see Eq. (3.129). The remaining, deviatoric part of the stress tensor t^{D} is related to the strain-rate deviator D^{D} via

$$\mathsf{t}^{\text{D}} = 2\eta\,\mathsf{D}^{\text{D}}\,. \tag{9.54}$$

For the description of the firn rheology, it is convenient to introduce the dimensionless *relative density*

$$\tilde{\rho} = \frac{\rho}{\hat{\rho}_{\text{i}}}\,, \tag{9.55}$$

where ρ is the density of firn and $\hat{\rho}_{\text{i}} = 910\,\text{kg}\,\text{m}^{-3}$ is the bulk density of pure ice. The following power laws for the shear and bulk viscosities, respectively, have been established (Gagliardini and Meyssonnier 1997, Zwinger et al. 2007),

$$\eta(T',\delta) = \frac{1}{2a(\tilde{\rho})} B(T')\,\delta^{-(1-1/n)}\,, \tag{9.56}$$

$$\zeta(T',\delta) = \frac{1}{2b(\tilde{\rho})} B(T')\,\delta^{-(1-1/n)}\,, \tag{9.57}$$

where

$$\delta = \left(\frac{\frac{1}{2}\,\text{tr}\,(\mathsf{D}^{\text{D}})^2}{a(\tilde{\rho})} + \frac{(\text{tr}\,\mathsf{D})^2}{4b(\tilde{\rho})} \right)^{1/2} = \left(\frac{\text{tr}\,(\mathsf{D}^{\text{D}})^2}{2a(\tilde{\rho})} + \frac{(\text{div}\,\mathbf{v})^2}{4b(\tilde{\rho})} \right)^{1/2} \tag{9.58}$$

is a strain invariant. The above viscosities fulfill the relation

$$\zeta(T',\delta) = \frac{a(\tilde{\rho})}{b(\tilde{\rho})}\,\eta(T',\delta)\,. \tag{9.59}$$

The dependency of $a(\tilde{\rho})$ and $b(\tilde{\rho})$ upon the relative density has been obtained from field data,

$$a(\tilde{\rho}) = \begin{cases} \exp(c_1 - c_2\,\tilde{\rho})\,, & \tilde{\rho} \le 0.81\,, \\ \left[1 + \dfrac{2}{3}(1-\tilde{\rho})\right] \tilde{\rho}^{-2n/(n+1)}\,, & \tilde{\rho} > 0.81\,, \end{cases}$$

$$b(\tilde{\rho}) = \begin{cases} \exp(c_3 - c_4\,\tilde{\rho})\,, & \tilde{\rho} \le 0.81\,, \\ \dfrac{3}{4} \left\{ \dfrac{(1-\tilde{\rho})^{1/n}}{n \cdot \left[1 - (1-\tilde{\rho})^{1/n}\right]} \right\}^{2n/(n+1)}\,, & \tilde{\rho} > 0.81\,. \end{cases} \tag{9.60}$$

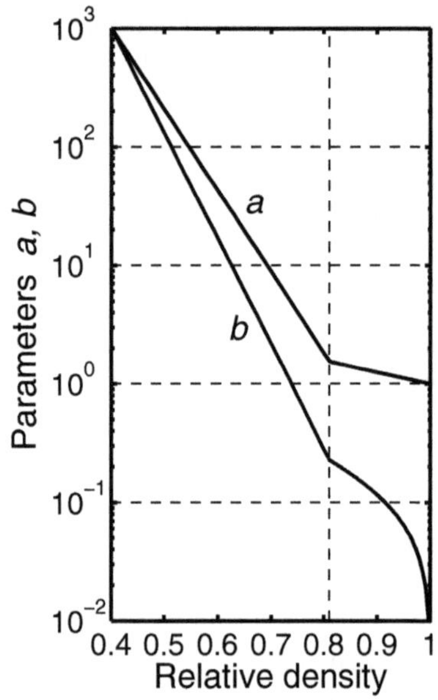

Fig. 9.17. Parameters a and b for the shear and bulk viscosities of compressible firn according to Eq. (9.60).

For $n = 3$, the constants are $c_1 = 13.22240$, $c_2 = 15.78652$, $c_3 = 15.09371$ and $c_4 = 20.46489$, see Fig. 9.17. In the limit of pure ice, $\tilde{\rho} \to 1$, we have $a \to 1$, $b \to 0$, $\operatorname{div}\mathbf{v} \to 0$ and $\delta \to d_\mathrm{e}$, and thus Glen's flow law in the form (4.21) with the shear viscosity (4.22) is restored.

9.2.4 Field Equations

Due to the compressibility of firn, the mass balance must be applied in the form of Eq. (3.58),

$$\frac{\mathrm{d}\rho}{\mathrm{d}t} + \rho \operatorname{div}\mathbf{v} = \frac{\partial \rho}{\partial t} + (\operatorname{grad}\rho)\cdot\mathbf{v} + \rho\operatorname{div}\mathbf{v} = 0\,. \tag{9.61}$$

The density ρ is now a prognostic variable, and it is necessary to incorporate the densification process. The stress tensor t for a compressible fluid is given by Eq. (3.127),

$$\mathsf{t} = -p(\rho, T)\,\mathsf{I} + (\zeta \operatorname{tr}\mathsf{D})\,\mathsf{I} + 2\eta\,\mathsf{D}^\mathrm{D}\,, \tag{9.62}$$

and with Eqs. (3.128), (3.129) and the negligibility of the thermodynamic pressure this yields

$$\begin{aligned} \mathsf{t} &= (\zeta \operatorname{tr}\mathsf{D})\,\mathsf{I} + 2\eta\,\mathsf{D}^\mathrm{D} \\ &= (\zeta \operatorname{tr}\mathsf{D})\,\mathsf{I} + 2\eta\left[\mathsf{D} - \frac{1}{3}(\operatorname{tr}\mathsf{D})\,\mathsf{I}\right] \\ &= (\lambda \operatorname{tr}\mathsf{D})\,\mathsf{I} + 2\eta\,\mathsf{D} \end{aligned} \tag{9.63}$$

(where $\lambda = \zeta - 2\eta/3$). For firn, the viscosities are functions of position, and thus, the divergence of the stress tensor is

$$\begin{aligned} \operatorname{div} \mathsf{t} &= \operatorname{grad}(\lambda \operatorname{div} \mathbf{v}) + \operatorname{div}(2\eta \mathsf{D}) \\ &= \operatorname{grad}(\lambda \operatorname{div} \mathbf{v}) + \operatorname{div}\left[\eta\left(\operatorname{grad} \mathbf{v} + (\operatorname{grad} \mathbf{v})^{\mathrm{T}}\right)\right]. \end{aligned} \tag{9.64}$$

Inserting this result in the momentum balance (3.72), neglecting the acceleration term and setting $\mathbf{f} = \rho\mathbf{g}$ yields the equation of motion

$$\operatorname{grad}(\lambda \operatorname{div} \mathbf{v}) + \operatorname{div}\left[\eta\left(\operatorname{grad} \mathbf{v} + (\operatorname{grad} \mathbf{v})^{\mathrm{T}}\right)\right] + \rho\mathbf{g} = \mathbf{0}. \tag{9.65}$$

The temperature evolution equation (5.14) for ice sheets can still be applied,

$$\rho c \frac{\mathrm{d}T}{\mathrm{d}t} = \operatorname{div}(\kappa \operatorname{grad} T) + 4\eta\, d_{\mathrm{e}}^2. \tag{9.66}$$

Expressions for the heat conductivity κ and specific heat c as functions of density and temperature have been suggested by Zwinger et al. (2007).

Based on measurements in the cold firn area of central Greenland, *Sorge's Law* was formulated (Bader 1954):

> At any given location in a dry firn area of a glacier, the density as a function of depth below the surface is invariant with time.

Since the surface change of glaciers and ice sheets is slowest in their highest reaches, Sorge's Law may be interpreted as

$$\frac{\partial \rho}{\partial t} = 0 \tag{9.67}$$

in an Eulerian description of a quasi-stationary field. Thus, Eq. (9.61) reduces to

$$(\operatorname{grad} \rho) \cdot \mathbf{v} + \rho \operatorname{div} \mathbf{v} = 0. \tag{9.68}$$

An advection equation of the form (9.68) is difficult to solve numerically. A possible solution strategy uses the fact that Eq. (9.61) is equivalent to an integral along the particle trajectories. Thus Eq. (9.68) can be transformed into

$$\rho \operatorname{div} \mathbf{v} + v_s \frac{\partial \rho}{\partial s} = 0, \tag{9.69}$$

where s and v_s are the arc length and component of the velocity vector along the particle trajectory, respectively. Equation (9.69) constitutes an ordinary differential equation for ρ that can be solved stably by integration along trajectories. With the additional assumption of a stationary flow field, the particle trajectories are identical to the flow lines. In this case, a solution strategy first solves Eq. (9.65) for the velocity with an initially assumed density field, and then solves Eq. (9.69) for the density with the resulting velocity field. This procedure offers the possibility of a fixed point iteration scheme to obtain consistent solutions for the velocity and density fields. To obtain a stable and smooth convergence, a relaxation scheme analogue to Eq. (6.118) may be employed for both fields.

Exploiting Sorge's Law simplifies the problem further if the density field can be prescribed based on field observations. This then reduces the problem to the computation of the velocity field for a given density field.

9.2.5 Parallel Sided Slab

Governing Equations

Similar to Sect. 7.2, a parallel sided slab with a given layering of the firn density can be reduced to a quadrature. We make the following assumptions:

- Plane strain approximation: purely two-dimensional flow in the vertical x-z plane, no dependencies on the transverse coordinate y.
- Constant inclination angle α.
- Uniformity in the downslope (x) direction: $\partial(\cdot)/\partial x = 0$ for all field quantities.
- Flat, rigid bed: $b(x,t) = 0$ (in the inclined coordinate system). The ice thickness may be time-dependent, $H(t)$, and thus the free surface is given by $h(x,t) = h(t) = H(t)$.
- No surface accumulation ($a_\mathrm{s}^\perp = 0$).
- No basal melting ($a_\mathrm{b}^\perp = 0$), no basal sliding ($C_\mathrm{b} = 0$).
- Flow law (9.56), (9.57) with exponent $n = 3$.
- Constant rate factor:
  $A(T') = A = \text{const} \;\Leftrightarrow\; B(T') = A(T')^{-1/n} = A^{-1/n} = B = \text{const}.$

Note that we do not make the steady-state assumption $\partial(\cdot)/\partial t = 0$ here. In particular, Sorge's Law (9.67) is not applicable to the parallel sided firn slab because there is no replenishment of fresh snow at the surface (due to $a_\mathrm{s}^\perp = 0$) which counteracts the compaction of the firn.

With the above assumptions, the x- and z-components of Eq. (9.65) reduce to the forms

$$\frac{\partial}{\partial z}\left(\eta\frac{\partial v_x}{\partial z}\right) + \rho g \sin\alpha = 0\,, \tag{9.70}$$

$$\frac{\partial}{\partial z}\left[\left(\zeta + \frac{4\eta}{3}\right)\frac{\partial v_z}{\partial z}\right] - \rho g \cos\alpha = 0\,, \tag{9.71}$$

and Eq. (9.61) becomes

$$\frac{\partial \rho}{\partial t} + \frac{\partial}{\partial z}(\rho v_z) = 0\,. \tag{9.72}$$

The first integrals of Eqs. (9.70) and (9.71) are

$$\eta\frac{\partial v_x}{\partial z} = g\sin\alpha \int_z^h \rho(z')\mathrm{d}z' \equiv g\sin\alpha\, D(z)\,, \tag{9.73}$$

$$\left(\zeta + \frac{4\eta}{3}\right) \frac{\partial v_z}{\partial z} = -g \cos\alpha \int_z^h \rho(z')\mathrm{d}z' \equiv -g \cos\alpha \, D(z) \, . \tag{9.74}$$

For the second integrals, the constitutive law for the viscosity of firn, Eqs. (9.56) and (9.57) must be substituted in Eqs. (9.73) and (9.74). With the symmetry of the parallel sided slab, the deviatoric strain rate tensor is

$$\mathsf{D}^{\mathrm{D}} = \frac{1}{2} \begin{pmatrix} -\dfrac{\partial v_z}{\partial z} & \dfrac{\partial v_x}{\partial z} \\ \dfrac{\partial v_x}{\partial z} & \dfrac{\partial v_z}{\partial z} \end{pmatrix} , \tag{9.75}$$

and thus, the invariant [Eq. (9.58)] becomes

$$\delta^2 = \frac{1}{4a} \left(\frac{\partial v_x}{\partial z}\right)^2 + \left(\frac{1}{4a} + \frac{1}{4b}\right) \left(\frac{\partial v_z}{\partial z}\right)^2 . \tag{9.76}$$

Through the invariant δ, the two equations (9.70) and (9.71) are coupled. The shear rate of the firn is influenced by the settling rate of the firn, and vice versa the settling rate is influenced by the shear rate. To illustrate the effects of shear and compaction in the firn layer, the problem is split into the two parts. In the first part, the shear flow of the firn layer is computed with the assumption that the firn does not settle ($\partial\rho/\partial t = 0$). In the second part, the parallel sided slab is set horizontal so that no shear flow occurs ($v_x = 0$).

Shear Flow of Firn

If compaction is disregarded, the second term in the invariant δ in Eq. (9.76) vanishes; thus Eq. (9.56) becomes

$$\eta = B \left(\frac{1}{2}\right)^{1/n} a^{-(n+1)/2n} \left(\frac{\partial v_x}{\partial z}\right)^{-(1-1/n)} . \tag{9.77}$$

With Eq. (9.73),

$$\eta \frac{\partial v_x}{\partial z} = B \left(\frac{1}{2}\right)^{1/n} a^{-(n+1)/2n} \left(\frac{\partial v_x}{\partial z}\right)^{1/n} = g \sin\alpha D(z) \, . \tag{9.78}$$

This can be solved for the shear strain rate,

$$\frac{\partial v_x}{\partial z} = 2A \, a^{(n+1)/2n} \left(g \sin\alpha D(z)\right)^n , \tag{9.79}$$

and, by inserting $n = 3$, integration yields the velocity v_x,

$$v_x = 2A \, (g \sin\alpha)^3 \int_0^z a^2 \, D^3 \, \mathrm{d}z' \, . \tag{9.80}$$

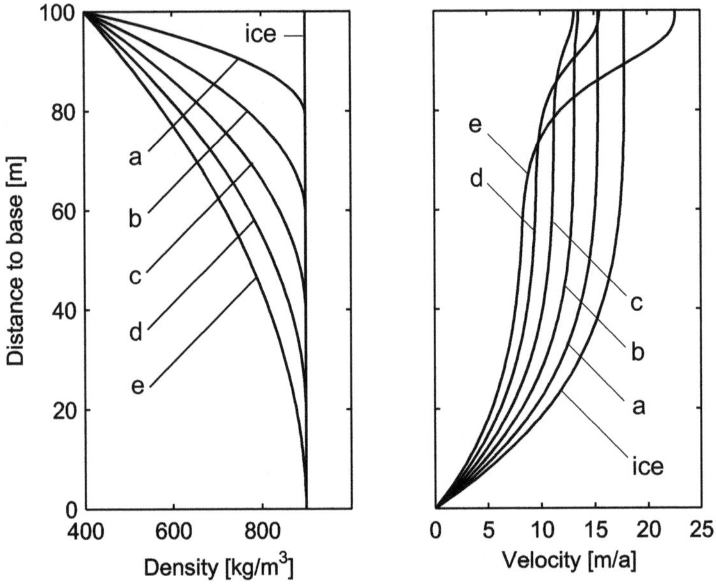

Fig. 9.18. Different firn density profiles (left panel) and resulting velocity profiles (right panel) for a slab with the parameters $H = 100\,\mathrm{m}$, $\alpha = 10°$, $n = 3$, $A = 10^{-16}\,\mathrm{a}^{-1}\,\mathrm{Pa}^{-3}$ and $g = 9.81\,\mathrm{m\,s}^{-2}$. The firn-ice transitions are (a) 20 m, (b) 40 m, (c) 60 m, (d) 80 m and (e) 100 m from the surface. The profiles labelled "ice" correspond to the solution for a homogeneous ice slab (Fig. 7.2, left panel).

Figure 9.18 shows velocity profiles for a slab of 100 m thickness and 10° inclination. The different density profiles have been prescribed by quadratic functions,

$$\rho(z) = (\rho_{z=h} - \hat{\rho}_\mathrm{i})\,\frac{(z - z_\mathrm{fi})^2}{(h - z_\mathrm{fi})^2} + \hat{\rho}_\mathrm{i}\,, \tag{9.81}$$

where $\rho_{z=h}$ is the firn density at the surface of the slab and z_fi the position of the firn-ice transition.

Settling of a Firn Layer

The bulk viscosity of firn grows asymptotically towards infinity while the relative density approaches unity. Thus, a layer of firn settles as long as parts of the layer have relative densities smaller than unity, the densities approach the density of ice, $\hat{\rho}_\mathrm{i}$, everywhere and the thickness of the layer decreases asymptotically towards

$$H_\infty = \frac{1}{\hat{\rho}_\mathrm{i}} \int_0^{H_0} \rho\,\mathrm{d}z\,, \tag{9.82}$$

where H_0 is the initial thickness of the slab. To simplify the ensuing equations, the following abbreviations are introduced,

$$Q \equiv \frac{1}{4a} + \frac{1}{4b}, \qquad \frac{1}{R} \equiv \frac{4}{3a} + \frac{1}{b}. \tag{9.83}$$

Then Eq. (9.71) becomes

$$\left(\zeta + \frac{4\eta}{3}\right) \frac{\partial v_z}{\partial z} = \frac{B}{2} \frac{1}{R} Q^{(1-n)/2n} \left(\frac{\partial v_z}{\partial z}\right)^{1/n} = -g \cos\alpha\, D. \tag{9.84}$$

With $n = 3$ and $B = A^{-1/3}$ we obtain

$$\frac{\partial v_z}{\partial z} = -8\, A\, (g\, \cos\alpha)^3\, D^3\, R^3\, Q, \tag{9.85}$$

and the profile of the vertical velocity component, v_z, is given by

$$v_z = -8\, A\, (g\, \cos\alpha)^3 \int_0^z D^3\, R^3\, Q\, \mathrm{d}z'. \tag{9.86}$$

The settling of the firn slab is a transient problem, where not only the density and the vertical velocity component change with time, but also the domain in consideration is changing. The handling of the discretised equations is made easier by mapping the variable layer thickness to unity at every time. To achieve this, we apply a terrain-following coordinate transformation according to Eq. (5.127), but for the simplified situation of the slab,

$$\tilde{z} = \frac{z}{H(t)}, \qquad \tilde{t} = t, \tag{9.87}$$

where $\tilde{z}$ is the transformed coordinate in the slab of unit thickness. The required coordinate differentials are

$$\frac{\partial \tilde{z}}{\partial z} = \frac{1}{H(t)}, \qquad \frac{\partial \tilde{z}}{\partial t} = -\frac{\tilde{z}}{H} \frac{\mathrm{d}H}{\mathrm{d}t}. \tag{9.88}$$

Equation (9.85) in the transformed coordinates, with $\alpha = 0°$ (horizontal slab), is

$$\frac{\partial v_z}{\partial \tilde{z}} = -8\, H^4\, A\, g^3\, D^3\, R^3\, Q, \tag{9.89}$$

where

$$D(\tilde{z}) \equiv \int_{\tilde{z}}^{H} \rho(\tilde{z}')\mathrm{d}\tilde{z}', \tag{9.90}$$

and the profile of the vertical velocity component, v_z, is given by

$$v_z = \int_0^{\tilde{z}} \frac{\partial v_z}{\partial \tilde{z}'} \mathrm{d}\tilde{z}'. \tag{9.91}$$

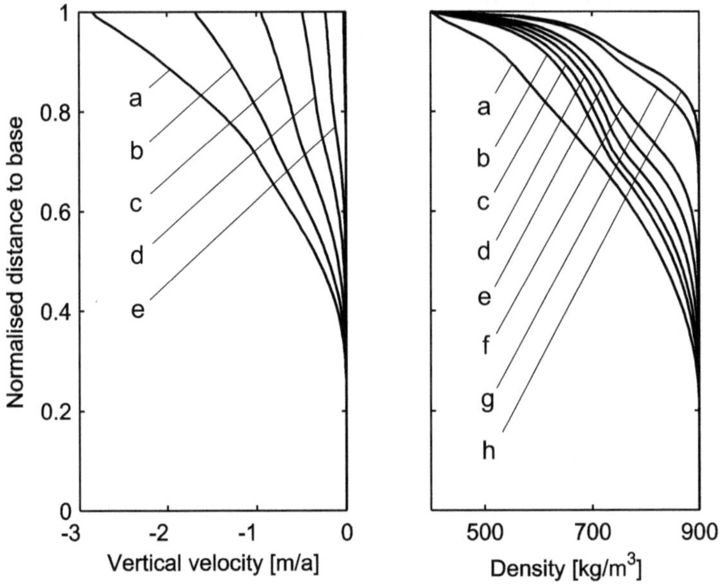

Fig. 9.19. Vertical velocity (left panel) and density profiles (right panel) computed for a settling firn slab with an initial thickness of $H_0 = 100\,\text{m}$ and no inclination ($\alpha = 0°$). (a) initial profiles, (b) profiles after 0.5 a, (c) after 1 a, (d) after 2 a, (e) after 4 a, (f) after 8 a, (g) after 50 a, (h) after 100 a.

Finally, the temporal evolution of the density profile, Eq. (9.72), is given by

$$\frac{\partial \rho}{\partial \tilde{t}} = \frac{1}{H}\left[\left(\tilde{z}\, v_z|_{\tilde{z}=1} - v_z\right)\frac{\partial \rho}{\partial \tilde{z}} - \rho\,\frac{\partial v_z}{\partial \tilde{z}}\right] \equiv S\,, \tag{9.92}$$

and the rate of change of the slab thickness is equal to the firn velocity at the surface of the slab,

$$\frac{\mathrm{d}H}{\mathrm{d}t} = v_z|_{\tilde{z}=1}\,. \tag{9.93}$$

To compute the evolution of the thickness of the slab and the density profiles, an explicit Euler scheme can be applied,

$$H_2 = H_1 + \Delta t\ v_z|_{\tilde{z}=1}\,, \tag{9.94}$$

$$\rho_2(\tilde{z}) = \rho_1(\tilde{z}) + \Delta t\ S_1(\tilde{z})\,, \tag{9.95}$$

where the subscripts 1 and 2 denote the corresponding values at the times t_1 and $t_2 = t_1 + \Delta t$, respectively. For a slab with an initial thickness $H_0 = 100\,\text{m}$, a time step $\Delta t = 0.01\,\text{a}$ is stable. Figure 9.19 shows the temporal evolution of the profiles of the vertical velocity and the density, and Fig. 9.20 shows the thinning of the slab. The results demonstrate that low densities cannot persist very long and settling of the firn rapidly increases the density, except near the surface.

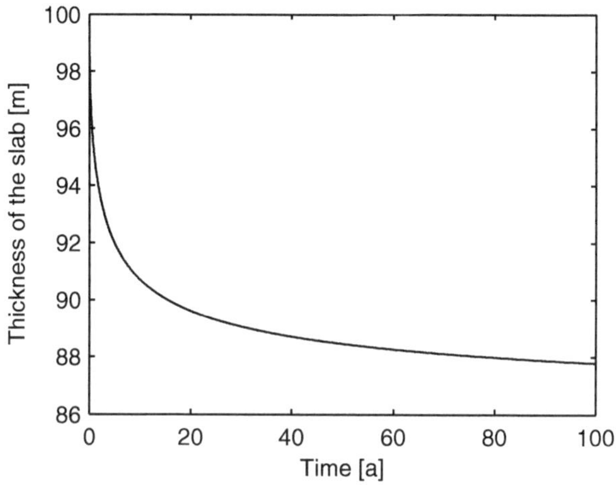

Fig. 9.20. Thinning of a horizontal firn slab (see Fig. 9.19). The initial thickness is $H_0 = 100\,\mathrm{m}$, and the final thickness is $H_\infty = 85.18\,\mathrm{m}$.

Coupling of Shear and Settling of a Firn Layer

The equations for shear flow and for settling of the firn, Eqs. (9.70) and (9.71), are coupled through the invariant δ, Eq. (9.76). Substitution of δ from Eq. (9.76) in the viscosity terms of Eqs. (9.70) and (9.71) yields a system of two non-linear first order partial differential equations. A solution strategy follows a similar idea as applied for the solution of the first order approximation of the ice flow problem, Eqs. (7.39) and (7.40). Assuming that η and ζ are known, Equations (9.70) and (9.71),

$$\frac{\partial v_x}{\partial z} = \frac{g \sin\alpha\, D}{\eta}\,, \tag{9.96}$$

$$\frac{\partial v_z}{\partial z} = -\frac{3g\cos\alpha\, D}{3\zeta + 4\eta}\,, \tag{9.97}$$

can be solved by integration. Assuming that η_i and ζ_i are known profiles of the viscosities, the velocity profiles $v_x|_{i+1}$ and $v_y|_{i+1}$ can be found by integration,

$$v_x|_{i+1} = g\sin\alpha \int_0^z \frac{D}{\eta_i}\mathrm{d}z'\,, \tag{9.98}$$

$$v_z|_{i+1} = -3g\cos\alpha \int_0^z \frac{D}{3\zeta_i + 4\eta_i}\mathrm{d}z'\,. \tag{9.99}$$

With the new $v_x|_{i+1}$ and $v_z|_{i+1}$, the viscosities can be updated by using Eqs. (9.56) and (9.57) with the updated δ from Eq. (9.76),

$$\eta_{i+1} = \frac{B}{2a}\,\delta^{-2/3}|_{i+1}\,, \tag{9.100}$$

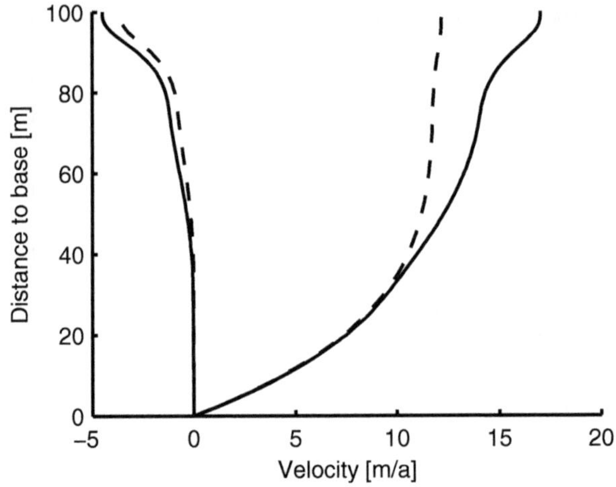

Fig. 9.21. Vertical (left curves) and horizontal (right curves) velocity components in a coupled shear and settling flow (solid lines) of a firn slab with a thickness of 100 m. The dashed lines represent the corresponding uncoupled case (b) in Fig. 9.18.

$$\zeta_{i+1} = \frac{B}{2b}\,\delta^{-2/3}|_{i+1}\,. \tag{9.101}$$

Equations (9.98) and (9.99), together with Eqs. (9.100) and (9.101) define a fixed point iteration. The iteration can be started from the uncoupled solutions for the velocity components, Eqs. (9.80) and (9.86), to compute the starting values for the viscosities.

Figure 9.21 shows the profiles of the velocity components in the uncoupled and in the coupled case. Since firn is treated as a strain softening fluid, the shearing additionally contributes to the decrease of the bulk viscosity and thus enhances the settling rate. In turn, the settling strain decreases the shear viscosity and thus enhances the shear rate.

9.3 Temperate and Polythermal Glaciers

9.3.1 Background

Most ice in large ice sheets and glaciers at high latitudes is cold (temperature below the pressure melting point). The thermodynamics of cold ice sheets is therefore described in Chapter 5. However, as mentioned in Sect. 7.1, many glaciers at lower latitudes contain significant amounts of temperate ice (temperature at the pressure melting point), or are entirely temperate except for a temporary cold surface layer in winter. Temperate ice may contain a small fraction of liquid water, and changes in heat content lead to changes in water content, whereas for cold ice changes in heat content lead to temperature changes. Recall that glaciers that consist entirely of cold ice are called cold glaciers, glaciers that consist entirely of temperate ice are called temperate glaciers, and glaciers that consist of both temperate and cold parts are called polythermal glaciers.

9.3.2 Temperate Ice

Flow properties of temperate ice depend on the content of water in the ice matrix; however, only one study (Lliboutry and Duval 1985) has attempted to quantify this relation. In order to comprehensively model the flow of a temperate or polythermal glacier, it is essential to know the spatial distribution of the water content in the temperate part of the glacier and the temperature in the cold part. This is particularly true near the bed in the ablation area where shear rates are high, and thus the impact of the water content on the flow behavior is expected to be significant.

As opposed to cold ice, the temperature of temperate ice is at the pressure melting point, so that it need not be calculated separately, but follows immediately from Eq. (4.13),

$$T = T_\mathrm{m} = T_0 - \beta p\,. \tag{9.102}$$

Temperate ice contains a certain amount of water, described as the mass fraction W [see Eq. (9.104) below], which takes the role as the main thermodynamic quantity instead of the temperature. Therefore, in contrast to cold ice, temperate ice must be regarded as a *binary mixture* of ice and water, and ρ denotes the density of the mixture. Because of this, it is necessary to apply some basic concepts of mixture theory (Müller 1985). Since measured water contents in temperate ice are generally less than 3% (Pettersson et al. 2004, and references therein) and thus small, temperate ice will be described by two mass balances, one for the mixture as a whole and one for the water component, but only one momentum and one energy balance for the mixture. That is, water is considered as a tracer component whose motion relative to the barycentre of the mixture is described as a diffusive process. Alternative concepts, not considered here, include two separate momentum balances with a Darcy-type interaction force between the two components (Fowler 1984, Hutter 1993).

For the formulation of the field equations for temperate ice, we closely follow the description by Greve (1997). First, some quantities from mixture theory are introduced. The *barycentric velocity* is defined as

$$\mathbf{v} \equiv \frac{1}{\rho}(\rho_\mathrm{i}\mathbf{v}_\mathrm{i} + \rho_\mathrm{w}\mathbf{v}_\mathrm{w})\,. \tag{9.103}$$

The indices i and w refer to the ice and water components, respectively, ρ_i and ρ_w denote the corresponding partial densities (mass of ice or water per unit volume of the mixture), and $\mathbf{v}_\mathrm{i}$ and $\mathbf{v}_\mathrm{w}$ are the corresponding velocities. The *water content* is introduced as the mass fraction, W, of water in the mixture,

$$W \equiv \frac{\rho_\mathrm{w}}{\rho}\,. \tag{9.104}$$

In addition, a *diffusive water mass flux* $\mathbf{j}$ is defined that describes the water motion relative to the motion of the barycentre,

$$\mathbf{j} \equiv \rho_\mathrm{w}(\mathbf{v}_\mathrm{w} - \mathbf{v}) = \rho W(\mathbf{v}_\mathrm{w} - \mathbf{v})\,. \tag{9.105}$$

As for pure (cold) ice, the mixture of ice and water is assumed to be incompressible. This is problematic in so far as the bulk densities of ice and water are distinctly different ($\hat{\rho}_\mathrm{i} = 910\,\mathrm{kg\,m^{-3}}$ vs. $\hat{\rho}_\mathrm{w} = 1000\,\mathrm{kg\,m^{-3}}$). However, as long as the water content is less than 3%, relative changes of the mixture density due to changes of the water content do not exceed 0.5%. This variability is negligible, and we set $\rho \approx \hat{\rho}_\mathrm{i} = 910\,\mathrm{kg\,m^{-3}}$ instead. As a consequence, the *mixture mass balance* and the *mixture momentum balance* have the same form as for cold ice,

$$\operatorname{div}\mathbf{v} = 0\,, \tag{9.106}$$

$$-\operatorname{grad}p + \operatorname{div}\mathsf{t}^\mathrm{D} + \rho\mathbf{g} = \mathbf{0}\,, \tag{9.107}$$

where the stress tensor t has again been decomposed as $\mathsf{t} = -p\mathsf{I} + \mathsf{t}^\mathrm{D}$, the acceleration term has been neglected and the volume force has been set to $\mathbf{f} = \rho\mathbf{g}$.

When formulating the *mass balance for the component water*, it must be noted that the partial density of water, ρ_w, is not constant, but depends on the water content itself. Furthermore, the mass of water is not conserved due to the possibility of melting and freezing processes. It is therefore necessary to use the general form (3.58) of the mass balance and include a production term M, the rate of water mass produced per unit mixture volume,

$$\frac{\partial\rho_\mathrm{w}}{\partial t} + \operatorname{div}(\rho_\mathrm{w}\mathbf{v}_\mathrm{w}) = M\,. \tag{9.108}$$

This is equivalent to

$$\rho \dot{W} = -\mathrm{div}\,\mathbf{j} + M\,. \tag{9.109}$$

As was the case for cold ice, constitutive relations are required to close the system,

$$\mathsf{D} = A_\mathrm{t}(W)\,\sigma_\mathrm{e}^{n-1}\,\mathsf{t}^\mathrm{D}\,, \tag{9.110}$$

$$\dot{u} = L\dot{W} + c(T_\mathrm{m})\,\dot{T}_\mathrm{m}\,, \tag{9.111}$$

$$\mathbf{j} = -\nu\,\mathrm{grad}\,W\,, \tag{9.112}$$

$$\mathbf{q}_\mathrm{s} = -\kappa(T_\mathrm{m})\,\mathrm{grad}\,T_\mathrm{m}\,. \tag{9.113}$$

The first equation is the counterpart of Glen's flow law (4.16) for cold ice; however, the temperature-dependent rate factor is replaced by a factor depending on the water content, $A_\mathrm{t}(W)$. Following Lliboutry and Duval (1985), it can be chosen as

$$A_\mathrm{t}(W) = A(T'\!=\!0^\circ\mathrm{C}) \times (1 + 1.8125\,W[\%])\,. \tag{9.114}$$

The second equation relates changes of the internal energy to changes of the water content and of the melting temperature. From a strict thermodynamical point of view this relation is merely approximate. The third equation is Fick's diffusion law for the motion of water, and the last equation is Fourier's law of heat conduction for the sensible heat flux $\mathbf{q}_\mathrm{s}$ [see the counterpart for cold ice, Eq. (4.36)]. The latent heat L has the value $L = 3.35 \times 10^5\,\mathrm{J\,kg^{-1}}$, while suitable values for the water diffusivity ν (assumed to be constant) are not well constrained.

Next, consider the *mixture energy balance*. In (9.111) the internal energy u depends on the water content W, so that a non-vanishing diffusive water mass flux $\mathbf{j}$ contributes to a flux of internal energy, the so-called latent heat flux $\mathbf{q}_\mathrm{l} = L\mathbf{j}$. Therefore, the total heat flux $\mathbf{q}$ can be expressed as

$$\mathbf{q} = \mathbf{q}_\mathrm{s} + \mathbf{q}_\mathrm{l} = \mathbf{q}_\mathrm{s} + L\,\mathbf{j}\,. \tag{9.115}$$

With this modified form of the energy flux, the mixture energy balance results from the general energy balance (3.92) as

$$\rho\dot{u} = -\mathrm{div}\,(\mathbf{q}_\mathrm{s} + L\mathbf{j}) + \mathrm{tr}\,(\mathsf{t}^\mathrm{D}\cdot\mathsf{D})\,, \tag{9.116}$$

where the radiation r has been neglected. Introducing the constitutive relations (9.110)–(9.113) into the water mass balance (9.109) and into the mixture energy balance (9.116) yields the respective relations

$$\rho \dot{W} = \nu\Delta W + M \tag{9.117}$$

and

$$\rho L\dot{W} + \rho c\dot{T}_\mathrm{m} = L\nu\Delta W + \mathrm{div}\,(\kappa\,\mathrm{grad}\,T_\mathrm{m}) + 2A_\mathrm{t}(W)\,\sigma_\mathrm{e}^{n+1}\,, \tag{9.118}$$

which are consistent, provided that the water production rate M is given by

$$M = \frac{1}{L}\left(2A_\mathrm{t}(W)\,\sigma_\mathrm{e}^{n+1} + \mathrm{div}\,(\kappa\,\mathrm{grad}\,T_\mathrm{m}) - \rho c \dot{T}_\mathrm{m}\right). \tag{9.119}$$

This has the physical interpretation that the energy available for melting is composed of three terms, (i) the heat dissipated by stress power, (ii) the heat conducted to the point under consideration and (iii) the heat stored by changes in the melting temperature, the latter term being negative for $\dot{T}_\mathrm{m} > 0$). The latter two effects contribute only little to M, so that the dominant effect is water production by heat dissipation, which is the expected behaviour in an environment with two coexisting, exchanging phases.

9.3.3 Temperate Ice Surface

At a temperate ice surface, the kinematic and dynamic boundary conditions described in Sect. 5.1.2 [Eqs. (5.15)–(5.23)] are applicable without any changes. However, the thermodynamic boundary condition (5.24) must be replaced by an analogous statement for the water content,

$$W = W_\mathrm{s}. \tag{9.120}$$

The surface water content W_s, which results from water entrapment in the upper firn layer, must be prescribed.

9.3.4 Temperate Ice Base

Next consider a temperate ice base, where the temperature at the ice-lithosphere interface is at pressure melting. However, the presence of a temperate ice base does not necessarily entail the occurrence of a temperate ice layer of non-vanishing thickness above it. It is equally possible that the temperature gradient at the ice base is below the Clausius-Clapeyron gradient, so that the ice becomes cold immediately above the base, even though the base itself is temperate. The latter case is fully covered by Sect. 5.1.2, Eqs. (5.25)–(5.40), whereas the former case (temperate base with temperate layer above) requires some modifications that will now be discussed.

We formulate a *mass jump relation for the component water.* In order to account for basal melting, a surface production term $\mathcal{P}_\mathrm{b}^\mathrm{w} = \rho a_\mathrm{b}^\perp$ must be introduced in the general form (3.61). This yields

$$[\![\rho_\mathrm{w}(\mathbf{v}_\mathrm{w} - \mathbf{w}) \cdot \mathbf{n}]\!] = \mathcal{P}_\mathrm{b}^\mathrm{w} = \rho a_\mathrm{b}^\perp\,, \tag{9.121}$$

where $a_\mathrm{b}^\perp$ is the basal melting rate. Equation (9.121) can be transformed to

$$\rho_\mathrm{w}(\mathbf{v}_\mathrm{w} - \mathbf{w}) \cdot \mathbf{n} = \rho W(\mathbf{v}_\mathrm{w} - \mathbf{w}) \cdot \mathbf{n} = \dot{m}_\mathrm{b}^\mathrm{w} - \rho a_\mathrm{b}^\perp\,, \tag{9.122}$$

where the *water mass flux into the base* has been defined as

$$\dot{m}_\mathrm{b}^\mathrm{w} \equiv \rho_\mathrm{w}^\mathrm{lith}(\mathbf{v}_\mathrm{w}^\mathrm{lith} - \mathbf{w}) \cdot \mathbf{n}\,. \tag{9.123}$$

This quantity must be prescribed in general. Analogous to Eq. (9.121), the *mass jump relation for the component ice* yields

$$[\![\rho_\mathrm{i}(\mathbf{v}_\mathrm{i}-\mathbf{w})\cdot\mathbf{n}]\!] = -\mathcal{P}_\mathrm{b}^\mathrm{w} = -\rho a_\mathrm{b}^\perp\,. \tag{9.124}$$

Since the lithosphere is impermeable to ice, and therefore does not contain any ice, $\rho_\mathrm{i}^\mathrm{lith}=0$, and thus Eq. (9.124) simplifies to

$$\rho_\mathrm{i}\,(\mathbf{v}_\mathrm{i}-\mathbf{w})\cdot\mathbf{n} = \rho(1-W)\,(\mathbf{v}_\mathrm{i}-\mathbf{w})\cdot\mathbf{n} = \rho a_\mathrm{b}^\perp\,. \tag{9.125}$$

With the definition (9.103) of the barycentric velocity, it follows that

$$\mathbf{v}-\mathbf{w} = W(\mathbf{v}_\mathrm{w}-\mathbf{w}) + (1-W)(\mathbf{v}_\mathrm{i}-\mathbf{w})\,, \tag{9.126}$$

which upon scalar multiplication by $\mathbf{n}$ and use of (9.122) and (9.125) becomes

$$(\mathbf{v}-\mathbf{w})\cdot\mathbf{n} = \frac{\dot{m}_\mathrm{b}^\mathrm{w}}{\rho}\,. \tag{9.127}$$

In the case of a negligible diffusive water mass flux $\mathbf{j}$ (i.e., $\mathbf{v}_\mathrm{w}=\mathbf{v}_\mathrm{i}=\mathbf{v}$), the water mass flux into the base $\dot{m}_\mathrm{b}^\mathrm{w}$ can be calculated by comparing (9.125) and (9.127),

$$\frac{\dot{m}_\mathrm{b}^\mathrm{w}}{\rho} = \frac{a_\mathrm{b}^\perp}{1-W} \approx a_\mathrm{b}^\perp\,, \tag{9.128}$$

and therefore, in contrast to the general case, does not need separate prescription. The approximation $1-W\approx 1$ is justified because of $W<0.03$ (water content less than 3%; see Sect. 9.3.2).

By applying (9.127), the *kinematic condition* (5.30) changes to

$$\frac{\partial F_\mathrm{b}}{\partial t} + (\operatorname{grad} F_\mathrm{b})\cdot\mathbf{v} = N_\mathrm{b}\frac{\dot{m}_\mathrm{b}^\mathrm{w}}{\rho}\,, \tag{9.129}$$

or, equivalently,

$$\frac{\partial b}{\partial t} + v_x\frac{\partial b}{\partial x} + v_y\frac{\partial b}{\partial y} - v_z = N_\mathrm{b}\frac{\dot{m}_\mathrm{b}^\mathrm{w}}{\rho}\,. \tag{9.130}$$

For the *diffusive water mass flux* $\mathbf{j}$, from (9.105), (9.122) and (9.127),

$$\begin{aligned}\mathbf{j}\cdot\mathbf{n} &= \rho W(\mathbf{v}_\mathrm{w}-\mathbf{v})\cdot\mathbf{n}\\ &= \rho W(\mathbf{v}_\mathrm{w}-\mathbf{w})\cdot\mathbf{n} - \rho W(\mathbf{v}-\mathbf{w})\cdot\mathbf{n}\\ &= \dot{m}_\mathrm{b}^\mathrm{w} - \rho a_\mathrm{b}^\perp - W\dot{m}_\mathrm{b}^\mathrm{w}\\ &= (1-W)\dot{m}_\mathrm{b}^\mathrm{w} - \rho a_\mathrm{b}^\perp \;\approx\; \dot{m}_\mathrm{b}^\mathrm{w} - \rho a_\mathrm{b}^\perp\,.\end{aligned} \tag{9.131}$$

Together with the diffusion law (9.112), Eq. (9.131) represents a Neumann-type boundary condition for the basal water content.

The continuity of the stress vector, Eq. (5.33), the basal sliding law, Eq. $(5.35)_2$, and the temperature condition, Eq. (5.39), remain valid. The *energy jump relation for the mixture* follows from the general form (3.94) with the extended energy flux (9.115),

$$[\![\mathbf{q}_\mathrm{s} \cdot \mathbf{n}]\!] + L\,[\![\mathbf{j} \cdot \mathbf{n}]\!] - [\![\mathbf{v}]\!] \cdot \mathbf{t} \cdot \mathbf{n} + \left[\!\left[\rho\left(u + \tfrac{1}{2}v^2\right)\left((\mathbf{v} - \mathbf{w}) \cdot \mathbf{n}\right)\right]\!\right] = 0 \qquad (9.132)$$

[compare also Eq. (5.36)]. Ignoring the contribution from the kinetic energy and setting $u = LW$ [due to Eq. (9.111)] yields

$$[\![\mathbf{q}_\mathrm{s} \cdot \mathbf{n}]\!] + L\,[\![\mathbf{j} \cdot \mathbf{n}]\!] - [\![\mathbf{v}]\!] \cdot \mathbf{t} \cdot \mathbf{n} + L\,[\![\rho W(\mathbf{v} - \mathbf{w}) \cdot \mathbf{n}]\!] = 0\,. \qquad (9.133)$$

By introducing Fourier's law of heat conduction (9.113), the geothermal heat flux $q_\mathrm{geo}^\perp = -\mathbf{q}_\mathrm{lith} \cdot \mathbf{n}$, the sliding velocity $\mathbf{v}_\mathrm{b} = -[\![\mathbf{v}]\!]$ and the definition (9.105) of $\mathbf{j}$,

$$\kappa\,(\operatorname{grad} T_\mathrm{m} \cdot \mathbf{n}) - q_\mathrm{geo}^\perp + \mathbf{v}_\mathrm{b} \cdot \mathbf{t} \cdot \mathbf{n} + L\,[\![\rho W(\mathbf{v}_\mathrm{w} - \mathbf{w}) \cdot \mathbf{n}]\!] = 0$$

$$\stackrel{(9.121)}{\Rightarrow} \quad \kappa\,(\operatorname{grad} T_\mathrm{m} \cdot \mathbf{n}) - q_\mathrm{geo}^\perp + \mathbf{v}_\mathrm{b} \cdot \mathbf{t} \cdot \mathbf{n} + L\,\rho a_\mathrm{b}^\perp = 0$$

$$\Rightarrow \quad a_\mathrm{b}^\perp = \frac{q_\mathrm{geo}^\perp - \kappa\,(\operatorname{grad} T_\mathrm{m} \cdot \mathbf{n}) - \mathbf{v}_\mathrm{b} \cdot \mathbf{t} \cdot \mathbf{n}}{\rho L}\,. \qquad (9.134)$$

This is the same result as Eq. (5.40), saying that the basal melting rate originates from the heat fluxes flowing toward the interface from the ice and lithosphere sides, and from the basal heat production due to sliding.

9.3.5 Transition Conditions at the CTS

The *cold-temperate transition surface*, henceforth called CTS, constitutes the phase-change surface between the cold and temperate regions of a glacier and is therefore a singular surface at which the physical quantities may suffer a jump. The CTS is described explicitly by

$$z = z_\mathrm{m}(x, y, t) \qquad (9.135)$$

or implicitly by

$$F_\mathrm{m}(\mathbf{x}, t) = z - z_\mathrm{m}(x, y, t) = 0\,. \qquad (9.136)$$

As indicated in Fig. 9.22, the positive side is by convention the cold ice side, the negative side the temperate ice side, and the normal unit vector

$$\mathbf{n} = \frac{\operatorname{grad} F_\mathrm{m}}{|\operatorname{grad} F_\mathrm{m}|} \qquad (9.137)$$

points into the cold ice.

Analogue to Eq. (5.20) for the ice surface, the *kinematic condition* for the CTS is

$$\frac{\partial F_\mathrm{m}}{\partial t} + (\operatorname{grad} F_\mathrm{m}) \cdot \mathbf{v} = -N_\mathrm{m} a_\mathrm{m}^\perp\,, \qquad (9.138)$$

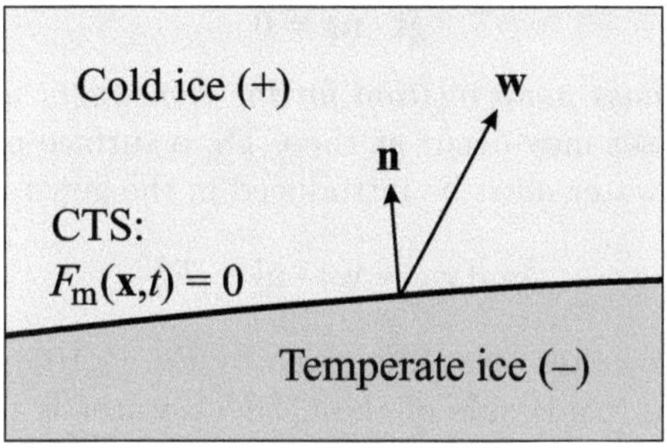

Fig. 9.22. Geometry of the CTS.

which becomes, with the above choice for F_{m},

$$\frac{\partial z_{\mathrm{m}}}{\partial t} + v_x \frac{\partial z_{\mathrm{m}}}{\partial x} + v_y \frac{\partial z_{\mathrm{m}}}{\partial y} - v_z = N_{\mathrm{m}} a_{\mathrm{m}}^{\perp}\,, \tag{9.139}$$

where

$$a_{\mathrm{m}}^{\perp} = (\mathbf{w} - \mathbf{v}) \cdot \mathbf{n} \tag{9.140}$$

is the ice volume flux through the CTS, and

$$N_{\mathrm{m}} = |\mathrm{grad}\, F_{\mathrm{m}}| = \left(1 + \left(\frac{\partial z_{\mathrm{m}}}{\partial x}\right)^2 + \left(\frac{\partial z_{\mathrm{m}}}{\partial y}\right)^2\right)^{1/2}. \tag{9.141}$$

The sign choice in Eq. (9.140) causes $a_{\mathrm{m}}^{\perp}$ to be positive when ice flows from the cold into the temperate region, and negative when ice flows from the temperate into the cold region. Because of the continuity of $\mathbf{v}$ across the CTS [see below, Eq. (9.144)], it is not necessary to distinguish between $\mathbf{v}^+$ and $\mathbf{v}^-$. In contrast to the accumulation-ablation function $a_{\mathrm{s}}^{\perp}$ on the free surface, $a_{\mathrm{m}}^{\perp}$ arises in the interior of the glacier, and in consequence it is part of the solution.

The temperature and the tangential velocity are assumed to be continuous across the CTS,

$$[\![T]\!] = 0, \quad [\![\mathbf{v} - (\mathbf{v} \cdot \mathbf{n})\mathbf{n}]\!] = \mathbf{0}\,. \tag{9.142}$$

The density difference between cold and temperate ice is at most 0.5% (see Sect. 9.3.2). If this small difference is ignored, the *mass jump relation* (3.61) yields continuity for the normal velocity at the CTS as well,

$$[\![\mathbf{v} \cdot \mathbf{n}]\!] = 0\,, \tag{9.143}$$

so that the entire velocity vector is continuous,

$$[\![\mathbf{v}]\!] = \mathbf{0}\,. \tag{9.144}$$

From this and from the *momentum jump relation* (3.73) follows the continuity of traction,

$$[\![\mathbf{t} \cdot \mathbf{n}]\!] = \mathbf{0}\,. \tag{9.145}$$

Now consider the *mass jump relation for the component water*. Since melting and freezing processes may occur at the CTS, a surface production term $\mathcal{P}_{\mathrm{m}}^{\mathrm{w}}$ for the component water must be introduced in the general form (3.61). Thus

$$[\![\rho_{\mathrm{w}}(\mathbf{v}_{\mathrm{w}} - \mathbf{w}) \cdot \mathbf{n}]\!] = \mathcal{P}_{\mathrm{m}}^{\mathrm{w}}\,, \tag{9.146}$$

or with the diffusive water mass flux given by Eq. (9.105), in view of the fact that at the positive (cold) side of the CTS no water is present, so that the quantities W^+ and $\mathbf{j}^+$ vanish,

$$-\mathbf{j}^- \cdot \mathbf{n} + \rho a_{\mathrm{m}}^{\perp} W^- = \mathcal{P}_{\mathrm{m}}^{\mathrm{w}}\,. \tag{9.147}$$

This relation can be interpreted in terms of a total water flux $\mathbf{j}_{\mathrm{tot}}$ relative to the CTS velocity $\mathbf{w}$, defined by

$$\mathbf{j}_{\mathrm{tot}} \equiv \rho_{\mathrm{w}}(\mathbf{v}_{\mathrm{w}} - \mathbf{w})\,, \tag{9.148}$$

with which Eq. (9.147) becomes

$$-\,\mathbf{j}_{\mathrm{tot}}^- \cdot \mathbf{n} = \mathcal{P}_{\mathrm{m}}^{\mathrm{w}}\,. \tag{9.149}$$

That is, the normal component of the total water flux relative to the CTS at the temperate side of the CTS equals the surface production of water.

In order to formulate the *energy jump relation* (3.94), as in the derivation of Eq. (9.133), the extended energy flux (9.115) for temperate ice is used, so that at the cold (positive) side $\mathbf{q} = \mathbf{q}_{\mathrm{s}}$, and at the negative (temperate) side $\mathbf{q} = \mathbf{q}_{\mathrm{s}} + L\mathbf{j}$. With Eqs. (9.111), (9.144) and (9.145),

$$\mathbf{q}_{\mathrm{s}}^+ \cdot \mathbf{n} - \mathbf{q}_{\mathrm{s}}^- \cdot \mathbf{n} - L\mathbf{j}^- \cdot \mathbf{n} = LW^-\rho(\mathbf{v} - \mathbf{w}) \cdot \mathbf{n} = -LW^-\rho a_{\mathrm{m}}^{\perp}\,, \tag{9.150}$$

or, with Fourier's law of heat conduction [Eqs. (4.36), (9.113)] and the definition (9.148) of $\mathbf{j}_{\mathrm{tot}}$,

$$\kappa\,(\mathrm{grad}\,T^+ - \mathrm{grad}\,T_{\mathrm{m}}^-) \cdot \mathbf{n} + L\,\mathbf{j}_{\mathrm{tot}}^- \cdot \mathbf{n} = 0\,. \tag{9.151}$$

At the cold side of the CTS the temperature $T = T_{\mathrm{m}}$ cannot increase from the CTS into the cold zone, otherwise the temperature would exceed the melting temperature, thus

$$\mathrm{grad}\,T^+ \cdot \mathbf{n} - \mathrm{grad}\,T_{\mathrm{m}}^- \cdot \mathbf{n} \le 0\,. \tag{9.152}$$

This and Eq. (9.151) imply $\mathbf{j}_{\mathrm{tot}}^- \cdot \mathbf{n} \ge 0$, so that

$$\mathcal{P}_{\mathrm{m}}^{\mathrm{w}} \quad (= -\mathbf{j}_{\mathrm{tot}}^- \cdot \mathbf{n}) \quad \le 0\,; \tag{9.153}$$

that is, the surface production of water $\mathcal{P}_{\mathrm{m}}^{\mathrm{w}}$ cannot be positive.

Because of this condition, for each point of the CTS three cases must be distinguished, depending on the sign of the quantity $(\mathbf{w} - \mathbf{v}_{\mathrm{w}}^-) \cdot \mathbf{n}$:

- $(\mathbf{w}-\mathbf{v}_\mathrm{w}^-)\cdot\mathbf{n}>0$ ("melting condition"):
  With the above definition (9.148) of $\mathbf{j}_\mathrm{tot}$ and $\rho_\mathrm{w}=\rho W$, Eq. (9.153) can only be fulfilled if
  $$W^- = 0\,, \tag{9.154}$$
  so the equality in Eq. (9.153) holds. Inserting this into Eq. (9.151) yields further
  $$\operatorname{grad} T^+ \cdot \mathbf{n} = \operatorname{grad} T_\mathrm{m}^- \cdot \mathbf{n}\,, \tag{9.155}$$
  which means that in case of melting conditions both the water content and the normal temperature derivative are continuous at the CTS, because W^+ is equal to zero anyway at the cold side of the CTS.

- $(\mathbf{w}-\mathbf{v}_\mathrm{w}^-)\cdot\mathbf{n}<0$ ("freezing condition"):
  In this case, Eq. (9.153) is compatible with
  $$W^- \ge 0\,, \tag{9.156}$$
  so that Eq. (9.152) can hold in its general form
  $$\operatorname{grad} T^+ \cdot \mathbf{n} \le \operatorname{grad} T_\mathrm{m}^- \cdot \mathbf{n}\,. \tag{9.157}$$
  As a consequence, in case of freezing conditions the water content and the normal temperature derivative can be discontinuous at the CTS; the jumps of these quantities are connected by Eq. (9.151).

- $(\mathbf{w}-\mathbf{v}_\mathrm{w}^-)\cdot\mathbf{n}=0$ ("parallel-flow condition"):
  For this case, too, Eq. (9.153) is compatible with Eq. (9.156), however, equality holds automatically in Eq. (9.153). Inserting this into (9.151) provides
  $$\operatorname{grad} T^+ \cdot \mathbf{n} = \operatorname{grad} T_\mathrm{m}^- \cdot \mathbf{n}\,. \tag{9.158}$$
  Hence, the parallel-flow condition is characterised by a continuous normal temperature derivative as for melting conditions, but the possibility of a jump in water content as for freezing conditions.

This behaviour can be understood as follows: if a non-vanishing total water flux $\mathbf{j}_\mathrm{tot}^-$ reaches the CTS from the temperate side (freezing condition), the transported water can freeze at the CTS. The latent heat released must be conducted away by a negative normal temperature derivative in the cold zone exceeding the small negative gradient in the temperate zone. Therefore, this entails a jump of both the normal temperature derivative and the water content.

However, the opposite situation cannot occur: it is impossible that cold ice flows toward the CTS, melts partly at the CTS and produces a non-vanishing total water flux at the temperate side. The reason for this is that the melting heat necessary for this process cannot be produced at or transported to the CTS. To achieve this the normal temperature derivative would have to be more positive at the cold side than at the temperate side, which is impossible because then the temperature in the cold zone would exceed the melting temperature of ice. Ice flow from the cold region through the CTS toward the temperate region is only possible without surface melting when passing the CTS, so that in this case $W^- = 0$ and $\operatorname{grad} T^+ \cdot \mathbf{n} = \operatorname{grad} T_\mathrm{m}^- \cdot \mathbf{n}$ hold, thus the water content and the normal temperature derivative are continuous.

In the case of a negligible diffusive water mass flux $\mathbf{j}$ in temperate ice, that is, a very small water diffusivity ν, the distinction between melting conditions, freezing conditions and parallel-flow conditions can simply be made by the sign of the ice volume flux through the CTS $a_\mathrm{m}^\perp$, because in this case $\mathbf{v} = \mathbf{v}_\mathrm{w}$ holds. Ice flow from cold to temperate ice, $a_\mathrm{m}^\perp > 0$, then corresponds to the melting condition, ice flow from temperate to cold ice, $a_\mathrm{m}^\perp < 0$, to the freezing condition and $a_\mathrm{m}^\perp = 0$ to the parallel-flow condition.

9.3.6 Parallel Sided Polythermal Slab

Let us once more consider a parallel sided slab (see Sects. 7.2 and 9.2.5), this time for polythermal conditions with a cold ice layer on top of a temperate layer (Fig. 9.23). The following assumptions are made:

- Plane strain approximation: purely two-dimensional flow in the vertical x-z plane, no dependencies on the transverse coordinate y.
- Constant thickness H and inclination angle α.
- Uniformity in the downslope (x) direction: $\partial(\cdot)/\partial x = 0$ for all field quantities.
- Steady-state conditions: $\partial(\cdot)/\partial t = 0$ for all field quantities.
- Flat, rigid bed: $b(x,t) = 0$ (in the inclined coordinate system). The free surface is therefore given by $h(x,t) = H$.
- Prescribed surface mass balance $a_\mathrm{s}^\perp$.
- Prescribed basal sliding velocity $v_{\mathrm{b}x}$.
- Glen's flow law (4.16), (9.110) with stress exponent $n = 3$.
- Constant rate factors: $A(T') = A_\mathrm{t}(W) = A = \mathrm{const.}$
- Constant heat conductivity: $\kappa(T) = \kappa = \mathrm{const.}$
- Neglect of the pressure dependence of the melting point of ice: $T_\mathrm{m} = 0°\mathrm{C}$.
- Neglect of water diffusion: $\nu = 0 \Rightarrow \mathbf{j} = \mathbf{0}$.

With these assumptions, the velocity field is not coupled to the thermodynamics, and the solutions (7.3) and (7.13) for the parallel sided slab in Sect. 7.2, modified by the prescribed basal sliding velocity $v_{\mathrm{b}x}$ and surface mass balance $a_\mathrm{s}^\perp$, can be applied:

$$v_x(z) = \frac{A(\rho g \sin\gamma)^3}{2}\,[H^4 - (H-z)^4] + v_{\mathrm{b}x}\,, \tag{9.159}$$

$$v_z(z) = -a_\mathrm{s}^\perp = -a_\mathrm{m}^\perp = \mathrm{const}\,. \tag{9.160}$$

The energy balance for the cold region, Eq. (5.14), reduces to

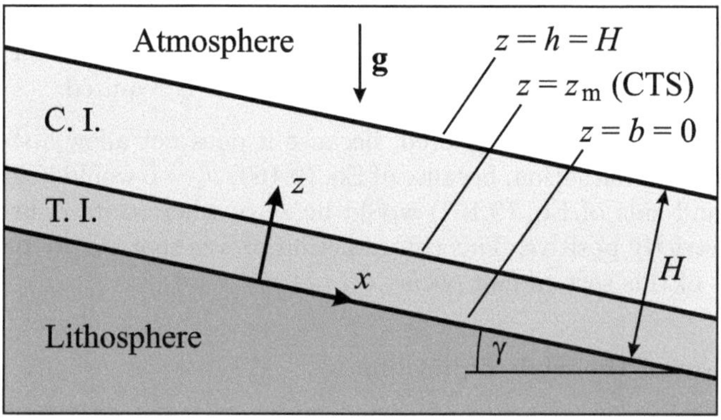

Fig. 9.23. Parallel sided polythermal slab: geometry and coordinate system. C. I.: cold ice, T. I.: temperate ice.

$$\rho c v_z \frac{\mathrm{d}T}{\mathrm{d}z} = \kappa \frac{\mathrm{d}^2 T}{\mathrm{d}z^2} + 2A\,\sigma_\mathrm{e}^4\,, \tag{9.161}$$

and for the temperate region, corresponding to the mass balance for the component water, Eqs. (9.117) and (9.118)) reduce to

$$\rho\, v_z \frac{\mathrm{d}W}{\mathrm{d}z} = 2\frac{A}{L}\,\sigma_\mathrm{e}^4\,. \tag{9.162}$$

The thermal boundary condition at the cold free surface is a prescribed surface temperature,

$$T = T_\mathrm{s}\,. \tag{9.163}$$

The boundary condition (9.131), which determines the normal diffusive water mass flux, is also redundant, since water diffusion is neglected. For simplicity, the sliding velocity $v_{\mathrm{b}x}$ is directly prescribed. The only impact of $v_{\mathrm{b}x}$ on the results is that it adds a constant to the velocity profile $v_x(z)$, whereas temperature and water contents are not affected at all.

The transition conditions at the CTS follow from Eqs. (9.142), (9.144), (9.145), (9.150) and (9.152),

$$\begin{aligned} & T^+ = T^-\,, \\ & v_x^+ = v_x^-\,, \quad v_z^+ = v_z^-\,, \\ & \kappa \frac{\mathrm{d}T^+}{\mathrm{d}z} = L\,W^- \rho\, a_\mathrm{m}^\perp \quad \text{with} \quad \frac{\mathrm{d}T^+}{\mathrm{d}z} \le 0\,. \end{aligned} \tag{9.164}$$

The secondary condition in the last equation entails that two different cases must be distinguished (see Sect. 9.3.5):

- Melting condition: $a_\mathrm{m}^\perp > 0$ and $\mathrm{d}T^+/\mathrm{d}z = 0$, $W^- = 0$.

- Freezing condition: $a_{\mathrm{m}}^{\perp} < 0$; Eq. (9.164) in its non-trivial form, i.e., $\mathrm{d}T^{+}/\mathrm{d}z$ can be strictly negative and W^{-} strictly positive; in this case an additional boundary condition for the basal water content is required.

The case of parallel flow is ignored, because it does not allow a steady-state solution. For this situation, because of Eq. (9.160) $v_z = 0$ would hold, and thus the left-hand side of Eq. (9.162) would be zero, whereas its right-hand side would be strictly positive. This contradiction proves that steady parallel-flow conditions of this sort cannot occur.

Integration of the Slab Equations

The equations derived above can be solved almost entirely analytically. Only for the CTS position $z = z_{\mathrm{m}}$ does an implicit algebraic equation remain, which can easily be solved by a numerical root finder.

The construction of the solution of Eqs. (9.161) and (9.162) for the temperature and the water content in the cold and the temperate region, respectively, and the associated determination of the CTS position, is outlined in the following. First, inserting Eqs. (7.9) for σ_{e} and (9.160) for v_z yields

$$\kappa \frac{\mathrm{d}^2 T}{\mathrm{d}z^2} + \rho c a_{\mathrm{s}}^{\perp} \frac{\mathrm{d}T}{\mathrm{d}z} = -2A\,(\rho g \sin\gamma)^4 (H-z)^4 \tag{9.165}$$

and

$$\rho a_{\mathrm{s}}^{\perp} \frac{\mathrm{d}W}{\mathrm{d}z} = -2\frac{A}{L}\,(\rho g \sin\gamma)^4 (H-z)^4\,. \tag{9.166}$$

For easier calculation, the vertical coordinate z is mapped onto the interval [0,1] by the transformation $z = H\zeta$,

$$D\frac{\mathrm{d}^2 T}{\mathrm{d}\zeta^2} + M\frac{\mathrm{d}T}{\mathrm{d}\zeta} = -K(1-\zeta)^4 \tag{9.167}$$

and

$$M\frac{\mathrm{d}W}{\mathrm{d}\zeta} = -K_t(1-\zeta)^4\,, \tag{9.168}$$

where

$$\begin{aligned} D &= \frac{\kappa}{\rho c}\,, \\ M &= H a_{\mathrm{s}}^{\perp}\,, \\ K &= \frac{2A}{\rho c} H^6 (\rho g \sin\gamma)^4\,, \\ K_{\mathrm{t}} &= \frac{2A}{\rho L} H^6 (\rho g \sin\gamma)^4\,. \end{aligned} \tag{9.169}$$

The solution of the homogeneous part of Eq. (9.167) is

$$T_{\mathrm{h}} = c_1\,\mathrm{e}^{-(M/D)\zeta} + c_2\,. \tag{9.170}$$

A particular integral of the inhomogeneous equation has the form

$$T_{\mathrm{p}} = a_1\zeta + a_2\zeta^2 + a_3\zeta^3 + a_4\zeta^4 + a_5\zeta^5\,, \tag{9.171}$$

where the coefficients a_1 to a_5 are calculated by balancing powers of ζ in Eq. (9.167), yielding

$$\begin{aligned}
a_5 &= -\frac{K}{5M}\,,\\
a_4 &= \frac{K}{M} + \frac{DK}{M^2}\,,\\
a_3 &= -2\frac{K}{M} - 4\frac{DK}{M^2} - 4\frac{D^2K}{M^3}\,,\\
a_2 &= 2\frac{K}{M} + 6\frac{DK}{M^2} + 12\frac{D^2K}{M^3} + 12\frac{D^3K}{M^4}\,,\\
a_1 &= -\frac{K}{M} - 4\frac{DK}{M^2} - 12\frac{D^2K}{M^3} - 24\frac{D^3K}{M^4} - 24\frac{D^4K}{M^5}\,.
\end{aligned} \tag{9.172}$$

With the above results, the general solution of the temperature equation (9.167) is

$$T = c_1\,\mathrm{e}^{-(M/D)\zeta} + c_2 + a_1\zeta + a_2\zeta^2 + a_3\zeta^3 + a_4\zeta^4 + a_5\zeta^5\,. \tag{9.173}$$

The constants c_1 and c_2 are still free and must be determined by boundary and transition conditions.

The water content equation (9.168) can be integrated directly,

$$W = \frac{K_{\mathrm{t}}}{5M}(1-\zeta)^5 + c_3\,, \tag{9.174}$$

leaving a further constant c_3 to be determined. The last step consists of calculating the integration constants c_1, c_2 and c_3, and the CTS position ζ_{m}. This must be performed separately for the cases of melting conditions ($a_{\mathrm{m}}^{\perp} > 0$) and freezing conditions ($a_{\mathrm{m}}^{\perp} < 0$) at the CTS.

Slab with Melting Condition at the CTS

In this case, according to Eq. (9.160), the velocity perpendicular to the bed is negative, i.e., the ice flows from the free surface toward the base. Owing to Eqs. (9.163) and (9.164) and the subsequent discussion,

$$T(1) = T_{\mathrm{s}}\,, \quad T^{+}(\zeta_{\mathrm{m}}) = 0\,, \quad (\mathrm{d}T^{+}/\mathrm{d}\zeta)_{\zeta_{\mathrm{m}}} = 0\,, \quad W^{-}(\zeta_{\mathrm{m}}) = 0\,. \tag{9.175}$$

Substitution of the first three of these equations into the general temperature solution (9.173) yields

$$T_\mathrm{s} = c_1\,\mathrm{e}^{-(M/D)} + c_2 + a_1 + a_2 + a_3 + a_4 + a_5\,, \tag{9.176}$$

$$0 = c_1\,\mathrm{e}^{-(M/D)\zeta_\mathrm{m}} + c_2 + a_1\zeta_\mathrm{m} + a_2\zeta_\mathrm{m}^2 + a_3\zeta_\mathrm{m}^3 + a_4\zeta_\mathrm{m}^4 + a_5\zeta_\mathrm{m}^5\,, \tag{9.177}$$

$$0 = -\frac{M}{D}c_1\,\mathrm{e}^{-(M/D)\zeta_\mathrm{m}} + a_1 + 2a_2\zeta_\mathrm{m} + 3a_3\zeta_\mathrm{m}^2 + 4a_4\zeta_\mathrm{m}^3 + 5a_5\zeta_\mathrm{m}^4\,, \tag{9.178}$$

which are three equations for the three unknowns c_1, c_2 and ζ_m. With c_2 from Eq. (9.177) and c_1 from Eq. (9.178), Eq. (9.176) becomes an implicit algebraic equation for the CTS position ζ_m,

$$\begin{aligned} 0 &= \frac{D}{M}\left(1 - \mathrm{e}^{(M/D)(\zeta_\mathrm{m}-1)}\right)(a_1 + 2a_2\zeta_\mathrm{m} + 3a_3\zeta_\mathrm{m}^2 + 4a_4\zeta_\mathrm{m}^3 + 5a_5\zeta_\mathrm{m}^4) + T_\mathrm{s} \\ &\quad + a_1(\zeta_\mathrm{m}-1) + a_2(\zeta_\mathrm{m}^2-1) + a_3(\zeta_\mathrm{m}^3-1) + a_4(\zeta_\mathrm{m}^4-1) + a_5(\zeta_\mathrm{m}^5-1) \\ &=: f(\zeta_\mathrm{m})\,. \end{aligned} \tag{9.179}$$

Equation (9.179) can be solved with a numerical root finder, which yields ζ_m with great accuracy. This is the only step in the whole solution procedure that must be performed numerically. Now c_1 follows from Eq. (9.178), and then c_2 from Eq. (9.176). The temperature in the cold region, given by Eq. (9.173), is therefore determined completely.

The forth boundary condition of Eq. (9.175) determines the coefficient c_3 in the expression (9.174) for the water content,

$$W = \frac{K_\mathrm{t}}{5M}[(1-\zeta)^5 - (1-\zeta_\mathrm{m})^5]\,. \tag{9.180}$$

Figure 9.24 shows computed profiles for the velocity v_x, temperature T and water content W for a slab with melting condition at the CTS.

Slab with Freezing Condition at the CTS

In this case, the velocity perpendicular to the bed is positive due to Eq. (9.160). The boundary conditions, by Eqs. (9.163), (9.164) and the subsequent discussion, are

$$T(1) = T_\mathrm{s}\,, \quad T^+(\zeta_\mathrm{m}) = 0\,, \quad \frac{\kappa}{H}\frac{\mathrm{d}T^+}{\mathrm{d}\zeta} = L\,W^-\rho a_\mathrm{m}^\perp\,, \quad W(0) = 0\,. \tag{9.181}$$

It is convenient to first determine c_3 by the last condition, when (9.174) becomes

$$W = \frac{K_\mathrm{t}}{5M}[(1-\zeta)^5 - 1]\,. \tag{9.182}$$

Thus the water content at the temperate side of the CTS is

$$W^- = W(\zeta_\mathrm{m}) = \frac{K_\mathrm{t}}{5M}[(1-\zeta_\mathrm{m})^5 - 1]\,, \tag{9.183}$$

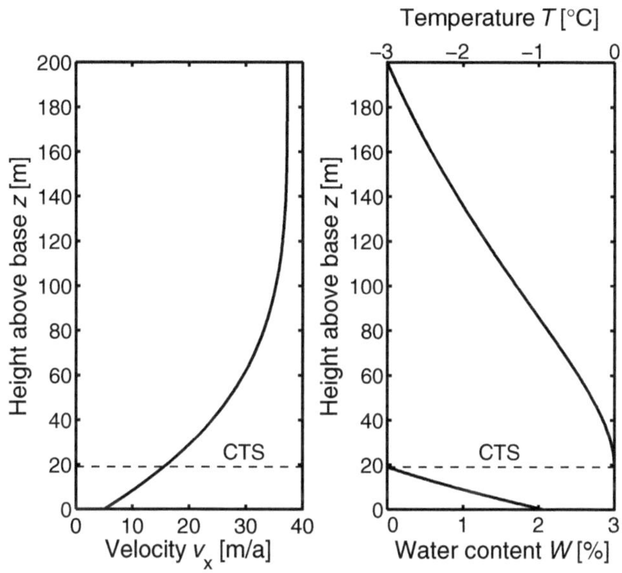

Fig. 9.24. Velocity v_x parallel to the bed, temperature T and water content W, for the slab with melting conditions at the CTS. $H = 200\,\mathrm{m}$, $\gamma = 4°$, $T_\mathrm{s} = -3°\mathrm{C}$, $a_\mathrm{s}^\perp = a_\mathrm{m}^\perp = 0.2\,\mathrm{m\,a^{-1}}$, $v_{\mathrm{b}x} = 5\,\mathrm{m\,a^{-1}}$.

where, however, ζ_m is still undetermined. Now, the temperature-gradient condition becomes

$$\begin{aligned}\frac{\mathrm{d}T^+}{\mathrm{d}\zeta} &= \frac{H}{\kappa} L\rho a_\mathrm{m}^\perp \frac{K_\mathrm{t}}{5M}[(1-\zeta_\mathrm{m})^5 - 1] \\ &= \frac{L\rho K_\mathrm{t}}{5\kappa}[(1-\zeta_\mathrm{m})^5 - 1], \quad \text{since} \quad M = Ha_\mathrm{s}^\perp = Ha_\mathrm{m}^\perp\,. \end{aligned} \tag{9.184}$$

The three temperature conditions, Eqs. $(9.181)_{1,2}$ and (9.184), now relate c_1, c_2 and ζ_m,

$$T_\mathrm{s} = c_1\,\mathrm{e}^{-(M/D)} + c_2 + a_1 + a_2 + a_3 + a_4 + a_5\,, \tag{9.185}$$

$$\begin{aligned}0 &= c_1\,\mathrm{e}^{-(M/D)\zeta_\mathrm{m}} + c_2 \\ &\quad + a_1\zeta_\mathrm{m} + a_2\zeta_\mathrm{m}^2 + a_3\zeta_\mathrm{m}^3 + a_4\zeta_\mathrm{m}^4 + a_5\zeta_\mathrm{m}^5\,,\end{aligned} \tag{9.186}$$

$$\begin{aligned}\frac{L\rho K_\mathrm{t}}{5\kappa}[(1-\zeta_\mathrm{m})^5 - 1] &= -\frac{M}{D} c_1\,\mathrm{e}^{-(M/D)\zeta_\mathrm{m}} \\ &\quad + a_1 + 2a_2\zeta_\mathrm{m} + 3a_3\zeta_\mathrm{m}^2 + 4a_4\zeta_\mathrm{m}^3 + 5a_5\zeta_\mathrm{m}^4\,.\end{aligned} \tag{9.187}$$

Again, with c_2 from Eq. (9.186) and c_1 from Eq. (9.187), Eq. (9.185) becomes an implicit algebraic equation for ζ_m,

$$0 = \frac{D}{M}(1 - \mathrm{e}^{(M/D)(\zeta_\mathrm{m}-1)})\Big(a_1 + 2a_2\zeta_\mathrm{m} + 3a_3\zeta_\mathrm{m}^2 + 4a_4\zeta_\mathrm{m}^3 + 5a_5\zeta_\mathrm{m}^4$$

$$
\begin{aligned}
&-\frac{L\rho K_{\mathrm{t}}}{5\kappa}[(1-\zeta_{\mathrm{m}})^5-1]\Big)+T_{\mathrm{s}}\\
&+a_1(\zeta_{\mathrm{m}}-1)+a_2(\zeta_{\mathrm{m}}^2-1)+a_3(\zeta_{\mathrm{m}}^3-1)+a_4(\zeta_{\mathrm{m}}^4-1)+a_5(\zeta_{\mathrm{m}}^5-1)\\
=:&\ g(\zeta_{\mathrm{m}})\,. \qquad (9.188)
\end{aligned}
$$

After solution by a Newtonian root finder, c_1 follows from Eq. (9.187), c_2 from Eq. (9.185), and the temperature distribution, Eq. (9.173), is fully determined.

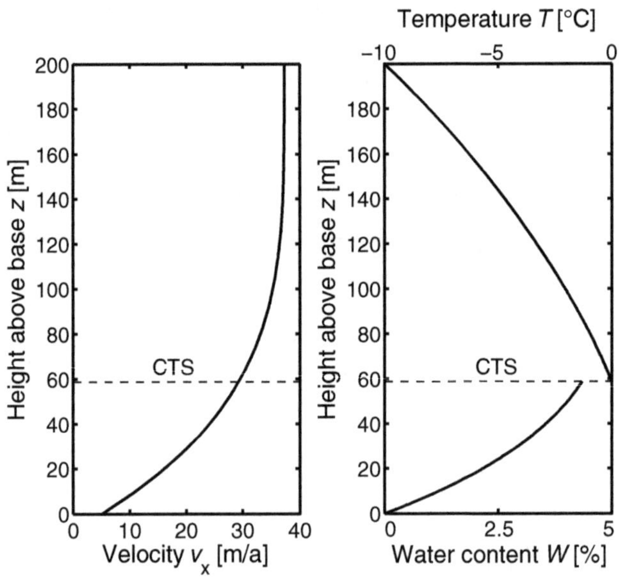

Fig. 9.25. Velocity v_x parallel to the bed, temperature T and water content W, for the slab with freezing conditions at the CTS. $H = 200\,\mathrm{m}$, $\gamma = 4°$, $T_{\mathrm{s}} = -10°\mathrm{C}$, $a_{\mathrm{s}}^{\perp} = a_{\mathrm{m}}^{\perp} = -0.2\,\mathrm{m\,a^{-1}}$, $v_{\mathrm{b}x} = 5\,\mathrm{m\,a^{-1}}$.

Figure 9.25 shows computed profiles for the velocity v_x, temperature T and water content W for a slab with freezing condition at the CTS. The distribution of the velocity v_x is identical to the slab with melting condition shown in Fig. 9.24. However, the behaviour of the temperature and the water content are entirely different. In the case of melting conditions the temperature gradient $\mathrm{d}T^+/\mathrm{d}z$ at the cold side of the CTS and the water content at the temperate side vanish. Since these quantities are zero on the corresponding opposite sides of the CTS anyway, they are continuous and therefore do not jump. On the other hand, in the case of freezing conditions a strictly negative temperature gradient $\mathrm{d}T^+/\mathrm{d}z$ and a strictly positive water content W^- appear, so that indeed there are discontinuities of these quantities.

The slab solutions demonstrate that the model equations are well posed. The solutions are analytic except for the determination of the CTS position by a numerical root finder, which can easily be done with great accuracy. Therefore, they can be used to check the performance of any numerical solution procedure.

9.3.7 Polythermal Glaciers

Polythermal glaciers occur in different climates and at different geographic locations. Given the winter cooling of the surface layer in temperate glaciers, or the summer warming of the surface layer in cold glaciers, most glaciers are seasonally polythermal. The only exceptions are perhaps cold glaciers in the extremely cold climate of Antarctica. Considering only perennial polythermal glaciers, the most frequent structures are the so-called Scandinavian-type and Canadian-type polythermal glaciers (Fig. 9.26).

The Scandinavian type occurs in Svalbard (Bamber 1988, Jania et al. 1996), Scandinavia (Holmlund and Eriksson 1989), the Rocky Mountains (Paterson 1971), Alaska and the Antarctic Peninsula (Breuer et al. 2006). Such glaciers are mostly temperate except for a cold surface layer in the ablation zone. This seemingly paradoxical situation may be explained by the summer heat input, which is stored due to percolation of meltwater into the upper firn area, whereas this heat is lost due to runoff of meltwater over the impermeable surface ice in the ablation zone. If winter temperatures increase, the cold layer thins and may eventually disappear, leaving an entirely temperate glacier.

The Canadian type occurs at high Arctic latitudes in Canada (Blatter 1987, Blatter and Kappenberger 1988) and Alaska, but the large ice sheets in Greenland and Antarctica also show this polythermal structure locally. Canadian-type glaciers are mostly cold except for a temperate layer at the bed in the ablation zone. Depending on the climate, ice thickness and ice flow, the thickness of this basal layer may shrink to zero, leaving a so-called basal hot spot (Classen and Clarke 1971). There may be additional polythermal structures, such as combinations of the above if glaciers span an extreme altitude range, or at confluences of glaciers with different thermal structures (Eisen et al. 2009).

In order to describe the thermal structures, the source of the englacial water must be identified. In this context, we do not consider the water in englacial channels and in the basal hydraulic system, which has a relatively short residence time of hours to days. The water content in consideration is given by water included in the ice matrix on length scales of the grain size.

Equation (9.118) determines changes of the water content by boundary conditions, internal transport processes (advection, diffusion) and sources (strain heating, pressure changes). Changes of the hydrostatic pressure contribute only 0.007 grams of water per kilogram ice-water mixture per metre change of depth, which makes this contribution to the water content negligibly small everywhere in the temperate part of a glacier (Pettersson et al. 2004). Thus, water entrapment in the firn zone and strain heating are the major sources of water in the ice.

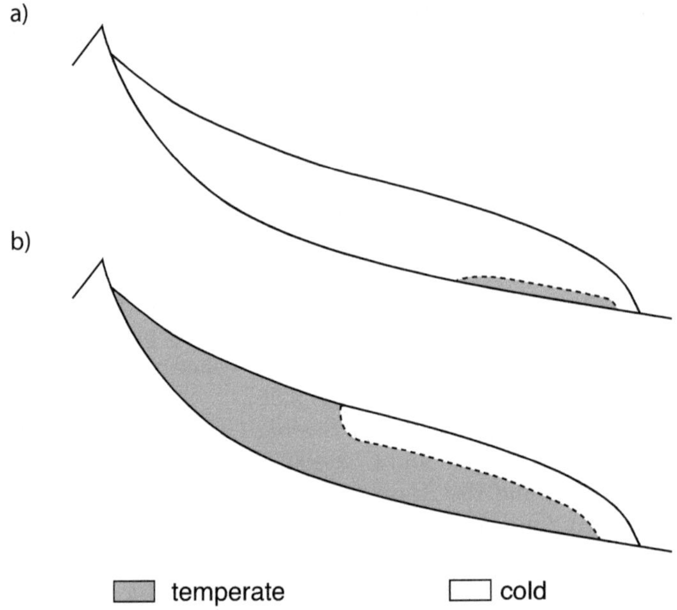

Fig. 9.26. Longitudinal section of glaciers with (a) Canadian-type and (b) Scandinavian-type polythermal structures. (Credit: A. Aschwanden, ETH Zurich, Switzerland.)

Water entrapment must be taken into account as a boundary condition at the firn surface in the accumulation area, if the firn is temperate [see Eq. (9.120)]. This leaves strain heating as the only significant source of water in the interior of the ice. If the diffusive water transport is neglected, Eq. (9.118) reduces to

$$\frac{\mathrm{d}W}{\mathrm{d}t} = \frac{2A_\mathrm{t}(W)}{\rho L}\,\sigma_\mathrm{e}^{n+1}\,. \tag{9.189}$$

This advection-production equation is difficult to handle since numerical solutions tend to be unstable. One way to solve it is to take advantage of the fact that water accumulates in an ice particle while it travels along its trajectory in the glacier. If the dependence of the rate factor on water content is ignored [$A_\mathrm{t}(W) = A = \text{const}$], Eq. (9.189) can readily be separated by variables, and integration along a trajectory from point O to point P yields

$$W_P = W_O + \int_O^P \frac{2A}{\rho L}\,\sigma_\mathrm{e}^{n+1}\,\mathrm{d}t\,. \tag{9.190}$$

In the case of a stationary flow field, trajectories are equivalent to flow lines. This simplifies the computation of the water-content field since the flow lines can be computed first and need not be updated during the integration.

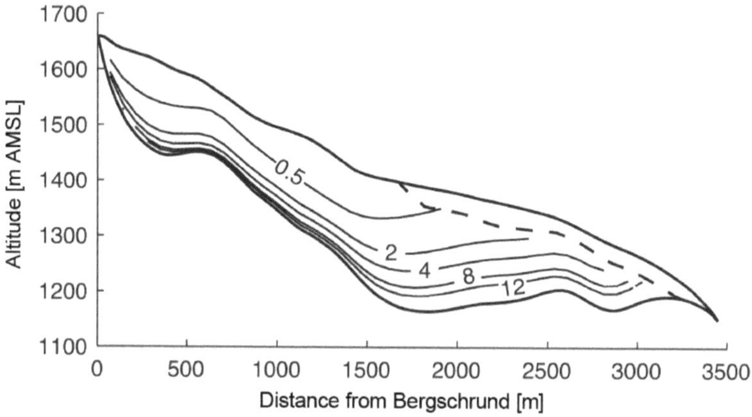

Fig. 9.27. Water content (in g water per kg ice-water mixture) due to strain heating in a longitudinal section of Storglaciären, Sweden, computed by integration of Eq. (9.190) along trajectories, starting with zero water content at the ice surface in the accumulation area. Adapted from Aschwanden and Blatter (2005), © American Geophysical Union.

Figure 9.27 shows the water content accumulated in the ice of Storglaciären, Sweden, by strain heating alone (Aschwanden and Blatter 2005). The flow and stress fields have been obtained by solving the full Stokes problem, and the water content has been computed by integrating Eq. (9.190) with the boundary condition $W_O = 0$ at the ice surface in the accumulation zone of the glacier.

The numerical model SICOPOLIS (see Sect. 5.7) can handle polythermal situations of the Canadian type in ice sheets. To this end, the temperature evolution equation (5.105) is solved in the cold domain and the evolution equation for the water content (9.118) in the temperate domain. The position and motion of the CTS are determined by exploiting the matching conditions described in Sect. 9.3.5. This *front tracking method* contrasts with enthalpy or enthalpy gradient methods (Nedjar 2002) which will be discussed in the following.

9.3.8 Enthalpy Formulation

Enthalpy Balance

The specific enthalpy, h, is commonly defined as (e.g., Moran and Shapiro 2000)

$$h = u + \frac{p}{\rho}, \qquad (9.191)$$

where u is the specific internal energy, p the pressure and ρ the density. If an incompressible material is heated under constant pressure, the enthalpy is equal to the internal energy (Alexiades and Solomon 1993). Since the term "enthalpy method" is widely used in the computational fluid dynamics literature, we refer to enthalpy instead of internal energy and rewrite Eq. (3.92) as

$$\rho\dot{h} = -\operatorname{div}\mathbf{q} + \operatorname{tr}(\mathsf{t}\cdot\mathsf{D}), \tag{9.192}$$

where the radiation r has again be neglected.

The enthalpy of ice is assumed to be uniquely related to temperature, $T = T(h)$, and water content, $W = W(h)$, in their respective domains (Aschwanden and Blatter 2009). In other words, enthalpy represents temperature in cold ice and water content in temperate ice.

We define h_i and h_l as the enthalpies of pure ice and pure water at the melting temperature T_m. Thus

$$h_\mathrm{i} = -L + h_\mathrm{l} = -L + c_\mathrm{w}T_\mathrm{m} \tag{9.193}$$

and

$$\begin{array}{llll}
\text{Cold ice:} & \left.\begin{array}{l} T = (h-h_\mathrm{i})/c + T_\mathrm{m} \\ W = 0 \end{array}\right\} & \text{if} & h < h_\mathrm{i}\,, \\[2ex]
\text{CTS:} & \left.\begin{array}{l} T = T_\mathrm{m} \\ W = 0 \end{array}\right\} & \text{if} & h = h_\mathrm{i}\,, \\[2ex]
\text{Temperate ice:} & \left.\begin{array}{l} T = T_\mathrm{m} \\ W = (h-h_\mathrm{i})/L \end{array}\right\} & \text{if} & h_\mathrm{i} < h < h_\mathrm{l}\,, \\[2ex]
\text{Liquid water:} & \left.\begin{array}{l} T = (h-h_\mathrm{l})/c + T_\mathrm{m} \\ W = 1 \end{array}\right\} & \text{if} & h \geq h_\mathrm{l}\,,
\end{array} \tag{9.194}$$

where L is the latent heat, c_i the specific heat of ice, c_w the specific heat of water and $c = (1-W)\,c_\mathrm{i} + W\,c_\mathrm{w}$ the specific heat of the ice-water mixture. Measured water contents in temperate ice are generally less than 3% (Pettersson et al. 2004, and references therein), and thus liquid water ($h \geq h_\mathrm{l}$) is disregarded. A schematic plot of the temperature and water content as functions of enthalpy is shown in Fig. 9.28.

Enthalpy Gradient Method

The enthalpy gradient method is suitable for smooth enthalpy distributions (Pham 1995). The enthalpy gradient method requires an enthalpy-gradient-driven energy flux in both cold and temperate ice. We assume that h_i and T_m vary negligibly with pressure and that thermodynamic properties such as c and κ are constant. To solve the enthalpy balance (9.192), the heat flux $\mathbf{q}$ is expressed in terms of enthalpy. In cold ice, the gradient of Eq. (9.194) gives

$$\operatorname{grad}T = \frac{1}{c}\operatorname{grad}h\,. \tag{9.195}$$

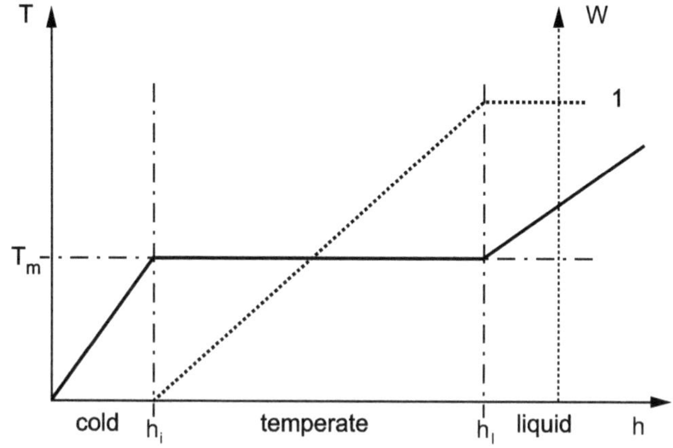

Fig. 9.28. Temperature-enthalpy relation (solid line) and water-content-enthalpy relation (dotted line). T is the temperature, W is the water content, T_{m} is the melting temperature, h_{i} and h_{l} are the enthalpies of pure ice and pure water at $T = T_{\mathrm{m}}$, respectively. (Credit: A. Aschwanden, ETH Zurich, Switzerland.)

Introducing Eq. (9.195) into Fourier's law of heat conduction (4.36) yields the heat flux in cold ice as a function of enthalpy,

$$\mathbf{q} = -\frac{\kappa}{c}\,\mathrm{grad}\,h\,. \tag{9.196}$$

In temperate ice, the heat flux is governed by Eq. (9.115). The sensible heat flux $\mathbf{q}_{\mathrm{s}}$ which arises from Eq. (9.113) is extremely small. As for the latent heat flux $\mathbf{q}_{\mathrm{l}} = L\mathbf{j}$, we assume a negligible diffusive water mass flux $\mathbf{j}$, so that $\mathbf{q}_{\mathrm{l}}$ is also negligible. Thus,

$$\mathbf{q} = \mathbf{0}\,. \tag{9.197}$$

By combining Eqs. (9.196) and (9.197), the heat flux $\mathbf{q}$ in both cold and temperate ice can be expressed in terms of the enthalpy gradient,

$$\mathbf{q} = -\rho k\,\mathrm{grad}\,h\,, \tag{9.198}$$

with the enthalpy diffusivity

$$k = \begin{cases} \kappa/(\rho c) & \text{for cold ice}\,, \\ 0 & \text{for temperate ice}\,. \end{cases} \tag{9.199}$$

Therefore, Eq. (9.192) can be rewritten as

$$\frac{\partial h}{\partial t} + \mathbf{v}\cdot\mathrm{grad}\,h = \mathrm{div}\,(k\,\mathrm{grad}\,h) + \frac{\mathrm{tr}\,(\mathsf{t}\cdot\mathsf{D})}{\rho}\,, \tag{9.200}$$

which holds for both cold and temperate ice. Equation (9.200) represents an advection-diffusion-production transport problem with the same mathematical form as the temperature evolution equation (5.14) for cold ice. The first

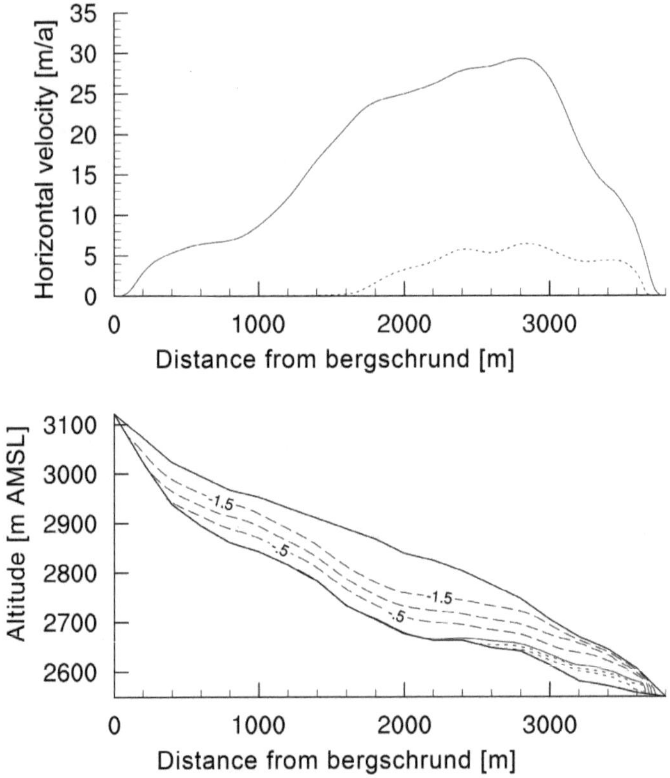

Fig. 9.29. Top: Horizontal velocity at the surface and the base of a Canadian-type polythermal glacier with a synthetically created geometry. Bottom: Contours of the ice temperature (dashed lines, in °C) and the water content (dotted lines, 0.3 and $1\,\mathrm{g\,kg^{-1}}$). The solid line in the ice domain represents the contour line $h = h_\mathrm{i}$ of the CTS. (Credit: A. Aschwanden, ETH Zurich, Switzerland.)

term is the local rate of change of enthalpy, the second term is the enthalpy advection, the third term is the enthalpy diffusion and $\mathrm{tr}\,(\mathsf{t}\cdot\mathsf{D})$ is the enthalpy production due to strain heating.

At the glacier surface, a Dirichlet condition, $h = h_\mathrm{s}$, with

$$\begin{aligned} h_\mathrm{s} &= h_\mathrm{i} + c\,(T_\mathrm{s} - T_\mathrm{m}) && \text{if} \quad h_\mathrm{s} < h_\mathrm{i}\,, \\ h_\mathrm{s} &= h_\mathrm{i} && \text{if} \quad h_\mathrm{s} = h_\mathrm{i}\,, \\ h_\mathrm{s} &= h_\mathrm{i} + LW_\mathrm{s} && \text{if} \quad h_\mathrm{i} < h_\mathrm{s} < h_\mathrm{l}\,, \end{aligned} \tag{9.201}$$

is applied, where T_s and W_s are the prescribed temperature at the cold surface and water content at the temperate surface, respectively. At a cold glacier bed, all geothermal heat $q_\mathrm{geo}^\perp$ enters the ice [see Eq. (5.38)]. At a temperate glacier bed, due to the neglect of water diffusion, no physical boundary condition is available although it is required by Eq. (9.200) (Griffiths 1997). Therefore, we choose a homogenous Neumann (zero flux) condition, because this boundary condition influences the result the least. This yields

$$\begin{aligned} (\mathbf{q}\cdot\mathbf{n})_\mathrm{b} &= -q_\mathrm{geo}^\perp \quad &\text{if} \quad & h_\mathrm{b} < h_\mathrm{i}\,, \\ (\mathbf{q}\cdot\mathbf{n})_\mathrm{b} &= 0 \quad &\text{if} \quad & h_\mathrm{i} \le h_\mathrm{b} < h_\mathrm{l}\,. \end{aligned} \tag{9.202}$$

The physical interpretation of the homogenous Neumann condition $(9.202)_2$ is that all heat from the geothermal heat flux and produced by basal friction is used for melting ice at the base.

Figure 9.29 shows results for a Canadian-type polythermal glacier computed with the enthalpy gradient method and a full Stokes solution for the velocity field. The enthalpy gradient method offers the advantage that the type of the polythermal structure emerges from the given boundary conditions. Furthermore, the CTS is given by the contour line $h = h_\mathrm{i}$ of the enthalpy field, thus no front tracking is necessary.

10

Conclusions, Summary and Outlook

In agreement with the scope of the series *Advances in Geophysical and Environmental Mechanics and Mathematics* (AGEM^2), it is our intention that this book serves the purposes

- to be a compact and up-to-date source of reference on the topic of ice sheet and glacier dynamics;
- to be an accessible introduction to this field for advanced students and researchers from related areas;
- to be a source of advanced teaching material for specialised seminars, courses and schools;

see http://www.springer.com/series/7540. We hope that it will be received in that sense. Naturally, the choice which topics are included and which ones are omitted was difficult. In order to stick to the above points, in particular the compactness and the introductory nature, we put the emphasis on the relatively well understood and established aspects of the field, and did not attempt to give an exhaustive overview of all directions of current research in a 600+ pages volume.

Regular flow of ice sheets and glaciers can be described well with the concepts presented in this book. This means that the dynamic and thermal response of the system depends smoothly on the boundary conditions. However, in certain situations small changes of the external forcings can trigger drastic reactions, such as surges, large calving events or break-off of large portions of ice. This lies partly in the non-linear nature of the problem (see, e.g., the mini-surges observed in the simulations of Sect. 5.7.4), but to a larger extent discontinuities are involved, which require a separate treatment.

The grand questions in future glaciology and glacier and ice sheet modelling concern possible rapid changes of ice masses in response to global warming. The conjectured processes are mostly related to discontinuities, which are not well understood (Sammonds 1999). Basal sliding (see Sect. 7.5) is strongly related to englacial and subglacial hydrology (e.g., the much discussed acceleration of basal sliding by surface meltwater percolating to the base; Zwally

R. Greve, H. Blatter, *Dynamics of Ice Sheets and Glaciers*, Advances in Geophysical and Environmental Mechanics and Mathematics,
DOI 10.1007/978-3-642-03415-2_10,

et al. 2002, Greve and Sugiyama 2009) and basal conditions (such as hard rocky beds with complex micro-topography or layers of granular material). This fact partly explains the difficulties encountered with a reliable handling of sliding in numerical glacier and ice sheet models. The often applied shallow ice approximation in conjunction with a Weertman-type sliding law falls short of modelling the details of basal sliding adequately, a fact that reduces the prognostic value of model calculations.

Fracturing of ice plays an important role in the dynamics of ice masses and may lead to instabilities. Glaciers that reach the water of lakes or oceans start to calve icebergs into the water, thus contributing to mass loss. The terminus of calving ice masses may remain grounded or become floating, two different conditions that are governed by different processes and require different treatment in numerical models.

Calving of grounded glaciers is mostly treated by simple parameterisations which must be tuned to observations (Vieli et al. 2001, Benn et al. 2007a,b); however, this does not allow a prognostic approach for glaciers without direct observations. A novel approach to predict fracturing, and thus calving, has been attempted with damage mechanics (Pralong and Funk 2005, 2006, Pralong et al. 2006). The state of the ice is described by a continuous quantity, called "damage", which has a source and a sink, and evolves in response to the state of stress. Fractures develop if damage values reach a threshold level, thus the basically discontinuous process of fracture is described by a continuous field.

Calving of floating ice shelves thus far resists successful treatment, both conceptually and in numerical models. The mechanisms of this process were discussed by Reeh (1968). In Sect. 6.1.3 we provided a continuous description based on a kinematic boundary condition, but the problem remains how to prescribe the parameter $c^{\perp}$ ("calving rate"). While sometimes a very simple, indirect approach is chosen by assuming a threshold value (e.g., 200 m) for the frontal ice thickness, in a recent study, Alley et al. (2007) suggested to relate the calving rate to the along-flow strain rate near the front of the ice shelf. In any case, it is still a long way to a description that allows prognostic simulations of rapid events like the recent disintegrations of the Larsen A, Larsen B and Wilkins Ice Shelves of the Antarctic Peninsula.

A related problem is to predict the position and migration of the grounding line of ice shelves (see Sect. 6.1.3), which is probably a key process affecting the stability of the West Antarctic Ice Sheet. Various model approaches have been tested to identify conditions for instabilities (Schoof 2007a,b); however, it is difficult to judge how conclusive they are (Vieli and Payne 2005). This leads to the question of the dynamic stability of ice sheets, ice shelves and glaciers in general, and whether there are critical thresholds ("points of no return") beyond which they evolve inevitably and irreversibly towards rapid disintegration.

ERRATA

Errata to: Dynamics of Ice Sheets and Glaciers

Ralf Greve and Heinz Blatter

R. Greve, H. Blatter, *Dynamics of Ice Sheets and Glaciers*, Advances in Geophysical and Environmental Mechanics and Mathematics,
DOI 10.1007/978-3-642-03415-2, © Springer-Verlag Berlin Heidelberg 2009

DOI 10.1007/978-3-642-03415-2_11

Chapter 7

- Page 154, caption of Fig. 7.5:
  The passage "dashed lines for $\Lambda = 0.25$, solid lines for $\Lambda = 1$, dash-dotted lines for $\Lambda = 100$" should be "dash-dotted lines for $\Lambda = 0.25$, solid lines for $\Lambda = 1$, dashed lines for $\Lambda = 10$".

Chapter 8

- Page 187, first line of subsection 8.2:
  "From seismic studies" should be "From seismological studies".

- Page 191, Eq. (8.12):
  The arguments of H should be $\check{x}$,$\check{y}$ rather than x, y. So the correct form is

$$w_{\mathrm{ss}}(x, y) = \int_{A_{\mathrm{ice}}} \rho g H(\check{x}, \check{y})\, G(x, \check{x}, y, \check{y})\, \mathrm{d}\check{x}\, \mathrm{d}\check{y}.$$

Chapter 9

- Page 204, second paragraph, second/third line:
  "Placidi et al. (2009)" should be "Placidi et al. (2010)".

- Page 205, Fig. 9.1:
  In some early copies of the book, this figure was rendered incorrectly. It should appear as follows:

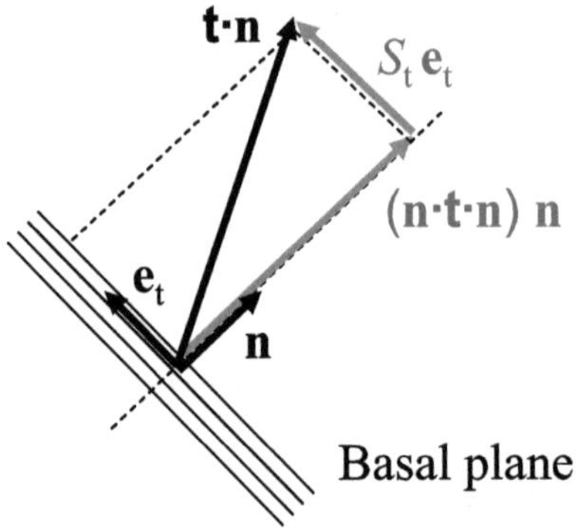

- Page 207, Fig. 9.2:
  In some early copies of the book, this figure was rendered incorrectly. It should appear as follows:

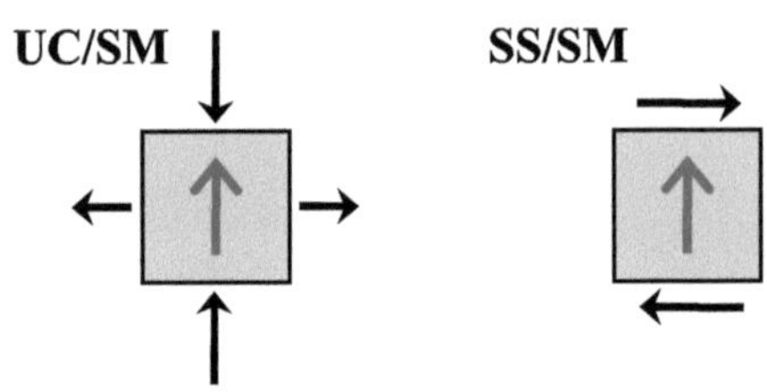

References Cited or Recommended

- Page 265, reference to Calov and Greve (2006):
  "`http://www.pik-potsdam.de/~{}calov/heino.html`" should be "`http://www.pik-potsdam.de/~calov/heino.html`".

- Page 265, reference to Durand, Gagliardini, Zwinger and Le Meur (2009):
  The list of authors should be "Durand, G., O. Gagliardini, T. Zwinger, E. Le Meur and R. C. A. Hindmarsh". The bibliographical information should be "*Annals of Glaciology*, **50** (52), 109–114".

- Page 266, reference to Gagliardini, Gillet-Chaulet and Montagnat (2009): The bibliographical information should be "*Low Temperature Science*, **68** (Supplement Issue "Physics of Ice Core Records II", Ed. T. Hondoh), 149–166".

- Page 266, reference to Greve, Placidi and Seddik (2009): The bibliographical information should be "*Low Temperature Science*, **68** (Supplement Issue "Physics of Ice Core Records II", Ed. T. Hondoh), 137–148".

- Page 270, reference to Placidi, Greve, Seddik and Faria (2009): The reference should be "Placidi, L., R. Greve, H. Seddik and S. H. Faria. 2010. Continuum-mechanical, Anisotropic Flow model for polar ice masses, based on an anisotropic Flow Enhancement factor. *Continuum Mechanics and Thermodynamics*, **22** (3), 221–237. doi:10.1007/s00161-009-0126-0."

The online version of the original book can be found at
http://dx.doi.org/10.1007/978-3-642-03415-2

References Cited or Recommended

Abe-Ouchi, A. 1993. *Ice sheet response to climatic changes: a modelling approach.* Zürcher Geographische Schriften No. 54. Geographical Institute, ETH Zurich, Switzerland.

Abramowitz, M. and I. A. Stegun. 1970. *Handbook of Mathematical Functions with Formulars, Graphs, and Mathematical Tables.* Dover Publications, New York, NY, USA.

Alexiades, V. and A. D. Solomon. 1993. *Mathematical Modeling of Melting and Freezing Processes.* Hemisphere Publishing Corporation, Washington, DC, USA.

Alley, R. B. 2000. *The Two-Mile Time Machine. Ice Cores, Abrupt Climate Change and Our Future.* Princeton University Press, Princeton, NJ, USA and Oxford, UK. ISBN 0-691-00493-5.

Alley, R. B., I. Joughin, H. J. Horgan, T. K. Dupont, B. R. Parizek, S. Anandakrishnan and K. M. Cuffey. 2007. A first calving law for ice shelves: spreading-rate control of calving rate. Abstract #C43A-01, American Geophysical Union Fall Meeting, San Francisco, CA, USA, 2007.12.10–14.

Aschwanden, A. and H. Blatter. 2005. Meltwater production due to strain heating in Storglaciären, Sweden. *Journal of Geophysical Research,* **110** (F4), F04024. doi:10.1029/2005JF000328.

Aschwanden, A. and H. Blatter. 2009. Mathematical modeling and numerical simulation of polythermal glaciers. *Journal of Geophysical Research,* **114** (F1), F01027. doi:10.1029/2008JF001028.

Azuma, N. 1995. A flow law for anisotropic polycrystalline ice under uniaxial compressive deformation. *Cold Regions Science and Technology,* **23** (2), 137–147.

Bader, H. 1954. Sorge's law of densification of snow on high polar glaciers. *Journal of Glaciology,* **2**, 319–323.

Bahr, D. B., W. T. Pfeffer and M. F. Meier. 1994. Theoretical limitations to englacial velocity calculations. *Journal of Glaciology,* **40** (136), 509–518.

Bamber, J. L. 1988. Internal reflecting horizons in Spitsbergen glaciers. *Annals of Glaciology,* **9**, 5–9.

R. Greve, H. Blatter, *Dynamics of Ice Sheets and Glaciers*, Advances in Geophysical and Environmental Mechanics and Mathematics,
DOI 10.1007/978-3-642-03415-2_BM2,

Benn, D. I., N. R. J. Hulton and R. H. Mottram. 2007a. 'Calving laws', 'sliding laws' and the stability of tidewater glaciers. *Annals of Glaciology*, **46**, 123–130.

Benn, D. I., C. R. Warren and R. H. Mottram. 2007b. Calving processes and the dynamics of calving glaciers. *Earth Science Reviews*, **82**, 143–179.

Bentley, C. R. 2004. Mass balance of the Antarctic ice sheet: observational aspects. In: J. L. Bamber and A. J. Payne (Eds.), *Mass Balance of the Cryosphere. Observations and Modelling of Contemporary and Future Changes*, pp. 459–489. Cambridge University Press, Cambridge, UK and New York, NY, USA.

Bindoff, N. L., J. Willebrand, V. Artale, A. Cazenave, J. Gregory, S. Gulev, K. Hanawa, C. Le Quéré, S. Levitus, Y. Nojiri, C. K. Shum, L. D. Talley and A. Unnikrishnan. 2007. Observations: Oceanic climate change and sea level. In: S. Solomon, D. Qin, M. Manning, Z. Chen, M. Marquis, K. B. Averyt, M. Tignor and H. L. Miller (Eds.), *Climate Change 2007: The Physical Science Basis. Contribution of Working Group I to the Fourth Assessment Report of the Intergovernmental Panel on Climate Change*, pp. 385–432. Cambridge University Press, Cambridge, UK, and New York, NY, USA. URL `http://ipcc-wg1.ucar.edu/wg1/wg1-report.html`.

Blatter, H. 1987. On the thermal regime of an arctic valley glacier, a study of the White Glacier, Axel Heiberg Island, N.W.T., Canada. *Journal of Glaciology*, **33** (114), 200–211.

Blatter, H. 1995. Velocity and stress fields in grounded glaciers: a simple algorithm for including deviatoric stress gradients. *Journal of Glaciology*, **41** (138), 333–344.

Blatter, H. and G. Kappenberger. 1988. Mass balance and thermal regime of the Laika Ice Cap, Coburg Island, N.W.T., Canada. *Journal of Glaciology*, **34** (116), 102–110.

Bond, G., H. Heinrich, S. Huon, W. Broecker, L. Labeyrie, J. Andrews, J. McManus, S. Clasen, K. Tedesco, R. Jantschik, C. Simet and M. Klas. 1992. Evidence for massive discharges of icebergs into the glacial Northern Atlantic. *Nature*, **360** (6401), 245–250.

Bond, G. C. and R. Lotti. 1995. Iceberg discharges into the North Atlantic on millennial time scales during the last glaciation. *Science*, **267** (5200), 1005–1010.

Breuer, B., M. A. Lange and N. Blindow. 2006. Sensitivity studies on model modifications to assess the dynamics of a temperate ice cap, such as that on King George Island, Antarctica. *Journal of Glaciology*, **52** (177), 235–247.

Bronshtein, I. N., K. A. Semendyayev, G. Musiol and H. Muehlig. 2004. *Handbook of Mathematics*. Springer, Berlin, Germany etc., 4th ed.

Brotchie, J. F. and R. Silvester. 1969. On crustal flexure. *Journal of Geophysical Research*, **74** (22), 5240–5252.

Buckingham, E. 1924. Dimensional analysis. *Philosophical Magazine*, **48** (283), 141–145.

Budd, W. F. and T. H. Jacka. 1989. A review of ice rheology for ice sheet modelling. *Cold Regions Science and Technology*, **16** (2), 107–144.

Bueler, E. 2003. Construction of steady state solutions for isothermal shallow ice sheets. Tech. Rep. 03-02, Department of Mathematics and Statistics, University of Alaska, Fairbanks.

Bueler, E., C. S. Lingle, J. A. Kallen-Brown, D. N. Covey and L. N. Bowman. 2005. Exact solutions and verification of numerical models for isothermal ice sheets. *Journal of Glaciology*, **51** (173), 291–306.

Calov, R. 1994. *Das thermomechanische Verhalten des Grönländischen Eisschildes unter der Wirkung verschiedener Klimaszenarien – Antworten eines theoretisch-numerischen Modells.* Doctoral thesis, Department of Mechanics, Darmstadt University of Technology, Germany.

Calov, R. and R. Greve. 2006. ISMIP HEINO. Ice Sheet Model Intercomparison Project – Heinrich Event INtercOmparison. Online publication. URL `http://www.pik-potsdam.de/~{}calov/heino.html`.

Church, J. A., J. M. Gregory, P. Huybrechts, M. Kuhn, K. Lambeck, M. T. Nhuan, D. Qin and P. L. Woodworth. 2001. Changes in sea level. In: J. T. Houghton, Y. Ding, D. J. Griggs, M. Noguer, P. J. van der Linden, X. Dai, K. Maskell and C. A. Johnson (Eds.), *Climate Change 2001: The Scientific Basis. Contribution of Working Group I to the Third Assessment Report of the Intergovernmental Panel on Climate Change*, pp. 639–693. Cambridge University Press, Cambridge, UK and New York, NY, USA.

Classen, D. F. and G. K. C. Clarke. 1971. Basal hot spot on a surge type glacier. *Nature*, **229**, 481–483.

Colinge, J. and H. Blatter. 1998. Stress and velocity fields in glaciers: Part I. Finite difference schemes for higher-order glacier models. *Journal of Glaciology*, **44**, 448–456.

Colinge, J. and J. Rappaz. 1999. A strongly nonlinear problem arising in glaciology. *Mathematical Modelling and Numerical Analysis*, **33** (2), 395–406.

Dansgaard, W. and S. J. Johnsen. 1969. A flow model and a time scale for the ice core from Camp Century, Greenland. *Journal of Glaciology*, **8** (53), 215–223.

de Berg, M., O. Cheong, M. van Kreveld and M. Overmars. 2008. *Computational Geometry. Algorithms and Applications.* Springer, Berlin, Germany etc., 3rd ed.

Debnath, L. and P. Mikusinski. 1993. *Introduction to Hilbert Spaces and Applications.* Academic Press, San Diego, CA, USA.

Durand, G., O. Gagliardini, T. Zwinger and E. Le Meur. 2009. Full-Stokes modeling of marine ice sheets: influence of the grid size. *Annals of Glaciology*, **52**. In press.

Eisen, O., A. Bauder, M. Lüthi, P. Riesen and M. Funk. 2009. Deducing the thermal structure in the tongue of Gornergletscher, Switzerland, from radar surveys and borehole measurements. *Annals of Glaciology*, **50** (51), 63–70.

Eisen, O., I. Hamann, S. Kipfstuhl, D. Steinhage and F. Wilhelms. 2007. Direct evidence for continuous radar reflector originating from changes in crystal-orientation fabric. *The Cryosphere*, **1** (1), 1–10. URL `http://www.the-cryosphere.net/1/1/2007/`.

Ekman, M. 1991. A concise history of postglacial land uplift research (from its beginning to 1950). *Terra Nova*, **3**, 358–365.

EPICA Community Members. 2006. One-to-one coupling of glacial climate variability in Greenland and Antarctica. *Nature*, **444** (7116), 195–198. doi: 10.1038/nature05301.

Faria, S. H. 2003. *Mechanics and thermodynamics of mixtures with continuous diversity.* Doctoral thesis, Department of Mechanics, Darmstadt University of Technology, Germany. URL `http://tuprints.ulb.tu-darmstadt.de/307/`.

Forsström, P.-L., O. Sallasmaa, R. Greve and T. Zwinger. 2003. Simulation of fast-flow features of the Fennoscandian ice sheet during the Last Glacial Maximum. *Annals of Glaciology*, **37**, 383–389.

Fowler, A. C. 1984. On the transport of moisture in polythermal glaciers. *Geophysical and Astrophysical Fluid Dynamics*, **24**, 99–140.

Gagliardini, O., F. Gillet-Chaulet and M. Montagnat. 2009. A review of anisotropic polar ice models: from crystal to ice-sheet flow models. In: T. Hondoh (Ed.), *Physics of Ice Core Records Vol. 2*. Yoshioka Publishing, Kyoto, Japan. In press.

Gagliardini, O. and J. Meyssonnier. 1997. Flow simulation of a firn-covered cold glacier. *Annals of Glaciology*, **24**, 242–248.

Glen, J. W. 1955. The creep of polycrystalline ice. *Proceedings of the Royal Society London A*, **228** (1175), 519–538. doi:10.1098/rspa.1955.0066.

Greve, R. 1994. Zwischenbericht zur Dissertation "Thermomechanisches Verhalten polythermer Eisschilde". Unpublished report, Department of Mechanics, Darmstadt University of Technology, Germany.

Greve, R. 1997. A continuum-mechanical formulation for shallow polythermal ice sheets. *Philosophical Transactions of the Royal Society London A*, **355** (1726), 921–974. doi:10.1098/rsta.1997.0050.

Greve, R. 2000. *Large-scale glaciation on Earth and on Mars.* Habilitation thesis, Department of Mechanics, Darmstadt University of Technology, Germany. URL `http://tuprints.ulb.tu-darmstadt.de/816/`.

Greve, R. 2005. Relation of measured basal temperatures and the spatial distribution of the geothermal heat flux for the Greenland ice sheet. *Annals of Glaciology*, **42**, 424–432.

Greve, R., L. Placidi and H. Seddik. 2009. A continuum-mechanical model for the flow of anisotropic polar ice. In: T. Hondoh (Ed.), *Physics of Ice Core Records Vol. 2*. Yoshioka Publishing, Kyoto, Japan. In press, preprint at arXiv:0903.3078 [physics.geo-ph].

Greve, R. and S. Sugiyama. 2009. Decay of the Greenland Ice Sheet due to surface-meltwater-induced acceleration of basal sliding. arXiv:0905.2027 [physics.geo-ph]. URL `http://arxiv.org/abs/0905.2027`.

Greve, R., R. Takahama and R. Calov. 2006. Simulation of large-scale ice-sheet surges: The ISMIP HEINO experiments. *Polar Meteorology and Glaciology*, **20**, 1–15.

Griffiths, D. F. 1997. The 'no boundary condition' outflow boundary condition. *International Journal for Numerical Methods in Fluids*, **24**, 393–411.

Gudmundsson, G. H. 1994. *Glacier sliding over sinusoidal bed and the characteristics of creeping flow over bedrock undulations.* Mitteilungen No. 130. Laboratory of Hydraulics, Hydrology and Glaciology (VAW), ETH Zurich, Switzerland.

Gudmundsson, G. H. 1997a. Basal-flow characteristics of a linear flow sliding frictionless over small bedrock undulations. *Journal of Glaciology*, **43** (143), 71–79.

Gudmundsson, G. H. 1997b. Basal-flow characteristics of a non-linear flow sliding frictionless over strongly undulating bedrock. *Journal of Glaciology*, **43** (143), 80–89.

Heinbockel, J. H. 1996. *Introduction to Tensor Calculus and Continuum Mechanics.* Trafford Publishing, Victoria, BC, Canada and Oxford, UK. ISBN 1-55369-133-4. Free online version available at http://www.math.odu.edu/~jhh/counter2.html (retrieved 2009-03-11).

Heinrich, H. 1988. Origin and consequences of cyclic ice rafting in the Northeast Atlantic Ocean during the past 130,000 years. *Quaternary Research*, **29** (2), 142–152.

Hindmarsh, R. C. A. 2004. A numerical comparison of approximations to the Stokes equations used in ice sheet and glacier modeling. *Journal of Geophysical Research*, **109** (F1), F01012. doi:10.1029/2003JF000065.

Hofmann, W. 1974. Die Internationale Glaziologische Grönland-Expedition EGIG. *Zeitschrift für Gletscherkunde und Glazialgeologie*, **5**, 217–224.

Holmlund, P. and M. Eriksson. 1989. The cold surface layer on Storglaciaren. *Geografiska Annaler*, **71A** (3-4), 241–244.

Hooke, R. L. 2005. *Principles of Glacier Mechanics.* Cambridge University Press, Cambridge, UK and New York, NY, USA, 2nd ed.

Hubbard, A., I. Willis, M. Sharp, D. Mair, P. Nienow, B. Hubbard and H. Blatter. 2000. Glacier mass-balance determination by remote sensing and high-resolution modelling. *Journal of Glaciology*, **46** (154), 491–498.

Humbert, A., R. Greve and K. Hutter. 2005. Parameter sensitivity studies for the ice flow of the Ross Ice Shelf, Antarctica. *Journal of Geophysical Research*, **110** (F4), F04022. doi:10.1029/2004JF000170.

Hundsdorfer, W. and J. G. Verwer. 2003. *Numerical Solution of Time-Dependent Advection-Diffusion-Reaction Equations.* Springer, Berlin, Germany etc.

Hutter, K. 1983. *Theoretical Glaciology; Material Science of Ice and the Mechanics of Glaciers and Ice Sheets.* D. Reidel Publishing Company, Dordrecht, The Netherlands.

Hutter, K. 1993. Thermo-mechanically coupled ice-sheet response – cold, polythermal, temperate. *Journal of Glaciology*, **39** (131), 65–86.

Hutter, K. and K. Jöhnk. 2004. *Continuum Methods of Physical Modeling.* Springer, Berlin, Germany etc.

Huybrechts, P., J. Gregory, I. Janssens and M. Wild. 2004. Modelling Antarctic and Greenland volume changes during the 20th and 21st centuries forced by GCM time slice integrations. *Global and Planetary Change*, **42** (1-4), 83–105. doi:10.1016/j.gloplacha.2003.11.011.

Huybrechts, P., A. J. Payne and EISMINT Intercomparison Group (including R. Greve). 1996. The EISMINT benchmarks for testing ice-sheet models. *Annals of Glaciology*, **23**, 1–12.

Huybrechts, P., O. Rybak, F. Pattyn, U. Ruth and D. Steinhage. 2007. Ice thinning, upstream advection, and non-climatic biases for the upper 89% of the EDML ice core from a nested model of the Antarctic ice sheet. *Climate of the Past*, **3** (4), 577–589. URL `http://www.clim-past.net/3/577/2007/`.

Jania, J., D. Mochnacki and B. Gadek. 1996. The thermal structure of Hansbreen, a tidewater glacier in southern Spitsbergen, Svalbard. *Polar Research*, **15** (1), 53–66.

Jänich, K. 1994. *Linear Algebra.* Springer, New York, NY, USA. ISBN 0-387-94128-2.

Kundu, P. K. and I. M. Cohen. 2004. *Fluid Mechanics.* Elsevier Academic Press, San Diego, CA, USA etc., 3rd ed.

Le Meur, E. 1996. Isostatic postglacial rebound over Fennoscandia with a self-gravitating spherical visco-elastic Earth model. *Annals of Glaciology*, **23**, 318–327.

Le Meur, E. and P. Huybrechts. 1996. A comparison of different ways of dealing with isostasy: examples from modelling the Antarctic ice sheet during the last glacial cycle. *Annals of Glaciology*, **23**, 309–317.

Liu, I.-S. 2002. *Continuum Mechanics.* Springer, Berlin, Germany etc.

Lliboutry, L. A. and P. Duval. 1985. Various isotropic and anisotropic ices found in glaciers and polar ice caps and their corresponding rheologies. *Annals of Geophysics*, **3**, 207–224.

Lunt, D. J., G. L. Foster, A. M. Haywood and E. J. Stone. 2008. Late Pliocene Greenland glaciation controlled by a decline in atmospheric CO_2 levels. *Nature*, **454** (7208), 1102–1105. doi:10.1038/nature07223.

Lüthi, M. and M. Funk. 2001. Modelling heat flow in a cold, high altitude glacier: interpretation of measurements from Colle Gnifetti, Swiss Alps. *Journal of Glaciology*, **47** (157), 314–324.

Marguerre, K. and H.-T. Woernle. 1969. *Elastic Plates.* Blaisdell Publishing Company, Waltham, MA, USA etc.

Meehl, G. A., T. F. Stocker, W. D. Collins, F. P., A. T. Gaye, J. M. Gregory, A. Kitoh, R. Knutti, J. M. Murphy, A. Noda, S. C. B. Raper, I. G. Watterson, A. J. Weaver and Z.-C. Zhao. 2007. Global climate projections. In: S. Solomon, D. Qin, M. Manning, Z. Chen, M. Marquis, K. B. Averyt, M. Tignor and H. L. Miller (Eds.), *Climate Change 2007: The Physical Science Basis. Contribution of Working Group I to the Fourth Assessment*

Report of the Intergovernmental Panel on Climate Change, pp. 747–845. Cambridge University Press, Cambridge, UK, and New York, NY, USA. URL `http://ipcc-wg1.ucar.edu/wg1/wg1-report.html`.

Miyamoto, A. 1999. *Mechanical properties and crystal textures of Greenland deep ice cores.* Doctoral thesis, Hokkaido University, Sapporo, Japan.

Moran, M. J. and W. N. Shapiro. 2000. *Fundamentals of Engineering Thermodynamics.* Wiley, New York, NY, USA.

Morland, L. W. 1984. Thermomechanical balances of ice sheet flows. *Geophysical and Astrophysical Fluid Dynamics*, **29**, 237–266.

Morland, L. W. 1987. Unconfined ice-shelf flow. In: C. J. van der Veen and J. Oerlemans (Eds.), *Dynamics of the West Antarctic Ice Sheet*, pp. 99–116. D. Reidel Publishing Company, Dordrecht, The Netherlands.

Morton, K. W. and D. F. Mayers. 1994. *Numerical Solution of Partial Differential Equations.* Cambridge University Press, Cambridge, UK and New York, NY, USA.

Müller, I. 1985. *Thermodynamics.* Pitman Advanced Publishing Program, Boston, MA, USA etc.

Nedjar, B. 2002. An enthalpy-based finite element method for nonlinear heat problems involving phase change. *Computers and Structures*, **80** (1), 9–21.

Nye, J. F. 1957. The distribution of stress and velocity in glaciers and ice sheets. *Proceedings of the Royal Society London A*, **239** (1216), 113–133. doi:10.1098/rspa.1957.0026.

Paterson, W. S. B. 1971. Temperature measurements in Athabasca Glacier, Alberta, Canada. *Journal of Glaciology*, **10** (60), 339–349.

Paterson, W. S. B. 1994. *The Physics of Glaciers.* Pergamon Press, Oxford, UK etc., 3rd ed.

Pattyn, F., L. Perichon, A. Aschwanden, B. Breuer, B. de Smedt, O. Gagliardini, G. H. Gudmundsson, R. Hindmarsh, A. Hubbard, J. V. Johnson, T. Kleiner, Y. Konovalov, C. Martin, A. J. Payne, D. Pollard, S. Price, M. Rückamp, F. Saito, O. Souček, S. Sugiyama and T. Zwinger. 2008. Benchmark experiments for higher-order and full-Stokes ice sheet models (ISMIP-HOM). *The Cryosphere*, **2** (2), 95–108. URL `http://www.the-cryosphere.net/2/95/2008/`.

Pedlosky, J. 1987. *Geophysical Fluid Dynamics.* Springer, New York, NY, USA etc., 2nd ed.

Petrenko, V. F. and R. W. Whitworth. 1999. *Physics of Ice.* Oxford University Press, Oxford, UK etc.

Pettersson, R., P. Jansson and H. Blatter. 2004. Spatial variability in water content at the cold-temperate transition surface of the polythermal Storglaciären, Sweden. *Journal of Geophysical Research*, **109** (F2), F02009. doi:10.1029/2003JF000110.

Pham, Q. T. 1995. Comparison of general-purpose finite-element methods for the Stefan problem. *Numerical Heat Transfer, Part B*, **27**, 417–435.

Pimienta, P., P. Duval and V. Y. Lipenkov. 1987. Mechanical behaviour of anisotropic polar ice. In: E. D. Waddington and J. S. Walder (Eds.), *The*

Physical Basis of Ice Sheet Modelling, IAHS Publication No. 170, pp. 57–66. IAHS Press, Wallingford, UK.

Placidi, L. 2004. *Thermodynamically consistent formulation of induced anisotropy in polar ice accounting for grain-rotation, grain-size evolution and recrystallization.* Doctoral thesis, Department of Mechanics, Darmstadt University of Technology, Germany. URL `http://tuprints.ulb.tu-darmstadt.de/614/`.

Placidi, L., R. Greve, H. Seddik and S. H. Faria. 2009. Continuum-mechanical, anisotropic flow model for polar ice masses, based on an anisotropic flow enhancement factor (CAFFE). *Continuum Mechanics and Thermodynamics.* Submitted, preprint at arXiv:0903.0688 [physics.geo-ph].

Pralong, A. and M. Funk. 2005. Dynamic damage model of crevasse opening and application to glacier calving. *Journal of Geophysical Research*, **110** (B1), B01309.

Pralong, A. and M. Funk. 2006. On the instability of avalanching glaciers. *Journal of Glaciology*, **52** (176), 31–48.

Pralong, A., K. Hutter and M. Funk. 2006. Anisotropic damage mechanics for viscoelastic ice. *Continuum Mechanics and Thermodynamics*, **17** (5), 387–408.

Press, W. H., S. A. Teukolsky, W. T. Vetterling and B. P. Flannery. 1996. *Numerical Recipes in Fortran 77.* Cambridge University Press, Cambridge, UK and New York, NY, USA, 2nd ed.

Reddy, J. N. 2006. *An Introduction to the Finite Element Method.* McGraw-Hill, Boston, MA, USA etc., 3rd ed. International edition.

Reeh, N. 1968. On the calving of ice from floating glaciers and ice shelves. *Journal of Glaciology*, **7** (50), 215–232.

Ridley, J. K., P. Huybrechts, J. M. Gregory and J. A. Lowe. 2005. Elimination of the Greenland ice sheet in a high CO_2 climate. *Journal of Climate*, **18** (17), 3409–3427.

Ritz, C. 1987. Time dependent boundary conditions for calculation of temperature fields in ice sheets. In: E. D. Waddington and J. S. Walder (Eds.), *The Physical Basis of Ice Sheet Modelling*, IAHS Publication No. 170, pp. 207–216. IAHS Press, Wallingford, UK.

Russell-Head, D. S. and W. F. Budd. 1979. Ice sheet flow properties derived from borehole shear measurements combined with ice core studies. *Journal of Glaciology*, **24** (90), 117–130.

Saito, F. and A. Abe-Ouchi. 2004. Thermal structure of Dome Fuji and east Dronning Maud Land, Antarctica, simulated by a three-dimensional ice-sheet model. *Annals of Glaciology*, **39**, 433–438.

Saito, F., A. Abe-Ouchi and H. Blatter. 2006. European Ice Sheet Modelling Initiative (EISMINT) model intercomparison experiments with first order mechanics. *Journal of Geophysical Research*, **111** (F2), F02012. doi:10.1029/2004JF000273.

Sammonds, P. R. 1999. Understanding the fundamental physics governing the evolution and dynamics of the earth's crust and ice sheets. *Philosophical Transactions of the Royal Society London A*, **357** (1763), 3377–3401.

Schoof, C. 2007a. Ice sheet grounding line dynamics: Steady states, stability, and hysteresis. *Journal of Geophysical Research*, **112** (F3), F03S28. doi: 10.1029/2006JF000664.

Schoof, C. 2007b. Marine ice-sheet dynamics. Part 1. The case of rapid sliding. *Journal of Fluid Mechanics*, **573**, 27–55. doi:10.1017/S0022112006003570.

Schwerzmann, A. A. 2006. *Borehole analysis and flow modeling of firn-covered glaciers*. Ph.D. thesis, ETH Zurich, Switzerland. URL `http://e-collection.ethbib.ethz.ch/view/eth:28327`.

Seddik, H., R. Greve, L. Placidi, I. Hamann and O. Gagliardini. 2008. Application of a continuum-mechanical model for the flow of anisotropic polar ice to the EDML core, Antarctica. *Journal of Glaciology*, **54** (187), 631–642.

Seibert, P. 1993. Convergence and accuracy of numerical methods for trajectory calculations. *Journal of Applied Meteorology*, **3**, 558–566.

Smith, G. D. and L. W. Morland. 1981. Viscous relations for the steady creep of polycrystalline ice. *Cold Regions Science and Technology*, **5**, 141–150.

Straughan, B., R. Greve, H. Ehrentraut and Y. Wang (Eds.). 2001. *Continuum Mechanics and Applications in Geophysics and the Environment*. Springer, Berlin, Germany etc. ISBN 3-540-41660-9.

Sugiyama, S., A. Bauder, M. Funk and C. Zahno. 2007. Evolution of Rhonegletscher, Switzerland, over the past 125 years and in the future: application of an improved flowline model. *Annals of Glaciology*, **46**, 268–274.

Thoma, M. and D. Wolf. 1999. Bestimmung der Mantelviskosität aus Beobachtungen der Landhebung und Schwere in Fennoskandien. Scientific Technical Report STR99/02, GeoForschungsZentrum Potsdam, Germany.

Thomas, R. H. 2004. Greenland: recent mass balance observations. In: J. L. Bamber and A. J. Payne (Eds.), *Mass Balance of the Cryosphere. Observations and Modelling of Contemporary and Future Changes*, pp. 393–436. Cambridge University Press, Cambridge, UK and New York, NY, USA.

Thomas, R. H., D. R. MacAyeal, D. H. Eilers and D. R. Gaylord. 1984. Glaciological studies on the Ross Ice Shelf, Antarctica, 1973-1978. In: C. R. Bentley and D. E. Hayes (Eds.), *The Ross Ice Shelf: Glaciology and Geophysics*, Antarctic Research Series No. 42, pp. 21–53. American Geophysical Union, Washington DC, USA.

Truffer, M. 2004. The basal speed of valley glaciers: an inverse approach. *Journal of Glaciology*, **50** (169), 236–242.

Turcotte, D. L. and G. Schubert. 2002. *Geodynamics*. Cambridge University Press, Cambridge, UK and New York, NY, USA, 2nd ed.

van der Veen, C. J. 1999. *Fundamentals of Glacier Dynamics*. A. A. Balkema, Rotterdam, The Netherlands.

Vialov, S. S. 1958. Regularities of glacial shields movement and the theory of plastic viscous flow. In: *Physics of the Motion of Ice*, IAHS Publication No. 47, pp. 266–275. IAHS Press, Wallingford, UK.

Vieli, A., M. Funk and H. Blatter. 2000. Tidewater glaciers: frontal flow acceleration and basal sliding. *Annals of Glaciology*, **31**, 217–221.

Vieli, A., M. Funk and H. Blatter. 2001. Flow dynamics of tidewater glaciers: a numerical modelling approach. *Journal of Glaciology*, **47** (159), 595–606.

Vieli, A. and A. J. Payne. 2005. Assessing the ability of numerical ice sheet models to simulate grounding line migration. *Journal of Geophysical Research*, **110**. doi:10.1029/2004JF000202.

Weis, M. 2001. *Theory and finite element analysis of shallow ice shelves.* Doctoral thesis, Department of Mechanics, Darmstadt University of Technology, Germany. URL `http://tuprints.ulb.tu-darmstadt.de/171/`.

Wilson, R. C. L., S. A. Drury and J. L. Chapman. 2000. *The Great Ice Age. Climate Change and Life.* Routledge, London, UK and New York, NY, USA. ISBN 0-415-19842-9.

Zotikov, I. A. 1986. *The Thermophysics of Glaciers.* D. Reidel Publishing Company, Dordrecht, The Netherlands.

Zwally, H. J., W. Abdalati, T. Herring, K. Larson, J. Saba and K. Steffen. 2002. Surface melt-induced acceleration of Greenland ice-sheet flow. *Science*, **297** (5579), 218–222. doi:10.1126/science.1072708.

Zwinger, T., R. Greve, O. Gagliardini, T. Shiraiwa and M. Lyly. 2007. A full Stokes-flow thermo-mechanical model for firn and ice applied to the Gorshkov crater glacier, Kamchatka. *Annals of Glaciology*, **45**, 29–37.

List of Symbols

Only the principal symbols are listed. A scalar is indicated by italics type, a vector by bold face upright type and a tensor by sans serif upright type.

$\mathbf{0}$	zero vector
I	unit tensor
$\mathbf{a}$	acceleration $(= \mathrm{d}\mathbf{v}/\mathrm{d}t = \mathrm{d}^2\mathbf{x}/\mathrm{d}t^2)$
a_b	basal melting (-freezing) rate in the vertical direction
$a_\mathrm{b}^\perp$	basal melting (-freezing) rate
$a_\mathrm{m}^\perp$	volume flux through the CTS
a_s	accumulation-ablation function (surface mass balance) in the vertical direction
$a_\mathrm{s}^\perp$	accumulation-ablation function (surface mass balance)
b	z-coordinate of the ice base
b_0	isostatically relaxed value for b without ice load
c	(1) specific heat in general (2) specific heat of ice $[= (146.3 + 7.253\,T[\mathrm{K}])\,\mathrm{J\,kg^{-1}K^{-1}}]$ (3) specific heat of firn (Sect. 9.2)
$c^\perp$	calving rate
d_e	effective strain rate
$\mathbf{e}_i$	set of orthonormal basis vectors (in the present configuration κ_t)
$\mathbf{e}_\mathrm{t}$	tangential unit vector
$\mathbf{f}$	volume force
$f(\sigma_\mathrm{e})$	creep function
$f^\star(\mathbf{n})$	orientation distribution function $[= \rho^\star(\mathbf{n})/\rho]$
$\mathbf{g}$, g	gravitational acceleration $(= 9.81\,\mathrm{m\,s^{-2}})$

g	density of arbitrary physical quantity
g_s	arbitrary physical quantity per unit mass ("specific ...")
h	(1) z-coordinate of the ice surface (2) specific enthalpy ($= u + p/\rho$; Sect. 9.3)
$\mathbf{j}$	diffusive water mass flux in temperate ice
k	enthalpy diffusivity
$\dot{m}_\mathrm{b}^\mathrm{w}$	water mass flux into the base
$\mathbf{n}$	(1) unit normal vector (2) orientation (direction of the c-axis; Sect. 9.1)
n	stress exponent ($= 3$)
p	(1) pressure (2) production density of arbitrary physical quantity (Chapter 3)
$p(\rho, T)$	thermodynamic pressure
p_visc	viscous pressure
p_tot	total pressure $[= p(\rho, T) + p_\mathrm{visc}]$
p_sw	hydrostatic pressure of sea water
p_sea	hydrostatic pressure of sea water at the ice-sea interface
p, q	basal sliding exponents
$\mathbf{q}$	heat flux
$\mathbf{q}^\star(\mathbf{n})$	orientation flux
$\mathbf{q}_\mathrm{l}$	latent heat flux in temperate ice
$\mathbf{q}_\mathrm{s}$	sensible heat flux in temperate ice
q	load per unit area
$q_\mathrm{geo}^\perp$	geothermal heat flux
$q_\mathrm{sea}^\perp$	heat flux on the sea side of the ice-sea interface
r	specific radiation power
s	supply density of arbitrary physical quantity
t	Cauchy stress tensor
t^D	Cauchy stress deviator $[= \mathsf{t} - (\frac{1}{3}\,\mathrm{tr}\,\mathsf{t})\,\mathsf{I}]$
t_lith	stress at the lithosphere side of the ice-lithosphere interface
t_sea	stress at the sea side of the ice-sea interface
$\mathbf{t}_\mathbf{n}$	stress vector
t	time
$[t]$	typical time-scale
$\mathbf{u}$	displacement
$\mathbf{u}^\star(\mathbf{n})$	orientation transition rate
u	specific internal energy
$\mathbf{v}$	velocity ($= \mathrm{d}\mathbf{x}/\mathrm{d}t$)

$\mathbf{v}^\star(\mathbf{n})$	orientation-dependent velocity
$\mathbf{v}_\mathrm{b}$	basal sliding velocity
$\mathbf{v}_\mathrm{gl}$	velocity at the grounding line
$\mathbf{v}_\mathrm{h}$	horizontal velocity
$\mathbf{v}_\mathrm{i}$	ice velocity in temperate ice
$\mathbf{v}_\mathrm{sea}$	velocity of subglacial sea water
$\mathbf{v}_\mathrm{w}$	water velocity in temperate ice
$\mathbf{w}$	velocity of a singular surface
w	vertical displacement of the lithosphere
w_ss	steady-state value of w
$\mathbf{x}$	position vector (in the present configuration κ_t)
x, y	horizontal Cartesian coordinates
z	vertical Cartesian coordinate
z_l	z-coordinate of the lithosphere surface
z_m	z-coordinate of the CTS
z_sl	z-coordinate of the mean sea level

$A(T')$	rate factor
$A_\mathrm{t}(W)$	rate factor of temperate ice
$\mathcal{A}$	deformability of polycrystalline ice
$\mathcal{A}^\star(\mathbf{n})$	crystallite deformability
A_0	pre-exponential constant $(= 3.985 \times 10^{-13}\,\mathrm{s}^{-1}\,\mathrm{Pa}^{-3}$ for $T' \le 263.15\,\mathrm{K}$, $= 1.916 \times 10^{3}\quad \mathrm{s}^{-1}\,\mathrm{Pa}^{-3}$ for $T' > 263.15\,\mathrm{K})$
B	left Cauchy Green tensor $(= \mathsf{V}^2)$
$B(T')$	associated rate factor
C	right Cauchy Green tensor $(= \mathsf{U}^2)$
C_b	basal sliding coefficient
C_wi	water-ice drag coefficient $(\approx 2.5 \times 10^{-3})$
D	strain-rate (stretching) tensor $[= (\mathsf{L} + \mathsf{L}^\mathrm{T})/2]$
D^D	strain-rate (stretching) deviator $[= \mathsf{D} - (\frac{1}{3}\,\mathrm{tr}\,\mathsf{D})\,\mathsf{I}]$
D	diffusivity of the ice surface
D_a	diffusivity of the asthenosphere
$\mathbf{E}_A$	set of orthonormal basis vectors (in the reference configuration κ_r)
E	(1) flow enhancement factor (2) Young's modulus
$E(\mathcal{A})$	anisotropic flow enhancement factor
E_s	stress enhancement factor

F	deformation gradient ($=\operatorname{Grad}\mathbf{x}$)
F_{b}	implicit representation of the ice base ($=b-z$)
F_{cf}	implicit representation of the calving front
F_{gl}	implicit representation of the grounding line
F_{m}	implicit representation of the cold-temperate transition surface ($=z-z_{\mathrm{m}}$)
F_{s}	implicit representation of the ice surface ($=z-h$)
Fr	Froude number $[=[U]^2/(g[H])]$
H	displacement gradient ($=\operatorname{Grad}\mathbf{u}=\mathsf{F}-\mathsf{I}$)
H	(1) thickness in general (2) ice thickness ($=h-b$)
H_{a}	thickness of the asthenosphere
H_{l}	thickness of the lithosphere
$[H]$	typical vertical extent
J	Jacobian of the deformation gradient ($=\det\mathsf{F}$)
K	flexural stiffness
K_{l}	flexural stiffness of the lithosphere ($\approx 10^{25}\,\mathrm{N\,m}$)
L	velocity gradient ($=\operatorname{grad}\mathbf{v}$)
L	latent heat of ice ($=3.35\times10^5\,\mathrm{J\,kg^{-1}}$)
L_{r}	radius of relative stiffness $[=(K_{\mathrm{l}}/(\rho_{\mathrm{a}}g))^{1/4}]$
$[L]$	typical horizontal extent
M	water mass production rate in temperate ice
N	membrane stress
$\mathbf{N}_{\mathrm{b}}$, N_{b}	basal normal stress
$[P]$	typical pressure
$\mathcal{P}_{\mathrm{m}}^{\mathrm{w}}$	water surface production rate at the CTS
$\mathbf{Q}$	horizontal volume flux
Q	activation energy ($=60\,\mathrm{kJ\,mol^{-1}}$ for $T'\le 263.15\,\mathrm{K}$, $=139\,\mathrm{kJ\,mol^{-1}}$ for $T'>263.15\,\mathrm{K}$)
R	rotation tensor ($=\mathsf{F}\cdot\mathsf{U}^{-1}=\mathsf{V}^{-1}\cdot\mathsf{F}$)
R	universal gas constant ($=8.314\,\mathrm{J\,mol^{-1}\,K^{-1}}$)
Ro	Rossby number $[=[U]/(2\Omega[L])]$
T	temperature
T_{m}	pressure melting temperature
T_0	melting temperature at standard pressure ($=273.15\,\mathrm{K}$)
T'	temperature relative to the pressure melting point ($=T-T_{\mathrm{m}}+T_0$)
T_{cf}	temperature at the calving front
T_{gl}	temperature at the grounding line

T_{s}	ice surface temperature
T_{sea}	temperature of sea water at the ice-sea interface ($\approx -2°\mathrm{C}$)
U	right stretch tensor $[= (\mathsf{F}^{\mathrm{T}} \cdot \mathsf{F})^{1/2}]$
$[U]$	typical horizontal velocity
V	left stretch tensor $[= (\mathsf{F} \cdot \mathsf{F}^{\mathrm{T}})^{1/2}]$
V	activation volume
W	spin tensor $[= (\mathsf{L} - \mathsf{L}^{\mathrm{T}})/2]$
W	water content of temperate ice (mass fraction)
W_{s}	water content at the ice surface
$[W]$	typical vertical velocity
$\mathbf{X}$	position vector (in the reference configuration κ_r)
α	inclination angle
β	Clausius-Clapeyron constant ($= 7.42 \times 10^{-8}\,\mathrm{K\,Pa^{-1}}$ for pure ice, $= 9.8 \times 10^{-8}\,\mathrm{K\,Pa^{-1}}$ for air-saturated glacier ice)
$\gamma_{ij}\ (i \neq j)$	shear angle
$\dot{\gamma}_{ij}\ (i \neq j)$	shear rate
$\delta(\cdot)$	Dirac's delta function
δ_{ij}	Kronecker symbol
δ_{sea}	frictional dissipation at the sea side of the ice-sea interface
ϵ	infinitesimal strain tensor $[= (\mathsf{H} + \mathsf{H}^{\mathrm{T}})/2]$
ϵ^{D}	infinitesimal strain deviator $[= \epsilon - (\frac{1}{3}\,\mathrm{tr}\,\epsilon)\,\mathsf{I}]$
ε	aspect ratio ($= [H]/[L] = [W]/[U]$)
ε_{ijk}	Levi-Civita symbol
ζ	(1) bulk viscosity ($= \lambda + 2\eta/3$) (2) vertical coordinate of the sigma transformation
η	shear viscosity
$\bar{\eta}$	depth-integrated shear viscosity
ι	constitutive parameter for the orientation transition rate
κ	(1) bulk modulus ($= \lambda + 2\mu/3$) (2) heat conductivity in general (3) heat conductivity of ice ($= 9.828\,e^{-0.0057\,T[\mathrm{K}]}\,\mathrm{W\,m^{-1}K^{-1}}$) (4) heat conductivity of firn (Sect. 9.2)
κ_r	reference configuration
κ_t	present configuration
λ	(1) 1st Lamé parameter (2) viscosity coefficient (3) orientation diffusivity (Sect. 9.1)

μ	2nd Lamé parameter (shear modulus)
ν	(1) Poisson's ratio (2) water diffusivity in temperate ice (Sect. 9.3)
ξ, φ	horizontal coordinates of the sigma transformation
ρ	(1) density in general (2) density of ice ($= 910\,\mathrm{kg\,m^{-3}}$) (3) density of firn (Sect. 9.2)
$\tilde{\rho}$	relative density of firn ($= \rho/\hat{\rho}_\mathrm{i}$)
$\rho^\star(\mathbf{n})$	orientation mass density
ρ_a	density of the asthenosphere ($= 3300\,\mathrm{kg\,m^{-3}}$)
ρ_i	partial density of ice in temperate ice
$\hat{\rho}_\mathrm{i}$	bulk density of ice ($= 910\,\mathrm{kg\,m^{-3}}$)
ρ_sw	density of sea water ($= 1028\,\mathrm{kg\,m^{-3}}$)
ρ_w	partial density of water in temperate ice
$\hat{\rho}_\mathrm{w}$	bulk density of water ($= 1000\,\mathrm{kg\,m^{-3}}$)
σ_e	effective stress
σ_0	residual stress
τ	time coordinate of the sigma transformation
τ_a	time lag of the relaxing asthenosphere ($\approx 3000\,\mathrm{a}$)
$\boldsymbol{\tau}_\mathrm{b}$, τ_b	basal drag (shear stress)
$\boldsymbol{\tau}_\mathrm{d}$	driving stress
$\boldsymbol{\tau}_\mathrm{sea}$	basal drag (shear stress) induced by circulating sea water
$\boldsymbol{\phi}$	flux density of arbitrary physical quantity
ψ	arbitrary scalar, vector or tensor field
ω	material volume in the present configuration κ_t
Γ	constitutive parameter for the orientation production rate
$\Gamma^\star(\mathbf{n})$	orientation production rate
$\Delta\zeta$	spacing of vertical coordinate ζ
$\Delta\xi$, $\Delta\varphi$	spacing of horizontal coordinates ξ and φ, respectively
$\Delta\tau$	spacing of time coordinate τ
$[\Delta T]$	typical temperature variation
Φ_n	set of basis functions
$\boldsymbol{\Omega}$, Ω	angular velocity of the Earth ($= 7.2921 \times 10^{-5}\,\mathrm{s^{-1}}$)
Ω	material volume in the reference configuration κ_r

List of Acronyms

AGCM	Atmosphere General Circulation Model
AMSL	Above Mean Sea Level
CAFFE	Continuum-mechanical, Anisotropic Flow model, based on an anisotropic Flow Enhancement factor
CTS	Cold-temperate Transition Surface,
DA	Diffusive Asthenosphere
EAIS	East Antarctic Ice Sheet
EDML	EPICA ice core in Dronning Maud Land
EGIG	Expédition Glaciologique International au Grœnland
EISMINT	European Ice Sheet Modeling INiTiative
EL	Elastic Lithosphere
ELDA	Elastic Lithosphere / Diffusive Asthenosphere
ELRA	Elastic Lithosphere / Relaxing Asthenosphere
EPICA	European Project for Ice Coring in Antarctica
ETH	Eidgenössische Technische Hochschule (Swiss Federal Institute of Technology) Zurich
FESSACODE	Finite Element Shallow Shelf Approximation Code
FOA	First Order Approximation
FS	Full Stokes
GCM	General Circulation Model
GPS	Global Positioning System
GRIP	Greenland Ice Core Project
HEINO	Heinrich Event INtercOmparison
IAI	International Antarctic Institute
IPCC	Intergovernmental Panel on Climate Change

ISMIP	Ice Sheet Model Intercomparison Project
LGM	Last Glacial Maximum
LL	Local Lithosphere
LLDA	Local Lithosphere / Diffusive Asthenosphere
LLRA	Local Lithosphere / Relaxing Asthenosphere
MODIS	MODerate-resolution Imaging Spectroradiometer
NADW	North Atlantic Deep Water
NASA	National Aeronautics and Space Administration
ODE	Ordinary Differential Equation
ODF	Orientation Distribution Function
OGCM	Ocean General Circulation Model
OMD	Orientation Mass Density
QED	Quod Erat Demonstrandum (which was to be demonstrated)
RA	Relaxing Asthenosphere
RIGGS	Ross Ice Shelf Geophysical and Glaciological Survey
SGVE	Self-Gravitating, spherical, Visco-Elastic multi-layer
SIA	Shallow Ice Approximation
SICOPOLIS	SImulation COde for POLythermal Ice Sheets
SSA	Shallow Shelf Approximation
WAIS	West Antarctic Ice Sheet

Index

MIX
Papier aus verantwortungsvollen Quellen
Paper from responsible sources
FSC® C105338

If you have any concerns about our products,
you can contact us on
ProductSafety@springernature.com

In case Publisher is established outside the EU,
the EU authorized representative is:
Springer Nature Customer Service Center GmbH
Europaplatz 3, 69115 Heidelberg, Germany

Printed by Libri Plureos GmbH
in Hamburg, Germany